LIBRARY/NEW ENGLAND INST. OF TECHNOLOGY

3 0147 1003 9507 1

```
TJ 1225 .S557 2010
Smid, Peter.
CNC control setup for
 milling and turning
```

NEW ENGLAND INSTITUTE OF TECHNOLOGY
LIBRARY

CNC Control Setup
for Milling
and Turning

Other books by Peter Smid:

CNC PROGRAMMING HANDBOOK
Comprehensive Guide to Practical CNC Programming
Third Edition - ISBN 9780831133474

CNC PROGRAMMING TECHNIQUES
An Insider's Guide to Effective Methods and Applications
ISBN 9780831131852

FANUC CNC CUSTOM MACROS
Practical Resources for Fanuc Custom Macro B Users
ISBN 9780831131579

CNC Control Setup for Milling and Turning

Mastering CNC Control Systems

Peter Smid

Industrial Press, Inc.
989 Avenue of the Americas
New York, NY 10018, USA
http://www.industrialpress.com

Library of Congress Cataloging-in-Publication Data

Smid, Peter.
  CNC control setup for milling and turning : mastering CNC control systems / Peter Smid.
  p. cm.
  Includes index
  ISBN 978-0-8311-3350-4 (hard cover)
  1. Milling-machines--Numerical control. 2. Turning (Lathe work)--Numerical control. 3. Machining--Automation. I. Title.

TJ1225.S557 2010

671.3'5--dc22

2010007023

Industrial Press, Inc.
989 Avenue of the Americas
New York, NY 10018, USA

Copyright © 2010
Printed in the United States of America

All Rights Reserved.

This book or parts thereof may not be reproduced, stored in a retrieval system, or transmitted in any form without the permission of the publishers.

Cover Design: Janet Romano
Editorial Director: John Carleo
Marketing & Sales Director: Patrick V. Hansard

3 4 5 6 7 8 9 10

About the Author

Peter Smid is a professional consultant, educator and speaker, with many years of practical, hands-on experience, in the industrial and educational fields. During his career, he has gathered an extensive experience with CNC and CAD/CAM applications on all levels. He consults to manufacturing industry and educational institutions on practical use of *Computerized Numerical Control* technology, part programming, *Autocad®*, *Mastercam®* and other CAD/CAM software, as well as advanced machining, tooling, setup, and many other related fields. His comprehensive industrial background in CNC programming, machining and company-oriented training has assisted several hundred companies to benefit from his wide-ranging knowledge.

Mr. Smid's long time association with advanced manufacturing companies and CNC machinery vendors, as well as his affiliation with a number of Community and Technical College industrial technology programs and machine shop skills training, have enabled him to broaden his professional and consulting skills in the areas of CNC and CAD/CAM training, computer applications and needs analysis, software evaluation, system benchmarking, programming, hardware selection, software customizing, and operations management.

Over the years, Mr. Smid has developed and delivered hundreds of customized educational programs to thousands of instructors and students at colleges and universities across United States, Canada and Europe, as well as to a large number of manufacturing companies and private sector organizations and individuals.

He has actively participated in many industrial trade shows, conferences, workshops and various seminars, including submission of papers, delivering presentations and a number of speaking engagements to professional organizations. He is also the author of articles and many in-house publications on the subject of CNC and CAD/CAM. For six years he had a monthly column in the ShopTalk Magazine related to CNC programming. During his many years as a highly respected professional in the CNC industrial and educational field, he has developed tens of thousands of pages of high quality training materials.

Peter Smid is also the author of three other best selling CNC books published by Industrial Press, Inc.:

CNC PROGRAMMING HANDBOOK
Comprehensive Guide to Practical CNC Programming
Third Edition - ISBN 9780831133474

CNC PROGRAMMING TECHNIQUES
An Insider's Guide to Effective Methods and Applications
ISBN 9780831131852

FANUC CNC CUSTOM MACROS
Practical Resources for Fanuc Custom Macro B Users
ISBN 9780831131579

All books are also available as e-books

The author welcomes comments, suggestions and other input from educators, students and industrial users.
You can contact him at *info@industrialpress.com*

Acknowledgments

I have made a number of references to several manufacturers and software developers in the book. It is only fair to acknowledge their names:

- **FANUC and CUSTOM MACRO or USER MACRO or MACRO B are registered trademarks of Fanuc Ltd, Japan**
- **Mastercam is a registered trademark of CNC Software Inc., Tolland, CT, USA**
- **Edgecam is a registered trademark of Pathtrace, Inc., UK**
- **NCPlot is a registered trademark of NCPlot LLC, Muskegon, MI, USA**
- **AutoCad is a registered trademark of Autodesk, Inc., San Rafael, CA, USA**
- **Kennametal is a registered trademark of Kennametal, Inc., Latrobe, PA, USA**
- **WINDOWS is a registered trademarks of Microsoft, Inc., Redmond, WA, USA**

There are other companies mentioned in this handbook ...

- **Caterpillar**
- **Fadal, Fagor**
- **Haas, Heidenhain**
- **Iscar**
- **Kitamura**
- **Makino, Matsuura, Mazak, Mitsubishi, Mori-Seiki**
- **Okuma, Osaka KiKo (OKK)**
- **Sandvik, Seco, Siemens, Sumitomo**
- **Valenite**
- **Widia**
- **Yasnac, Yamazaki**
- **...** *and some others that may have escaped me*

Table of Contents

Chapter 1
CONCEPTS OF CNC MACHINING 1
    CNC PROCESS 1
        Drawing Evaluation 1
        Material Identification 1
        Part Holding 2
        Tooling Selection 2
        Cutting Conditions 2
        Program Writing 2
        Program Verification 2
        Program Documentation 3
        Program Transfer 3
    WORK COMPLETION 3
        Program Evaluation 3
        Material Check 3
        Tooling Preparation 3
        Tooling Setup 3
        Fixture Setup 4
        Program Loading 4
        Setting Offsets 4
        First Part Run 4
        Program Optimization 4
        Production Run 4
        Part Inspection 4
        Safety Issues 4

Chapter 2
CNC MACHINE SPECIFICATIONS 5
    MACHINE SPECIFICATIONS 5
        Useful Information 5
    VERTICAL MACHINING CENTER 5
        Typical Specifications - VMC 6
    HORIZONTAL MACHINING CENTER 6
        Typical Specifications - HMC 6
    LATHES AND TURNING CENTERS 7
        Typical Specifications 7
    FANUC SYSTEM SPECIFICATIONS 7
        Controlled Axes 7
        Operation Functions 8
        Interpolation Functions 8
        Feedrate Functions 8
        Spindle Functions 8
        Tool Functions 9
        Program Input 9
        Editing Functions 9
        Setting and Display Functions 9

TABLE OF CONTENTS

        Data Input / Output 10
        Options 10

Chapter 3
PROGRAM INTERPRETATION 11
    PROGRAM STRUCTURE 11
      Logical Structure 11
      Program Structure Consistency 14
    PROGRAM INTERPRETATION 14
      Units 15
      Type of Operation 15
      Speeds and Feeds 15
      Tool Numbers 15
      Dwell Time 16
      Other Considerations 16
    PROGRAM ERRORS 17
      Syntax Errors 17
      Logical Errors 17
      Offset Errors 18
      Incorrect Units 18

Chapter 4
CONTROL SYSTEM 19
    CONTROL SYSTEM OVERVIEW 19
    SYSTEM MEMORY 20
      Paper Tape 20
      Memory Capacity 21
    CONTROL PANEL 22
      ON-OFF Buttons 23
      Display Screen and Soft Keys 23
      Selection Keys 23
      Address Keyboard 25
      Shift Key 25
      Numeric Keyboard 25
      Edit Keys 26
      Page and Cursor Keys 26
      EOB - CAN - INPUT Keys 26
      Reset Key 26
      Help Key 26

Chapter 5
OPERATION PANEL 27
    OPERATION PANEL LAYOUT 28
    MAIN BUTTONS AND SWITCHES 28
      Power ON / Ready Light Indicator 28
      Cycle Start / Feedhold - Auto Operation 28
    STATUS INDICATOR LIGHTS 29
      Indicator Lights - Three Groups 29
    ON/OFF SWITCHES 32
      Toggle and Button Switches 33

TABLE OF CONTENTS

Program Testing and Setup Mode 37
ROTARY SWITCHES 37
   Mode Selection 37
   Axis Selection 41
OVERRIDES 41
   Rapid Override 41
   Feedrate Override 42
   Spindle Override 44
   Jog Feedrate Selection 44
EDIT KEY 45
EMERGENCY STOP SWITCH 45
OTHER FEATURES 45
CONTROL PANEL SYMBOLS 45
   Switches 46
   Modes 46
   Operations 47

Chapter 6
SETUP HANDLE 49
PURPOSE 49
DESIGN 49
HANDLE FEATURES 50
   Mode Selection 50
   Axis Selection 50
   Range Selection 50
   Dial Wheel 50
APPLICATIONS 52
   Manual Absolute Switch 52
   Safety Issues 52

Chapter 7
MILLING TOOLS - SETUP 53
MILLING TOOLS 53
   Tool Holders 53
   Cutting Tools 54
   Tooling System 54
ATC - AUTOMATIC TOOL CHANGE 55
   Manual Tool Change 55
   Tool Magazine 56
   Fixed Type ATC 57
   Random Memory Type ATC 57
REGISTERING TOOLS 58
TOOL SPECIFICATIONS 58
   Maximum Tool Length 58
   Maximum Tool Diameter 59
   Maximum Tool Weight 59
ATC PROCESS 60
TOOL REPETITION 62
   Repeating the Same Tool 62

Repeating a Different Tool 62
Solving the Problem 63
Tool Change - One or Two Blocks 64
DUMMY TOOL 64
SUMMARY 64

Chapter 8
SETTING PART ZERO 65
SYMBOLS 65
WHAT IS PART ZERO ? 65
Part Zero Location 65
VISES FOR CNC WORK 67
CNC Vise - Basic Design 67
CNC Vise - Setup 67
APPLICATION EXAMPLE 68
Part Orientation 68
Vise Orientation 69
FINDING A CENTER 71
PART ZERO ON CNC LATHES 71
X-axis Zero 71
Z-axis Zero 71
Work Shift 72

Chapter 9
WORK OFFSET SETTINGS 73
REFERENCE POINTS 73
Machine Zero 73
Part Zero 73
Machine Zero to Part Zero Connection 73
PART SETUP EVALUATION 73
USING WORK OFFSETS 74
Preparatory Commands G54-G59 75
Establishing the Connection 75
USING EDGE FINDERS 77
Magnetic Single Edge Finder 77
Magnetic Center Finder 77
DISTANCE-TO-GO 78
SHAFT TYPE EDGE FINDERS 78
Prerequisites 78
Individual Steps 79
Work Offset Calculations 79
Original Example Revisited 81
WORK OFFSET Z-SETTING 81
Where is the Z-setting? 81
Setting the Z-axis 82
MULTIPLE WORK OFFSETS 82
Program Entry 83
WORK OFFSET ADJUSTMENTS 83
Mixing Units of Measurement 83

TABLE OF CONTENTS

Height of Part 84
EXT WORK OFFSET 85
  EXT or COM ? 85
  Offset Adjustment Options 85
WORK OFFSETS - DATA INPUT 86
  Data Change Example 86

Chapter 10
TOOL LENGTH OFFSET 87

TOOL LENGTH 87
  Tool Length Offset Register 87
TOOL LENGTH OFFSET COMMANDS 87
  Limitations of Work Offsets 88
  G43 and G44 Commands 88
  Applying Tool Length Offset 89
  G49 Command 89
OFFSET MEMORY TYPE 89
  Offset Memory - Type A 90
  Offset Memory - Type B 90
  Offset Memory - Type C 90
  INPUT and +INPUT Key Selection 90
  Adjustment Examples 91
MACHINE GEOMETRY 92
SETTING THE TOOL LENGTH 92
  Tool Assembly Fixture 93
  Preset Tool Method 93
  Touch-Off Method 95
  Reference Tool Method 97
EXT OFFSET AND WORK OFFSET 100
WORK AND TOOL LENGTH OFFSET 100

Chapter 11
MACHINING A PART 101

MACHINING OBJECTIVES 101
PROCESS OF MACHINING 101
DRAWING EVALUATION 101
  Tolerances 101
  Radius from I and J 102
  Surface Finish 103
  Stock Allowance 103
MATERIAL INSPECTION 103
PROGRAM EVALUATION 103
  Programs Generated Manually 104
  Programs Generated by Software 106
  Part Setup and Tooling 106
PROGRAM INPUT 106
  Memory Mode 106
  DNC Mode 106
PROGRAM VERIFICATION 106

- Manual Verification 107
- Software Verification 107

PRODUCTION MACHINING 107
- Responsibilities 107

PART INSPECTION 108

PROGRAM SOURCES 108
- CAM Programs and Machining 108
- Conversational and Macro Programming 108

PREVENTING A SCRAP 109
- Causes of Scrap 109
- Steps to Eliminate Scrap 109
- Let Offsets Work for You 109
- Offset Adjustment - Example 110

PROGRAM MODIFICATION 111

MAINTAINING SURFACE FINISH 111
- Surface Finish in a Drawing 111
- Effect of Cutting Tools 111
- Selection of Cutting Data 112
- Machining Direction 112
- Chatter / Vibration 113

COOLANTS 113
- Benefits of Coolants 114
- Working with Coolant 114

Chapter 12
MACHINING HOLES 115

FIXED CYCLES - OVERVIEW 115

DRILLING HOLES 115
- Drilling Example 116
- Terminology 116
- Initial Level and R-Level 117
- Depth 117
- Spot Drilling + Counterbore + Countersink 117
- Peck Drilling 118
- Changing Q-depth 119

TAPPING OPERATIONS 120
- Tap Drill 120
- Feedrate Calculation 120
- Troubleshooting 121
- Rigid Tapping 122

BORING OPERATIONS 123
- G85 Cycle 123
- G86 Cycle 123
- G87 Cycle 123
- G88 Cycle 123
- G89 Cycle 124
- G76 Cycle 124

DRILLING FORMULAS 125
- Center Drills 125
- Spot Drills 126

Twist Drills 126

Chapter 13
MULTIPART SETUP 127
MULTIPLE PARTS 127
Setup Conditions 127
ORIGIN DISTANCES UNKNOWN 127
Program Data 128
Setup Process 129
ORIGIN DISTANCES KNOWN 129
Program Data 130
Using Work Offsets 130
Using Local Coordinate System 130
Using G10 Setting 130
Using Custom Macro 131
EXTENDED WORK OFFSETS 132
Available Range 132
Standard vs. Extended Work Offset Range 132
Z-AXIS AND MULTIPART SETUP 132
Variable Height 132
Tool Length Offsets 132
PROGRESSIVE SETUP 134
SUMMARY 134

Chapter 14
CUTTER RADIUS OFFSET 135
BENEFITS 135
PROGRAM DATA 135
Address D 135
Offset Memory Types 135
Equidistant Toolpath 136
Cutting Direction 136
WORKING WITH RADIUS OFFSET 137
D-offset Number 137
Look-Ahead Type Offset 138
INITIAL CONSIDERATIONS 139
Cutter Diameter 139
Effects of Machining 140
Scrap Prevention 140
ACTUAL OFFSET SETTING 140
Programmed Tool Size 141
External Offset Settings 141
Internal Offset Settings 142
Conclusion 142
USING OFFSET MEMORY TYPES 142
Geometry and Wear Offsets 142
Offset Memory - Type A 142
Offset Memory - Type B 143
Offset Memory - Type C 144

PROGRAMS FROM COMPUTER	144
Mastercam Example	144
SOLVING PROBLEMS	145
Alarm Number 041	145
Insufficient Clearance	146
Cutter Radius Too Large	147
Contour Start / End Error	148
CHANGE of TOOL DIAMETER	149
Units Conversion	150

Chapter 15
TOOLS FOR CNC LATHES ... 151

PROGRAMMMING LATHE TOOLS	151
Zero Wear Offset	151
Feedrate Error	151
Prevention	152
TOOL NUMBERS	152
TOOL TURRET	152
TOOL HOLDERS	153
Standards	153
Identification	153
ISO and ANSI	154
Gage Insert	154
Other Tool Holders	154
CUTTING INSERTS	156
Insert Shape Identification	156
ISO Unique Identification	157
ANSI Unique Identification	158
Grades, Coating and Chip Breakers	158
CHANGING AN INSERT	158
Corner Radius Change	158
TOOL CHANGING	159
LATHE CHUCK	160
Chuck Clamp and Unclamp	160
Chuck Pressure	160
SELECTION OF JAWS	160
Hard Jaws	160
Soft Jaws	160

Chapter 16
LATHE OFFSETS ... 161

Lathe Geometry	161
Machine Zero	161
Part Zero	162
Tool Reference Point	162
WORK SETUP	163
Definitions	163
Geometry Offset	163
The R and T Columns	164

TABLE OF CONTENTS

Wear Offset 165
DRAWING AND PROGRAM 165
Changing Tool Numbers 166
Bar Extension 166
MANAGING TOLERANCES 166
Tolerances and Program Data 167
Tolerances on Diameters - Part A 167
Tolerances on Shoulders - Part B 169
Tolerances on Diameters and Shoulders 170
WORKING WITH WEAR OFFSETS 171
Common Offset Change Example 171
WEAR OFFSET PARAMETERS 172

Chapter 17
TURNING AND BORING 173
LATHE CYCLES G70 - G71 - G72 173
Concepts of Turning and Boring Cycles 174
Program Listing 175
Contour Shape - External 176
G71 Cycle Format for Two Blocks 177
Contour Shape - Internal 177
PROGRAM FEATURES TO WATCH 178
Start Point Location 178
Clearances 179
Tool Change Clearance 180
Program after Changes 181
PROGRAM DATA TO WATCH 182
Start Point Too High or Too Low 182
Positive and Negative Stock Allowance 182
W-Stock Amount Too Heavy 182
Return to Start Point in the Cycle 183
P and/or Q Reference Defined Incorrectly 183
Feedrates for Roughing and Finishing 184
Cutter Radius Offset 184
Small Inside Radius 184
G72 AND G73 TURNING CYCLES 185
G72 Lathe Cycle 185
G72 Cycle Format for Two Blocks 185
G73 Pattern Repeating Cycle 186
G73 Cycle Format For Two Blocks 186
ONE-BLOCK FORMAT CYCLES 186
G71 Cycle Format for One Block 186
G72 Cycle Format for One Block 187
G73 Cycle Format for One Block 187
G71 TYPE i AND TYPE II CYCLES 187
Type I 187
Type II 187
SUMMARY 188

Chapter 18
TOOL NOSE RADIUS OFFSET ... 189
BRIEF REVIEW ... 189
TOOLPATH EVALUATION ... 189
Program Listing - No Offset ... 190
Program Listing - With Offset ... 191
Offset Screen ... 191
FACING TO CENTERLINE ... 192
Tool Height Error ... 192
Incorrect Tool Position ... 192
G41 - Yes or No? ... 193
OFFSET CHANGE ... 193
CLEARANCE AMOUNT ... 194

Chapter 19
GROOVING ON LATHES ... 195
GROOVE MACHINING ... 195
The Simplest Groove ... 195
Standard Grooves ... 195
Typical Groove ... 196
Grooving Tool ... 196
Grooving Insert ... 196
Control Settings ... 197
Program Evaluation ... 197
Basic Groove Dimensions ... 198
TOLERANCES NOT SPECIFIED ... 198
Dimensional Problems ... 199
DEPTH CONTROL ... 199
External and Internal Grooves ... 200
POSITION CONTROL ... 201
Left Side Reference Point ... 201
Right Side Reference Point ... 201
Reference Point Set in the Middle ... 201
Offset Adjustment ... 201
WIDTH CONTROL ... 203
Program Requirements ... 203
POSITION AND WIDTH CONTROL ... 204
GEOMETRY OFFSET ... 204
Change of Tool Width ... 204
Program Requirements - Summary ... 205
GROOVING CYCLES ... 205
G74 Cycle ... 205
G75 Cycle ... 205

Chapter 20
THREAD CUTTING ... 207
THREADING PROCESS ... 207
THREADING TOOLS ... 207
Threading Tool Holders and Inserts ... 207

TABLE OF CONTENTS

    Hand of Tool Selection 208
THREADING INFEED 210
    Radial Infeed 210
    Flank Infeed 210
    Modified Flank Infeed 210
    Alternating Flank Infeed 211
THREADING INSERT 211
    Full Profile 211
    Partial Profile 211
    Clearances 212
    Multitooth Inserts 212
THREAD RELATED DEFINITIONS 212
    Pitch 212
    Lead 212
    Other Definitions 212
METHODS OF PROGRAMMING 213
    Longhand Threading - G32 213
    Box Threading - G92 213
    Repetitive Cycle Threading - G76 214
THREADING DATA 214
    Depth of Thread 214
    Depth of the First Pass 214
    Threading Feedrate 214
    Angle of Thread 214
    Spring Pass 214
G76 THREAD CUTTING CYCLE 215
    One-Block Format 215
    Two-Block Format 215
    G76 Cycle Examples 216
MULTI START THREADS 218
SPINDLE SPEED LIMITATION 218
    Machine Specifications 218
THREADING DEFAULTS 218

Chapter 21
WORKING WITH BLOCK SKIP 219
BLOCK SKIP FUNCTION 219
    Block Skip Switch 219
    Using Block Skip 220
    Block Skip within a Block 220
    Variable Stock 220
TRIAL CUTS 221
    Maintaining Tolerances 221
    Trial Cut Concepts 221
    Trial Cut Methods 221
MILLING APPLICATIONS 222
    Program Sample 222
    Program Evaluation 222
    Finishing Tool - Program Details 223
    Conclusion 223

TURNING APPLICATIONS 223
Test Cut Addition 224

Chapter 22
OFFSET CHANGE BY PROGRAM 225
LOCAL COORDINATE SYSTEM 225
Work Offset Change 226
DATUM SHIFT 226
G10 Format - Milling Applications 226
G10 For Roughing And Finishing 229
G10 Format - Lathe Applications 230
SETTING PARAMETERS BY G10 232

Chapter 23
SYSTEM PARAMETERS 233
WHAT ARE PARAMETERS? 233
Parameters for Operators 233
Fanuc Parameter Classification 233
Access to Parameters 233
Backing Up Parameters 235
PARAMETER TYPES 235
Bit Type Description 236
Practical Interpretation 236
INPUT OF REAL NUMBERS 238
Increment System 239
Units of Input 239
Data Range 239
VALID DATA RANGES 240
PRACTICAL APPLICATIONS 240
PARAMETERS OF SETTING 240
Setting (Handy) 240
PARAMETERS OF FEEDRATE 241
PARAMETERS OF DISPLAY 243
Programs Registration and Editing 243
Password Protection 243
Program Display 244
Automatic Block Numbering 244
Other Parameters 244
PARAMETERS OF PROGRAMS 245
Decimal Point 245
Power-On Defaults 245
End of Program Functions 245
Number of M-Functions in a Block 246
Automatic Corner Rounding 247
PARAMETERS OF SPINDLE CONTROL 247
Maximum Spindle Speed 247
Spindle Orientation Code 247
PARAMETERS OF TOOL OFFSET 248
Offset Address 248

TABLE OF CONTENTS

    Lathe Offsets 248
    Clearing Offsets 248
    Radius vs. Diameter 248
    Wear Offset 249
PARAMETERS OF CYCLES 250
    Fanuc 15 Mode - Compatibility 250
    Spindle Rotation Direction 250
    Clearances in G73 / G83 Cycles 251
PARAMETERS OF THREADING CYCLE 251
PARAMETERS OF MULTIPLE REPETITIVE CYCLES 252
    Available Cycles 252
    Parameters for G71 / G72 Cycles 252
    Parameters for G73 Cycle 253
    Parameters for G74 / G75 Cycles 253
    Parameters for G76 Cycle 253
BATTERY BACKUP 254
PARAMETERS AND G10 254
NO DECIMAL POINT ENTRY 255
    Historical Background 255
    Equivalent Numbers 256
    Input Without a Decimal Point 256
SUMMARY 256

Chapter 24
PROGRAM OPTIMIZATION 257
PROGRAM REVIEW 257
PROGRAM RELATED CHANGES 257
CHANGES AT THE CONTROL 258
    Spindle Speeds and Feedrates 258
    Clearances 258
    Optimized Program 259
    Dwells 259
    Combined Tool Motions 259
SETUP RELATED CHANGES 260
SUMMARY 260

Chapter 25
REFERENCES 261
G-CODES AND M-FUNCTIONS 261
    Milling - G-codes 261
    Milling - M-functions 262
G-CODES AND M-FUNCTIONS 262
    Turning - G-codes 262
    Turning - M-functions 263
DECIMAL EQUIVALENTS 263
PARAMETERS CLASSIFICATIONS 267
COMMON ABBREVIATIONS 269
TAP DRILL SIZES 271
    Metric Threads - Coarse 271

    Metric Threads - Fine 271
    Imperial Threads - UNC/UNF 272
    Straight Pipe Taps - NPS 273
    Taper Pipe Taps - NPT 273
60-DEGREE THREAD FORMS 274
USEFUL FORMULAS 274
    Speeds and Feeds 274
    Machining Holes - Spot Drilling 275
    Tapers 276
    Control Memory Capacity 277
    Trigonometric Chart 278

Index 279

1 CONCEPTS OF CNC MACHINING

CNC PROCESS

Making a certain part (also called a *workpiece*) does not normally start at the CNC machine - it starts much earlier, at the design engineer's desk. Engineering design means developing an intended part that is economical to make, of high quality, as well as a part that does what it is supposed to do - simply, *to design a part that works*. This process takes place in various offices and laboratories, research centers, and other places, including engineer's imagination. Manufacturing process - CNC process included - is always a cooperative effort. Modern part design requires professionals from different disciplines, aided by a powerful computer installed with suitable design software, for example, SolidWorks®, Autodesk Inventor®, and many others, as well the venerable AutoCad® - one of the oldest and still very popular of the design group of application software. In simplified terms, engineering design starts with an idea and ends with the development of a drawing - or a series of drawings - that can be used in manufacturing at various stages.

For the CNC programmer as well as the CNC operator, this engineering drawing is the first source, and often the only source, of information about what the final part is to be. Typically, CNC programmer follows a certain process - or *workflow* - that can be summarized into a several critical points or steps:

- Evaluate drawing
- Identify material of the part
- Determine part holding method
- Select suitable tools
- Decide on cutting conditions
- Write the program
- Verify the program
- Complete documentation
- Send program to machine shop

Keep in mind that this is not always the step-by-step method as it may appear to be. Often, a decision made in one step influences a decision made in another step, which often leads to revisiting earlier stages of the process and making necessary changes.

Drawing Evaluation

This initial stage is very important because it will influence all activities that are necessary to machine a part. Evaluating a drawing means finding a solution to a single question:

What is the best way to machine the part?

This single question will be the foundation for solutions of all other items in the workflow. One of the biggest challenges modern machining faces is not the lack of people who can master a CAD/CAM system but lack of people who know how to machine a part.

> Knowledge of how to machine a part
> is the most important quality of a CNC programmer

An engineering drawing does not offer solutions - it only provides goals and objectives. Studying the features of the part, dimensions, tolerances, various relationships between features, quality requirements, etc., will largely determine the method of machining. All requirements of the drawing have to be met.

A single drawing is often not sufficient to provide all answers. Also having a drawing of a matching part or an assembly drawing, may be necessary in some cases.

Drawing - or even a set of drawings - offers the main source of information for both CNC programmers and CNC operators.

Material Identification

Part is made from a blank stock material specified either in the drawing itself or in another source. For programming and machining purposes, material should be identified by its type, size, shape, and condition.

Type of material is important when selecting tools, setup, and cutting conditions. Soft materials such as brass and aluminum will require a different method of machining than steels and space age materials. Size of the stock provides information about how much material has to be removed by roughing. Stock shape is important primarily for selection of the holding device (fixture).

Material condition may require special cuts, if necessary. Machining a forged steel requires a different approach than machining the same steel type as a round bar.

Shape of material comes in many different forms. Ideally, all material supplied for a particular batch of parts should be the same. That is not always the case; for example, bars that come cut-off into individual pieces should have but may not have the same rough length.

Part Holding

Part holding - or work holding - is a topic that can have its own book. From CNC perspective, the decision a CNC programmer has to make is to select such a work holding method that provides fixed, safe, and stable location for the mounted material.

Tooling Selection

Selecting tools is also part of the programming process, and experienced programmers often discuss the tooling possibilities with CNC operators. Selecting a tool for CNC machining means selection of:

- Cutting tool holder
- Cutting tool

For machining centers, cutting tool holder is the connection between the cutting tool and the machine spindle. For lathes, the tool holder is mounted in the turret.

Cutting Conditions

What is often called 'speeds and feeds' is only part of overall cutting conditions. Cutting conditions are influenced by many fixed factors, such as material being machined, its shape and condition, machine capabilities, etc. They are also influenced by conditions that are with the power of CNC programmer - for example, tooling selection, setup method, depth of cut, width of cut, and spindle speeds and cutting feedrates.

Program Writing

Part program can be developed by several methods:

- Manual programming
- Macro programming
- Computer programming
- Conversational programming

Manual programming means manual calculations and manual writing of the program. A computer is often used, but only as a text editor or a toolpath simulator. In manual programming, the computer does not generate the program code.

Enhanced methods of manual programming use so called macros, such as a *Fanuc Custom Macro B*. In simple terms, a macro is a type of a subprogram, but it can handle features that no subprogram can. Those features include variable data, conditional testing (**IF**), iteration (**WHILE**), arithmetic, algebraic and trigonometric calculations, and many other features.

In a true computer program development, a special software is used to generate complete program. Typical software used for such purpose includes Mastercam®, Edgecam®, and many others.

Programming directly at the machine control is called *conversation programming*. Generally, lathe programming can benefit from conversational programming a little more than programs for milling.

Regardless of how the program is developed, it has to be written in such a format that the controls system of the CNC machine 'understands' it. Part programs generated manually or by software should have the same format (some minor inconsistencies should be expected).

Program Verification

A program that contains even a single error is not desirable. One of the programmer's responsibilities is to check the completed program before it is used. There are several ways of verifying the program. One proven method is to use a toolpath simulator software, such as NCPlot®, another is to employ a fairly extensive variety of manual checks.

Simulation software varies quite a bit in features and cost. Price of the software by itself does not necessarily reflect its quality and features. Most simulation software lack support for some high end control features, such as coordinate rotation, polar coordinates and macros. Before purchasing software that simulates the toolpath, make sure to do some research.

Numerous manual checks can also be used to find program errors. Even a brief scan of the program may reveal some obvious errors. If you know what you are looking for, the program check is that much faster. Many errors in the program happen at the beginning of the program. Errors in the middle of a program are of a different kind. An experienced operator will discover an error before it does any damage.

Program Documentation

The final part program is the result of many small, progressive steps, some quite straightforward, others more complex. The program can be generated by a CAM software or it can be written completely manually. In any case, the programmer had to go through certain steps that required calculations, for example, as well as other steps. Documenting these calculations or processes can prove to be very valuable if the program has be modified for any reasons. Program modification can range from a simple error correction to an engineering change.

It is an unpleasant reality, that many CNC programs, regardless of how they were developed, lack any background information that can help the machine operator. At best, the part program may include some basic data regarding the setup and even some special instructions. What is often missing are details of individual steps. The operator needs to know what fixture has been used, how the part is oriented, what tools have been selected, and where the part X0Y0Z0 is located.

Program Transfer

Program transfer includes several methods of making the program available for machining. Commonly, this process is called *Program Loading*. There are different methods of loading program - the most common is to store it in the memory of the CNC unit. Other methods include DNC (processing the program from an external computer), or a program stored on a flash drive.

WORK COMPLETION

Once the drawing and program reach the machine shop, it is up to the CNC operator to continue with actual production. Production cannot start right away, and certain workflow is followed at the machine as well:

- Evaluate the part program
- Check supplied material
- Prepare required tools
- Setup and register tools
- Setup part in a fixture
- Load program
- Set various offsets
- Run first part
- Optimize program if necessary
- Run production
- Inspect frequently

Program Evaluation

Evaluating a part program serves the purpose of knowing what it will do and how it will do it. Careful evaluation provides means of organizing the part setup, tool preparation, and relates activities in an efficient manner. Program evaluation is typically combined with the drawing and material provided.

Material Check

Ideally, any material used for CNC work should be qualified as acceptable. Qualified material guarantees consistency in size, shape and condition from one blank piece to another. Unfortunately, this 'ideal' condition does not always exist, for different reasons.

A prudent CNC programmer or operator will make a check of at least a sample of the material delivered, and identify the differences, if any. Always watch for material inconsistency, not only in size, but also in shape and type. Harder material requires different cutting conditions than a softer material. Such a situation is common when getting material for the same part from two different suppliers.

Tooling Preparation

Tooling generally refers to cutting tools. For CNC machines, cutting tools have two parts - the holder and the cutting tool itself. The tool holder is the part of the assembly that allows the cutting tool to be mounted in the spindle (milling) or in the turret (turning).

Tool holders are standard, and from the setup point of view, the taper that is to be located in the spindle should always be clean. The same applies to the internal area of the spindle, where the holder will be located.

Tooling Setup

Setting up any CNC machine is a multi-step process. Some of these steps have to be performed in a particular order, others have a certain degree of flexibility. Most of the steps are common to all CNC machines, others may vary from one machine manufacturer to another. The first step is the selection of tools. Tools should always be set using an external tool fixture, never in the machine spindle. Use tools that are sharp, preferably new or at least lightly used. If possible, keep frequently used tools in the magazine all the time. This may require cooperation with the programmer, but shortens the unproductive setup time, often significantly.

Fixture Setup

Fixturing is part of work holding. This is the engineering area that studies and develops the most suitable methods of holding a part for safe and consistent machining. The main purpose of a fixture is to safely hold all parts of the batch in a fixed location.

Program Loading

A part program is typically developed off-machine, either manually or with the assistance of computer software. In order to use such a program, it has be made available to the CNC system. There are two methods to achieve this objective:

- Load the program to the control system
- Run the program from a remote computer

The very early method was to use a paper tape for running the program, but this method is now considered obsolete. Loading the completed program to the system memory can be done via control panel keyboard or via a cable from a remote computer. Both methods will result in the program loaded into the CNC memory. Once the program is loaded, it will run in the memory mode. Programs stored in the memory can also be edited there. Keep in mind that the control system memory has limited capacity and is not designed to store all your programs. Viewing an appropriate screen will show the remaining capacity.

Setting Offsets

Offsets are adjustments between fixed positions and actual positions. Although offsets can be set through the program, they are normally set at the control by the CNC operator. The reason why a program cannot be used for all offset setup is simple - the programmer has no way of knowing what the dimensions will be during actual part setup. It is quite normal to have different offsets the next time the same part is machined. There are three groups of offsets (adjustments) available:

- Work offset
- Tool length offset
- Cutter radius offset

The word offset is synonymous with the word compensation. It is the are of offsets that the operator's skills are most required. Each offset group has its own chapter in this handbook.

First Part Run

A lot can be said about running the first part of a given batch. When a certain job is assigned to the CNC machine, production people in the company decide (among other things) on the required output - the number of parts to be machined together in one setup. This number is often called the 'run', the 'order', the 'batch', and may even include some unique local terms as well.

Program Optimization

During the first part run, or even during the run of the next few parts, some program adjustments may be necessary - they generally belong to the features of the program that can be easily edited, such as spindle speeds, cutting feedrates, clearances, etc. program optimization make the current program run more efficiently within the given setup.

Production Run

When the optimization process is completed, the production run is more than just loading, running, and unloading parts. The operator has to monitor the progress, check dimensions, check tools, replace tools if necessary, adjust offsets, watch the coolant, deburr parts, and do many other tasks.

Part Inspection

Inspecting a finished part is one of standard manufacturing processes. Depending on the job, not every part in the batch has be to be inspected, but a certain number of parts in any batch are inspected as a norm. CNC operator may be called upon part inspection as part of everyday duties, or a QI (Quality Improvement) department may be responsible.

Safety Issues

In all stages of the machining process, safety should always be paramount.

> **Always obey all safety rules**

Study individual chapters in this handbook. Many of the items briefly described here are presented in great detail, including examples and applications.

2 CNC MACHINE SPECIFICATIONS

All CNC machine tool manufacturers provide very comprehensive information about each of their products they ship to their customers, mainly distributors, dealers, service providers, and even to the end users. Typically, this information is a collection of various data related to the particular CNC machine tool.

Machine specifications contain data relating to important dimensions, capacities, ranges, restrictions and limitations of the machine tool, *not* the CNC system. These specifications come in many versions, each being focused to a different purpose of usage. They can be in digital form stored on a disk, or in a reference book that is shipped with the CNC machine.

Most commonly available methods of machine specifications data can be summed up into several distinct categories:

- **Advertising material**
- **Promotional brochures - overview**
- **Promotional brochures - comprehensive**
- **Sales oriented information**
- **Service oriented information**
- **Machine tool manuals**
- **Control manuals**
- **Parameters manuals**
- **Training material with examples**
- **Information specific to the CNC machine**

How does all this information that may or may not be included in such diverse groups assist the CNC programmer during programming or the CNC operator during part setup and machining? The answer is *'selectivity'*.

Both CNC programmers and operators need only a very small portion of the overall information in the documentation provided by the control or machine manufacturers. This chapter tries to identify what data and other information are important to the CNC operator working at the machine, and to some extent, to the CNC programmer as well.

No specific machine specifications have been used. All information has been adapted from actual machine tools and is representative of the important features that directly affect CNC programming and CNC machining.

MACHINE SPECIFICATIONS

Every manufacturer of CNC machines provides the potential buyer with descriptions of their products, for three reasons - sales, information, and training. Sales definitely dominate. Machine specifications are a collection of data regarding a particular machine tool, without the control system. Control unit specifications, on the other hand, provide information about the control itself.

Useful Information

The following sections will concentrate on the three major groups of metal-cutting CNC machine tools:

- **Vertical Machining Centers**   VMC
- **Horizontal Machining Centers**   HMC
- **Lathes and Turning Centers**

Each group will cover specifications that are common to all machines, such as maximum and minimum axis motion or the range of spindle speeds. Each group will also cover specifications that are unique to that particular group of machines, such as indexing axis for horizontal machining centers.

In all examples, only those features that directly affect the work of CNC programmer and/or CNC machine operator will be described.

To make the specifications easier to read, the data will be collected in the form of tables.

VERTICAL MACHINING CENTER

This group of machining centers is often identified by the abbreviation **VMC** - *Vertical Machining Center*. VMC is the most widely used machine group in industry and many models, sizes and configurations are available to the end users.

Vertical machining centers are designed mainly for machining as the XY motions at a given depth. This is called 2-1/2 axis machining. Most of these machines also allow simultaneous XYZ motions. This feature is called 3-axis machining. Because of their design, cutting on sides is limited, unless special attachments are used.

Typical Specifications - VMC

	Metric	Imperial
Table length	900 mm	35.4 in
Table width	480 mm	18.9 in
Maximum table load	900 Kg	1984 lb
X-axis motion travel	750 mm	29.5 in
Y-axis motion travel	410 mm	16.1 in
Z-axis motion travel	490 mm	19.3 in
X-axis rapid rate Y-axis rapid rate	35 000 mm/min	1378.0 in/min
Z-axis rapid rate	30 000 mm/min	1181.0 in/min
Maximum cutting feedrate	15 000 mm/min	590.5 in/min
Spindle taper	BT-40	
Spindle speed	200 - 10 000 r/min	
Spindle motor (maximum)	15 kW	20 HP
Tool magazine capacity	20	
Maximum tool diameter (normal)	100 mm	3.94 in.
Maximum tool diameter (adjacent pockets empty)	130 mm	5.12 in.
Maximum tool length	250 mm	9.8 in.
Maximum tool weight	6 Kg	13.2 lb
Tool change time (chip-to-chip)	3 seconds	

For the CNC programmer, all features are equally important. For the CNC operator, some features are more important than others.

HORIZONTAL MACHINING CENTER

This group of machining centers is often identified by the abbreviation **HMC** - *Horizontal Machining Center*.

Typical Specifications - HMC

	Metric	Imperial
Table size	400 x 400 mm	15.7x15.7 in
Number of pallets	2	
Maximum table load	250 Kg	551 lb
X-axis motion travel	500 mm	19.7 in
Y-axis motion travel	500 mm	19.7 in
Z-axis motion travel	500 mm	19.7 in
X-axis rapid rate Y-axis rapid rate Z-axis rapid rate	35 000 mm/min	1378.0 in/min
Indexing increment	1° (0.001° optional)	
Pallet changing time	7 seconds	
Maximum cutting feedrate	1 - 20 000 mm/min	0.1 - 787.0 in/min
Spindle taper	BT-40	
Spindle speed	40 - 12 000 r/min	
Spindle motor (maximum)	15 kW	20 HP
Tool magazine capacity	30	
Maximum tool diameter	90 mm	3.5 in
Maximum tool diameter (adjacent pockets empty)	140 mm	5.5 in
Maximum tool length	275 mm	10.8 in
Maximum tool weight	7 Kg	15.4 lb
Tool change time (chip-to-chip)	3 seconds	

LATHES AND TURNING CENTERS

Only basic data for a 2-axes CNC lathe are included.

Typical Specifications

While the traditional and standard workhorse two-axis CNC lathe has been greatly improved by additional features, such as extra axes, sub-spindle, milling attachments, live tools, etc., the most important specifications still apply, regardless of the innovations.

	Metric	Imperial
Swing over bed	⌀ 500 mm	⌀ 19.7 in
Swing over cross-slides	⌀ 245 mm	⌀ 9.6 in
Normal turning diameter	⌀ 210 mm	⌀ 8.3 in
Maximum turning diameter	⌀ 280 mm	⌀ 11.0 in
Maximum turning length	485 mm	19.0 in
X-axis motion travel (stroke)	210 mm	8.3 in
Z-axis motion travel (stroke)	470 mm	18.5 in
X-axis rapid rate / Z-axis rapid rate	30 000 mm/min	1181 in/min
Chuck size	⌀ 200 mm	⌀ 8.0 in
Bar capacity	⌀ 50 mm	⌀ 2.0 in
Tool shank (external)	25.4 mm	1.0 in
Maximum cutting feedrate	1 265 mm/min	50.0 in/min
Maximum spindle speed	4 500 r/min	
Spindle motor (maximum)	15 kW	20 HP
Number of tool	12	
Tailstock travel	445 mm	17.5 in
Chip conveyor	Yes	

FANUC SYSTEM SPECIFICATIONS

The following tables contain typical control specifications of Fanuc CNC system and are based on the *0i, 21i* and *18i* models. Keep in mind that control specifications generally show the maximum capabilities which are not always present for a particular machine tool.

Most of the specifications listed are common to both milling and turning controls, although each control type has features specific to the particular machine tool.

Use the tables as general reference only and consult machine documentation for exact specifications.

Controlled Axes

Programmable axes	3 - X ,Y, Z 4 - X, Y, Z, B 2 - X, Z (turning)
Simultaneously controlled axes	3 or 4
Least input increment	0.001mm 0.0001 inch 0.001 degree
Max. command value	±99999.999 mm ±9999.9999 in
Fine acceleration & deceleration control	Standard
Inch-Metric conversion	G20 (imperial) G21 (metric)
Interlock	All axes
Machine lock	All axes / Z-axis
Emergency Stop	Standard
Overt ravel	Standard
Stored stroke check 1	Standard
Mirror image (setting)	Each axis
Programmable mirror Image XY axes	M-functions
Backlash compensation	Standard
Pitch error compensation	Standard

Operation Functions

Automatic operation	Memory
MDI operation	Standard
DNC operation	Reader-Puncher interface required
DNC operation with memory card	PCMCIA card attachment required
Program number search	Standard
Sequence number search	Standard
Buffer register	Standard
Dry run	Standard
Single block	Standard
JOG feed	Standard
Manual reference position return	Standard
Manual handle feedrate	x1, x10, x100
Z Axis neglect	Standard

Interpolation Functions

Positioning	G00
Exact Stop Mode	G61
Exact Stop	G09
Linear Interpolation	G01
Circular interpolation	G02 (CW) - G03 (CCW) (multi-quadrant standard)
Dwell function	G04
Skip function	G31
Reference position return	G28
Reference position return check	G27
2nd reference position return	G30
Polar coordination interpolation	Optional
Cylindrical interpolation	Optional
Helical interpolation	Optional
Thread cutting	G32, G90, G76 (turning only)

Feedrate Functions

Rapid traverse rate	Standard
Rapid override	F0, 25%, 50%, 100%
Feedrate per minute	G94
Cutting feedrate clamp	Standard
Automatic acceleration and deceleration	Rapid traverse = linear Cutting feed = exponential
Feedrate override	0-150% or 0-200% (10% incr.)
Jog override	0-100%
Feedhold	Standard

Spindle Functions

Auxiliary function lock	Standard
High speed MST interface	Standard
Spindle speed function	Standard
Spindle override	50 - 120%
Spindle orientation	M19
M-code function	3 digits
S-code function	5 digits
T-code function	3 digits
Rigid tapping	Standard or Optional

Tool Functions

Tool offset pairs	±6-digit 32 pairs
Tool length offset	G43, G44, G49
Cutter radius offset Type C	G40, G41, G42
Maximum number of tool offsets	400 sets (0i *and* 21i) 999 sets (18i)
Direct input of offset value	Standard

Program Input

EIA- ISO automatic recognition	Standard
Label skip and Parity check	Standard
Control In / Out (comments)	( )
Optional block skip	/
Maximum dimensional input	±8 digits
Program number	O (4 digits)
Sequence number	N (5 digits)
Absolute programming Incremental programming	G90 (XZ for turning) G91 (UW for turning)
Decimal point input	Standard
Calculator type input	Available
Plane selection	G17, G18, G19
Rotary axis designation	Standard
Coordinate system setting	Standard
Auto coordinate system setting	Standard
Work coordinate system	G52, G53, G54-G59
Manual absolute ON and OFF	Standard
Subprogram call	M98 - 4 nested levels
Circular interpolation with R	Standard
Program format	FANUC
Program stop functions Program end functions	M00 - M01 M02 - M30

Reset	Standard
Programmable data input	G10, G11
Custom Macro B	G65 (option)
Fixed cycles	G73, G74, G76 G80 - G89 G98, G99
Multiple repetitive cycles (turning controls only)	G70, G71, G72, G73, G74, G75, G76
Corner rounding (turning only)	Standard

Editing Functions

Program storage capacity (max.)	640 m (0i / 21i) 1280m (18i)
Maximum number of registered programs	400 (0i / 21i) 1000 (18i)
Program editing	Standard
Program protection	Standard
Background editing	Standard

Setting and Display Functions

Status and current position	Standard
Program name	31 characters max.
Parameter setting and display	Standard
Self diagnosis function	Standard
Alarm and alarm history	Standard
Operation history display	Standard
Help function	Standard
Run time and parts count	Standard
Actual cutting feedrate display	Standard
Display of S and T code	Standard (all screens)
Servo setting screen	Standard
Multi-language display	English (default)

Spindle speed indicator	Standard
Spindle load indicator	Standard
Data protection key	Standard
Graphic function	Standard
Clock function	Standard
Dynamic graphic display	Standard
Display unit - Color LCD	8.4" (0i) 10.4" (21i / 18i)

Increment system multiplied by 1/10	0.0001 mm 0.00001 inch
Scaling function	G50, G51
Small diameter peck drilling cycle	
Smooth interpolation	
Circular threading	
3D coordinate conversion	
3D cutter radius offset	
Load monitoring	
Position compensation (for backward compatibility only)	G45-G48
Tool retract and recover	

Data Input / Output

Reader / Puncher Interface	RS-232 interface
Memory card interface	Standard
External part number search	9999

Options

Only some software options are listed in the table:

Additional work coordinate systems	48 sets or 300 sets
Extra custom macro common variables	up to 999
Tool life management	
Additional tool life management sets	512
Automatic corner override	
Automatic corner deceleration	
Coordinate system rotation	
Feedrate clamp by arc radius	
Hypothetical axis interpolation	
Custom macro interruption	
Jerk control	
Polar coordinate interpolation	
Program restart	

3 PROGRAM INTERPRETATION

When working with CNC machining centers or CNC lathes programs, the machine operator monitors not only the numerous tool motions, but also the actual program flow. This is especially true - *and a very important observation* - for the beginning of the actual production run. During full production, the CNC operator will frequently check the program as it is displayed on the control monitor (screen), to keep an eye on the machining process. This initial evaluation usually takes place at the control system of the machine during part setup. Many machine shops also issue a printed copy of the part program to make the monitoring a bit easier, particularly for programs that are only a page or two.

Regardless of the extent a CNC operator is involved with a particular part program (on *screen* or on *paper*), it is important that he or she fully understands its contents - what *exact information* does the program really provide? There is a lot more information provided even in a short program than a brief simple look can absorb. The first step in the process of understanding the program itself is to understand the way it was written - to understand the *program structure*.

PROGRAM STRUCTURE

A CNC program can be written in so many different ways that is makes it impossible to find any degree of consistency between individual part programmers, even within the same company. Operators can greatly influence any programming style, based on their experience of using many programs to machine various parts. Operators can also communicate their preferences to the CNC programmer, who should use such information in the best way to improve program structure.

Logical Structure

Every part program has a single and very specific purpose that will always be the same - *to machine a part*. The way it is written has to be logical in many ways, some of them quite obvious. For example, the order of tools in the program is the order of tools for actual machining. If coolant is required for a certain operation or for the whole part, it is logical to include coolant ON function (usually M08) in the program. Machining order must also be logical - roughing will always be done before finishing, heavy operations before light operations, etc. All these structural features are generally followed by all CNC programmers, otherwise their programs would not work.

The area of the 'not-so-obvious' program structure flaws can be quite large. Poor program structure will increase lead time and also can make a life of the operator a bit unpleasant or even difficult. These flaws of structure belong to the logical program features and mainly relate to G-codes and M-functions, particularly to their application within a program.

One very common break of logical structure is using the right G-codes in the wrong place. Typical example are G-codes used *before* tool change, for example:

```
N1 G21 G17 G40 G80 G90 G54 G00 G49
N2 T01 M06
N3 ...
```

Although technically correct, this structure has several *potential* problems. One is having **G21** (or **G20**) in the same block with other G-codes, which is not allowed on some controls, especially those that can convert metric and imperial mode between each other. That may not be important for one machine, but for another, it may be very important. Placing **G20**/**G21** in a block by itself will make the program available to more similar machines and prevents problems on those machines that require it.

Is G20/G21 Really Needed ?

Some programmers do not use **G20** or **G21** unit selection in the program at all. They often based their reasoning on the fact that *only* metric jobs or *only* jobs in imperial units are ever done on that machine. While that may be true today, it may not be true tomorrow. Programmers who do *not* use the proper G-code for units selection count on the following control feature:

> **Units set *before* machine shut-off remain in effect *after* machine power-on**

In absolute terms, command **G20** or **G21** is *not* needed in the program, providing the units are set at the control. However, including the units selection in the program will *guarantee* that that only those units will be in effect - now or in the future. There is no practical difference between units set through the program and units set at the control.

This observation slightly changes the earlier example:

```
N1 G21
N2 G17 G40 G80 G90 G54 G00 G49
N3 T01 M06
N4 ...
```

By isolating the **G20** or **G21** command, the program is one step closer to being portable between machines with similar control systems.

G17 is Good for ... ?

Preparatory command **G17** is often part of the program initial blocks, and surprisingly many CNC programmers (and operators) have no idea why it is there in the first place or what it is for.

By definition, **G17** is the first of three similar preparatory commands - each define a unique working plane:

- G17 - XY plane
- G18 - ZX plane
- G19 - YZ plane

Plane selection is important only in two major areas:

- Circular interpolation   ... G02 and G03
- Cutter radius offset   ... G41 and G42

➔ *In circular interpolation, it supplies the MISSING AXIS:*

Consider this simple example ...

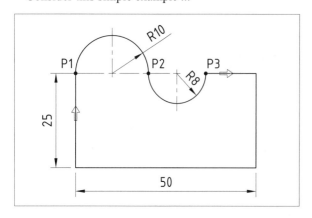

... and a section of the part program:

```
N1 G21
N2 G17 G40 G80 G90 G54 G00 G49
...
N21 G01 X0 Y25.0 F200.0   (P1)
N22 G02 X20.0 I20.0 J0    (P2)
N23 G03 X36.0 I8.0 J0     (P3)
N24 G01 ...
```

Note that in block N21, the linear motion ends and arc motion begins (P1). Both X and Y axes are programmed in this block. The following two blocks are both circular motions. As per drawing, the Y-axis is identical for both start points P1 and P2. Since axis commands are modal, they do not have to be repeated in every block. That accounts for only X being programmed for both arcs.

Considering that arcs always require two axes, and also considering that they can be machined in ZX and YZ planes, the control system has to replace the axis that is *not* included. This is where the plane selection command comes in. In the example, **G17** in block N2 provides the missing axis as the Y-axis. Control system will interpret the program as:

```
N1 G21
N2 G17 G40 G80 G90 G54 G00 G49
...
N21 G01 X0 Y25.0 F200.0    (P1)
N22 G02 X20.0 Y25.0 I20.0 J0 (P2)
N23 G03 X36.0 Y25.0 I8.0 J0  (P3)
N24 G01 ...
```

➔ *In cutter radius offset mode, it supplies the DIRECTION:*

Cutter radius offset is programmed by two preparatory commands:

- G41   ... cutter radius offset to the *LEFT*
- G42   ... cutter radius offset to the *RIGHT*

The *left* and *right* descriptions indicate position of the cutter when viewed into the cutting direction. As this direction is always relative to the three available planes, the correct plane selection command has to be included in the program.

Initial Settings

In addition to the plane selection command, such as **G17**, the initial block N2 contains several other G-codes:

`N2 G17 G40 G80 G90 G54 G00 G49`

Note that all are included *before* tool change, which is an important consideration.

Many CNC programmers include these G-codes at the program beginning, mainly for two reasons:

- **Cancellation of commands that may be in effect**
- **Initialization of the program mode**

The block that contains these codes is often called the *safety block*. Although well meant, it also presents a sense of false security. There is no guarantee that the program will run smoothly, regardless of this block.

There are three common *cancellations* in block N2:

- G40   ... Cancel cutter radius offset
- G80   ... Cancel fixed cycle
- G49   ... Cancel tool length offset

Are these cancellations necessary? In one word - *no*. They are mainly a leftover from early days of NC and even CNC and have gained a permanency of their own. What do they represent at the program beginning?

The answer uses three words *'just in case'*. Just in case the cutter radius is active when cutting was interrupted; just in case the fixed cycle has not been canceled; just in case the tool length offset is still active. Are these situations possible? They may be in some very extreme cases, but practically, they are not likely. The main reason is that when you press the **RESET** key on the control panel, these cancellations are automatically performed. On the other hand, it does no harm to include **G40** and **G80** in the program, *'just in case'*.

G49 cancellation is different. This command cancels the tool length compensation. There is no reason to program it at all - in fact, by using **G49** in the program, you may create a situation that could create *more* problems than it supposed to solve.

That leaves three other G-codes that need to be addressed. They are the *initiations* commands:

- G90   ... Absolute mode of programming
- G54   ... Work offset
- G00   ... Rapid motion

Again, these are good commands in a bad place:

```
N1 G21
N2 G17 G40 G80 G90 G54 G00 G49
N3 T01 M06
N4 X50.0 Y100.0 S900 M03 T02
N5 G43 Z10.0 H01 M08
... <G90 IS IN EFFECT>
...
N17 G91 X-10.0
N18 Y23.0        (... UNWANTED INTERRUPTION !)
...
```

If the program flows from the beginning to the end without a problem, it may appear that it is a good program in terms of its structure. Now consider that for some reason, the machining has to be interrupted in block N18. Perhaps because the tool may have broken or some other mishap. The typical course of action in this case is to remove the broken tool, replace it, set a new tool length offset and - *start machining from block N4* - this is the block after tool change. The proper holder with the tool is already in the spindle, so it makes sense to start from block N4. Here is where the real problem lies - the control is still in **G91** mode, although it should be in **G90** mode. The same is to be said about **G54** work offset setting and **G00** rapid motion. A change from a possible **G55** command or **G01** command may have serious consequences. A simple solution is to program these codes *after* tool change, not before:

```
N1 G21
N2 G17 G40 G80
N3 T01 M06
N4 G90 G54 G00 X50.0 Y100.0 S900 M03 T02
N5 G43 Z10.0 H01 M08
... <G90 IS IN EFFECT>
...
N17 G91 X-10.0
N18 Y23.0
...
N.. G90
...
```

Note that G90 is (and should be) programmed later in the program, to bring the control system to its 'normal' mode (typical mode).

Tool Change Block

Block N3 makes the actual tool change. It is the block where the *Automatic Tool Change* (ATC) takes place:

```
N3 T01 M06
```

On most machining centers, the T01 is a function that makes the tool magazine to rotate and place tool one (T01) into a special magazine position, to be ready for tool change. T-word in this case does not make any tool change, it only makes the selected tool to be *ready* for tool change. This method shortens the tool change time.

There is nothing wrong with this block in terms of programming, but for CNC operators, it may cause some inconvenience. As both tool call T01 and tool change function **M06** are in the same block, they will be processed simultaneously. Tool search will always be first, followed by the actual tool change. That is not the problem. Think how it will affect running the first part.

During the first part run, the operator generally selects *Dry Run* mode along with *Single Block* mode. Other settings may also be in effect. In single block mode, when the block N3 is activated, tool T01 should be placed into the spindle. What happens when the wrong tool had been registered and is placed into the spindle instead? At best, it forces the operator to remove the tool from the spindle, place it back to the magazine and re-register it. On the other hand, if the commands currently in block N3 were in individual blocks, the operator can look at the magazine a see if the right tool is ready *before* the tool change. An improvement to the program structure is simple:

Instead of **N3 T01 M06**, use:

```
N3 T01            (MAKE TOOL 1 READY FOR ATC)
N4 M06            (MAKE ACTUAL TOOL CHANGE)
```

Even a small program change like this can go a long way. Similar changes may be done in other blocks. Compare the original version with the modified version:

```
(ORIGINAL VERSION)
N1 G21 G17 G40 G80 G90 G54 G00 G49
N2 T01 M06
N3 ...

(MODIFIED VERSION)
N1 G21
N2 G17 G40 G80 T01
N3 M06
N4 G90 G54 G00 X.. Y.. S.. M03 T02
N5 ...
```

Even the **G17** could be placed after the tool change.

This section, as many others in the book, relates to the program development, which is the responsibility of the CNC programmer. The machine operator works with the resulting program - or programs - and has to understand the program contents. Programs reaching the machine shop are often written by more then one programmer, and even if only one programmer is responsible, the final programs may lack one important feature - *consistency*.

Program Structure Consistency

Many small machine shops usually employ only one CNC programmer, or - *more likely* - only one CNC programmer/operator, who is responsible for both programming *and* machine operation. Larger shops and large companies may employ several programmers who do not operate machine tools, as their main responsibility is only program development. In these cases, most CNC operators will agree that programs developed by different programmers *'look and feel'* quite different.

By itself, the difference does not present any major problem, unless is very significant. If those programs are good, they will be usable and present no problems.

In spite of all that, the CNC operator will have no choice but to interpret each program in a different way, depending on who he programmer was. The fact that all programs may be correct is not the issue here - the real issue is the *efficiency* of the CNC operator.

On the other hand, if all programs written by different programmers *'look and feel'* the same, the operator has a much easier - *and more efficient* - approach to program interpretation and part setup. The keyword that describes a program following a certain structure is ***consistency***.

> **Consistency in program development means using the same approach to program development by all programmers**

Typical problems related to consistency are usually found in the program structure *before* and *after* the actual machining. Consider just a small example:

```
N5 G43 Z25.0 H01 M08
```

The initial level was set to 25 mm before machining, for example, using a fixed cycle. When the cycle is finished, the programmer retracted the tool to 10 mm:

```
N18 G80 Z10.0 M09
```

There is absolutely nothing wrong with either block. The intent here is not to look for minute details like this one. The real issue is that the CNC operator make try to find if there was a reason. This particular example may not present the strongest argument for the importance of programmimng consistency, but it illustrates the idea.

PROGRAM INTERPRETATION

The ability of a CNC operator to interpret a part program is one of the most important skills in CNC setup and machining. It is unfortunate that not every part program that reaches the machine tool is flawless, but it does present a very real situation. Programming errors do happen, and in many cases, it is up to the CNC operator to find these errors and 'debug' the program.

> **Program interpretation requires at least some basic knowledge of CNC programming**

As an example, consider the following program listing. You are at the machine, drawing is not available and the CNC display shows this program:

```
O1001 (ELEVATION PLATE)
(T7 - 12.7 MM DIA SPOT DRILL)
N1 G21
N2 G17 G40 G80 T07
N3 M06
N4 G90 G54 G00 X20.0 Y60.0 S900 M03 T10
N5 G43 Z25.0 H07 M08
N6 G99 G82 R2.0 Z-2.6 P250 F200.0
N7 X30.0 Y100.0
N8 X100.0
N9 Y60.0
N10 G80 Z25.0 M09
N11 G28 Z25.0 M05
N12 M01
...
```

The above program is very simple, but it may present a surprise or two.

When evaluating a part program, some internal information is readily available, such as program number (if available), block numbers, messages and comments, and some others. In the example above, it is obvious that the part name is *Elevation Plate*. It is a bit less obvious that the tool used for this operation is a half-inch spot drill, converted to millimeters (12.7 mm = 0.5 inch). This is a common method in North American machine shops, where a metric part uses inch type cutting tools (imperial measures). What else can be gathered from the program?

Units

For any CNC machine setup, it is always important to know - *and understand* - the dimensional units. In an environment where both metric and imperial units are used, there are two methods to find out which one is active - by the program code - **G20** (imperial) or **G21** (metric) sets the units through the program. The other method is to find out directly at the control unit, usually under *Settings*. As the example contains **G21** command (N1), we know the program is in metric units.

Type of Operation

Even without an elaborate tool specification or detailed messages, the CNC operator should be able to find out what type of operation is taking place from the program itself. Prevailing use of **G01**, **G02** and **G03** commands suggests a roughing or finishing of a contour or a pocket. When XYZ axes are used frequently together with **G01** motion, the chances are that a 3D machining is taking place, typically collaborated by the tool used, often a spherical end mill (or similar). Machining holes uses various fixed cycles, typically in the range of **G81** to **G89**, along with **G73**, **G74** and **G76**. Also associated with fixed cycles are retract commands **G98** and **G99** and cycle cancellation command G80. In the sample program, the existence of both **G99** and **G82** (as well as **G80** cancellation) means the type of machining is a *fixed cycle* that requires a pause at the bottom of the hole, such as a spot drill in the example. Spot drilling can be further identified by a rather shallow cutting depth, dwell (P), and the existence of a drill as the next tool in the part program.

Speeds and Feeds

For a safe machining operation, identifying the spindle speeds and cutting feedrates in the program helps to make intelligent decisions about the use of spindle and feedrate overrides, if necessary, particularly for the first part trial. As a CNC operator, keep in mind that even the most experienced and reliable programmer may not be always exact in the area of speeds and feeds. In fact, optimizing speeds and feeds at the machine control is often one of the most important functions of a CNC operator.

The example above has only one spindle speed and one feedrate programmed for the tool. While it is common to have only one spindle speed for the tool, it is very common to have two or even more feedrates for the same tool. Typical example would be in milling, where a plunging feedrate along the Z-axis is lighter than the cutting feedrate in XY axes. The program example shows spindle speed as 900 r/min (S900) and the cutting feedrate as 200 mm/min (F200.0) or about 8 inches per minute.

Not to be missed is the spindle rotational direction. When interpreting a part program, look for miscellaneous functions relating to spindle rotation - **M03** (rotation CW) and **M04** (rotation CCW). There is also **M05** available as the spindle stop function. **M03** is the standard rotation for right hand tools, **M04** is used for left hand tools, for example, to make a left hand tap.

Tool Numbers

Many severe errors at the machine happen because of a wrong tool selection. In the program, the tool number is identified by the address *T*. Its exact meaning is determined by the type of automatic tool changer (ATC). In case of a fixed type tool changer, where the tool returns to the same pocket it came from, the address *T* indicates both the pocket number of the tool changer as well as the tool number. In case of the more common and efficient random type tool changer, the address *T* indicates the next tool to be used.

There are actually *two* T addresses in the above example - one in block N2 and the other in block N4. Block N2 contains the first *T* function - T07. Note that the following block N3 contains the actual tool change function **M06**. Once the tool change had been completed, another *T* address appears (block N4), this time it is T10. To interpret these blocks, read block N2 as 'get tool T07 ready for tool change'. The actual tool change follows in block N3 (**M06**), which caused T07 to be in the spindle. Block N4 should be interpreted as 'get the next tool T10 ready for tool change'. Having the next tool waiting is one of the greatest advantages of the random type tool changer. Calling the next tool (T10) is not required for CNC machines where the tool returns to its original pocket of the tool magazine.

Dwell Time

Programmers place a dwell function (a pause) in strategic places of the program structure, to force a delay in machining, when justified. In addition to optimizing speeds and feeds at the machine, the programmed dwell time is another function to be optimized. Unfortunately, some programmers use a dwell of unnecessarily long duration. Multiplied by a large number of parts, long dwell time can increase the cost per part quite significantly. A qualified CNC operator can change the dwell time to be both functional and efficient. The first requirement is the identification of dwell in the program.

There are two methods of programming a dwell:

- As a single block entry, using the G04 command
- Using the P address in certain fixed cycles

In both cases, the dwell address is followed by the dwell time. Single block entry for dwell can have three basic forms:

G04 X.. (mills and lathes)

... for example, **G04 X0.5** is 0.5 seconds dwell

G04 P.. (mills and lathes)

... for example, **G04 P500** is 500 ms dwell or 0.5 sec.

G04 U.. (lathes only)

... for example, **G04 U0.5** is 0.5 sec. dwell

Milliseconds can also be used with the X and U addresses, but seconds (with a decimal point) cannot be used with the P address. Typical use of the P address is in fixed cycles **G82** and **G89**. **G04** command is not used with fixed cycles (it is built into the cycle internally).

Once a dwell is identified in the program, make sure you understand the reason for its inclusion. As a rule of thumb, pay special attention to dwells longer than one half of a second. A longer dwell than that is not necessarily a waste of time. A long dwell is often programmed to make sure a certain mechanical function of the CNC machine is fully completed before program processing continues. That may apply to activities such as clamping and unclamping, automatic door closing, delay due to acceleration, tailstock application, bar feeder, and many others. In these cases, you should not change the dwell time. For machining, the dwell time change (usually a reduction) is quite often justified.

In order to optimize dwell time, you have to understand what a minimum dwell is:

> **Minimum dwell is the time required to complete ONE spindle revolution**

Based on the definition, a formula can be established to calculate minimum dwell:

- **Minimum dwell = 60 / rpm**

As an example, the minimum dwell is **60/900 = 0.067** seconds or 67 ms (milliseconds). Programmers often increase the minimum dwell to allow for more than one revolution, and also for the possibility that spindle speed override is set to 50% (normal minimum override). P250 in block N6 allows fewer than four revolutions while the dwell is in effect.

> **Dwell should never be shorter than twice the minimum dwell calculation**

Other Considerations

There is a lot more that can be found in the above program or any other program. An experienced CNC operator will be able to visualize various tool motions before testing the program or running the first part. One of the more important skills of being a first class CNC operator is to be able to identify program errors.

PROGRAM ERRORS

Mistakes are part of the human condition and CNC programmers are not immune in this area. Manual programming is prone to errors, but even a computer generated program may contain errors if the post processor is not properly configured. Entering program codes in MDI (Manual Data Input) mode can also contain errors.

Program errors can be very costly in production, causing expensive downtime. When evaluating a program before actual machining, looking for errors is often a time well spent. There are generally two groups of errors in the program:

- Syntax errors
- Logical errors

In addition, errors caused by equipment failure can be also added, but these belong to the area outside of the operator's influence.

Syntax Errors

Syntax errors are easy to detect. In fact, there is no real need for manual detection, because the control system will identify all syntax errors anyway. When evaluating a program, it is always a good idea to look for syntax errors before actual program run.

As the name suggests, syntax errors are equivalent to misspelled words in typical regular text. In CNC, syntax errors include all illegal characters, characters not available on a particular control unit, wrong order of characters, letter O instead of numeral 0 (or vice versa), and similar errors. All following examples are syntax errors:

```
MO8         instead of M08
            ... letter O instead of numeral 0
            (only one letter per word is allowed)

-X5.0       instead of X-5.0
            ... minus sign in front of an address
            instead of the number

1000S       instead of S1000
            ... address (letter) must always be
            the first character

Y 5.0       instead of Y5.0
            ... no spaces allowed within a word
```

These and other similar errors will be detected by the control, even if you miss them during program evaluation. By far, the most common syntax error is the letter O instead of numeral 0 or numeral 0 instead of letter O.

> Zeros are visually narrow ... 00000
> Letter O is visually wide ... OOOOO

Logical Errors

Logical errors present a much more serious group of errors, as they can lead to a scrap or even a collision. In short, a logical error can be defined as a program entry that is correct in its syntax but wrong in context. A very common example is a misplaced decimal point, an error in programming and program editing at the machine. The control has no way of selecting the correct dimension, as both are perfectly legitimate:

```
X100.0    instead of X10.0
```

Another common error is the change of G-code mode, for example:

```
G00 Z2.0       ... rapid to 2 mm above top of part
Z-5.0 F150.0   ... feed to 5 mm deep at 150 mm/min
...
```

> Feedrate is *NOT* effective in rapid mode

Missing Decimal Point

Also a very common error is a missing decimal point. This is an area that requires a little deeper explanation. Try to answer this question - *what is the decimal equivalent of a program entry of* X34720 ?

Don't rush with the answer yet - ask yourself another question first - what units are used in the program? Machine shop programmers and operators who work with both metric and imperial units have to think twice. Here is some background for better understanding.

In the very early stages of numerical control, the decision was made to limit every dimensional entry into the maximum of eight numerals (8-bit format), ranging from 00000000 to 99999999. At the same time, virtually all machine NC (hard wired) controls have been set to accept leading zero suppression as default. Its definition is quite clear:

> Leading zero suppression allows all zeros
> *before* the first significant digit to be dropped

The example of **X34720** is in reality equivalent to **X00034720**, with insignificant zeros dropped. Decimal point programming did not become widespread in part programming until the middle or even late 1980s (!). Until then, the decimal point had to be implied.

The rule was as simple as it was logical:

> For METRIC units, the implied decimal point is THREE decimal places left of the last of 8 digits

> For IMPERIAL units, the implied decimal point is FOUR decimal places left of the last of 8 digits

In both cases, the number of decimal places indicates the smallest resolution - the smallest amount of motion possible for each system of units. That means the example of **X34720** is equivalent to:

- Metric units:

    X34720   *is equivalent to*   **X34.720**   (= X34.72) mm

- Imperial units:

    X34720   *is equivalent to*   **X3.4720**   (= X3.472) inches

For those interested in the trailing zero suppression (not a required knowledge), the same program word **X34720** would have to be written as **X0003472** with the following results:

- Metric units:

    X0003472   *is equivalent to*   **X34.720**   (= X34.72) mm

- Imperial units:

    X0003472   *is equivalent to*   **X3.4720**   (= X3.472) inches

Trailing zero suppression is not used on CNC machines, because of its impracticality.

Offset Errors

As this section covers the subject of logical errors that cannot be normally detected by the control system, there is at least one notable exception:

> In G41/G42 cutter radius offset mode, the control system will issue an alarm when the cutter radius is *greater* than the inside arc radius

More on the subject of *Cutter Radius Offset* is covered in a separate comprehensive chapter (*page 135*).

Entering the wrong offset amount into the control system is also an error that can have serious consequences. For every dimensional problem, check not only the program but check the offset settings as well.

Incorrect Units

If parts to be machined frequently change from one unit system to another (metric to imperial or vice versa), a common error is to use a frequently used dimension is one format and using it in the other format.

For example,

G99 G81 R2.0 Z-10.0 F200.0

... is a normal *metric* example. In imperial units, the feedrate start point in R-level will be 2 inches, and - *even worse* - the depth of cut will be 10 inches. Feedrate will also be extremely fast, at 200 inches per minute.

Many other programming errors can be listed, but the most common have been covered.

4

CONTROL SYSTEM

The control system is the heart and brain of all CNC machines - it is their main component. In fact, this whole handbook is about the control system of a CNC machine. The abbreviation 'CNC' stands for *Computer Numerical Control*, and is commonly referred to as the **control** or the **control system** or the **control unit** and the even more common expression - **CNC Control,** in short. Although the expression **CNC control** does not follow any exact rules of the English language because of literal redundancy (*C*omputer *N*umerical *C*ontrol *control*), it is still a commonly accepted form of expression, at least in informal communications, in some cases including this publication.

Regardless of any specific description applied, the control system of any CNC machine can be a complex entity and often intimidating for new users. There are many aspects to the control system that are used by skilled professionals in different fields. For example, an electronic expert will look for different features of the system than a mechanical expert. Features relating to the CNC programmers and CNC operators will share many features, but each area of interest will always be unique. From a more practical perspective, both CNC operators and CNC programmers look only for those control features that are required in their respective fields.

CONTROL SYSTEM OVERVIEW

Looking at a CNC machine (mill, lathe or other), the control system is identified by its visual components, in the form of a large panel, where its most noticeable features are the display screen and a keyboard or a keypad. Various switches, buttons, lights, etc., are also part of the control system. A short summary lists the most important visual features of any control system:

- **Display screen (monitor) with soft keys**
- **Power ON/Off switches**
- **Keyboard**
- **Handle**
- **Push buttons - Toggle switches - Rotary switches**
- **Confirmation lights**
- **Alarms and Errors**
- *... any several others*

Each of these features is covered in this handbook in sufficient detail. While the overall control system panel may include many overwhelming features (particularly for a beginner), it is easier to understand individual parts by a simple separation into two distinct sections:

- **Control Panel**
- **Operation Panel**

To make such distinction or such separation, it is important to know some background of CNC machine manufacturing.

CNC machine tools are manufactured by companies that have expertise in building machine tools. Without suggesting any preferences and their origin, well known companies that manufacture CNC machining centers and have a large presence in North America are OKK®, Mori-Seiki®, Matsuura®, Makino®, Haas®, Fadal®, Mitsubishi®, Kitamura®, Okuma®, and others. European and Asian markets will be different in the brand names, but these sample names represent *machine tool manufacturers* regardless of the actual machine origin.

Except for some rare exceptions (Haas, Fadal, Fagor, etc.), most machine tool builders do *not* manufacture their own control system (the computer part of the machine) - they concentrate at what they do best, which is building machine tools. On the other hand, each CNC machine tool requires a CNC control system.

Just as there are CNC machine tool manufacturers that do not make their own controls, there are CNC control system manufacturers that do not make their own machine tools (except some minor exceptions). The reasoning is quite simple - *'we know what we're good at and specialize in doing it'*.

Those machine tool manufacturers that make no control systems of their own do only one thing - they *purchase* the control system that supports their machines.

That also means there is a field of manufacturing that specializes in design and manufacturing of the computer portion of the CNC machine only - the actual *control manufacturer*. Names such as Fanuc®, Yasnac®, Mitsubishi®, Siemens®, Heidenhain®, and many others are all part of this large field.

Let's sort all this out.

As stated earlier, any control system of a CNC machine can be identified by its two main components:

- Control Panel
- Operation Panel

For both operators and programmers, these two control system features form the main area of interest.

Control panel represents the computer part of the CNC machine operations. It is mainly used to make changes in the software part of the system whether in the part program itself or in various software settings, such as offsets and parameters.

Operation Panel represents the hardware part of the CNC machine operations. It is mainly used to change various settings and modes, overrides, and allows manual motions of the machine axes.

The distinction of software/hardware as separate entities is rather simplistic, as both are always connected. At the same time the distinction helps to illustrate the many functions offered by any CNC system.

SYSTEM MEMORY

Every CNC system provides a physical area where a number of part programs can be stored for a period of time. This area is called the *system memory*.

A program that exceeds the memory capacity cannot be loaded to or run from CNC memory. Even a large memory capacity is always limited, but can be increased by hardware updates. For a CNC operator, it is important to know how much memory is remaining before loading a program. Ideally, all memory capacity should be measured in characters of text, both used and remaining.

> The latest controls system DO show memory capacity in characters

Unfortunately, many older controls do not have this facility and use one of two other methods:

- Memory available as number of pages
- Memory available as meters or feet

In the first case, the number of pages refers to the number of full screen displays of the program. This is a rather hard way to establish the capacity remaining, and requires some practice. The method of measuring memory capacity in meters or feet seems a bit strange at first, until you understand what meters and feet refer to.

Yes, they refer to the equivalent length of paper tape that used to be the original source of program data in the early stages of numerical control. Although NC or CNC machines using a paper tape today are rather rare, the method of memory capacity measurement has prevailed.

Most current CNC programmers and CNC operators are not familiar with paper tape as the data source and probably have never even seen one. The one reason that the paper tape is even mentioned in a modern handbook such as this is not nostalgia. The reason for its inclusion is the fact that there are still remnants of tape references used on even the most modern CNC systems.

Paper Tape

Paper tape was a one inch wide tape, made of special paper (or other material) to exact specifications. Across the tape was a set of odd or even number of holes, punched in different configurations, where each set of holes represented a single character, according to EIA or ISO standard. The separation of one set of punched holes from the next was exactly 0.1000 inches (2.54 mm). This dimension - this 0.1000 inches - is the key to understand memory capacity of a control system, identified in feet or meters.

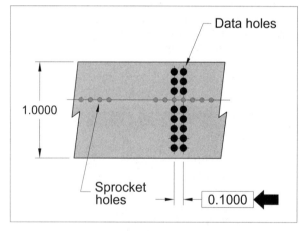

Keep in mind that this standard has been used around the world, although the base unit was an inch - after all, numerical control technology originated in the United States, where imperial measurements have been predominant (although that is now changing as well). Also keep in mind that most current controls are of either Japanese or German origin, two countries where metric system has been standard for centuries.

Memory Capacity

The following two tables provide conversions of several common memory capacities, as installed in the control system - the first one converts meters to feet:

Meters	Feet	Characters
15	49.21	5 906
20	65.62	7 874
40	131.23	15 748
60	196.85	23 622
80	262.47	31 496
320	1049.97	125 984

The second table converts feet to meters:

Feet	Meters	Characters
50	15.24	6 000
66	20.12	7 920
130	39.62	15 600
200	60.96	24 000
300	91.44	36 000
1000	304.80	120 000

As you may expect, it is impossible to list all possible options, but a simple calculation or two will provide the answers required.

First, some conversions of various units, in case they are necessary for additional calculations:

1 inch	=	25.4 mm
	=	0.0254 m
1 foot	=	(25.4 × 12 inches) / 1000
	=	0.3048 m
1 mm	=	1 inch / 25.4
	=	0.039370 inches
1 m	=	1000 mm / 25.4
	=	39.370079 inches
1 foot	=	12 inches / 39.370079
	=	0.3048 m
1 meter	=	1000 mm / 25.4 / 12
	=	3.28084 feet

Conversion of METERS to FEET:

Meters are the units of input to be converted into equivalent number of *feet*:

Feet = Meters / 0.3048

Conversion of FEET to METERS:

Feet are the units of input to be converted into equivalent number of *meters*:

Meters = Feet × 0.3048

Conversion of METERS to CHARACTERS:

Meters are the units of input to be converted into equivalent number of *characters*:

Characters = (Meters × 1000) / 2.54

... or ...

Characters = Meters / 0.00254

Conversion of FEET to CHARACTERS:

Feet are the units of input to be converted into equivalent number of *characters*:

Characters = Feet × 120

Conversion of CHARACTERS to METERS:

Characters are the units of input to be converted into equivalent number of *meters*:

Meters = Characters × 0.00254

... or ...

Meters = (Characters × 2.54) / 1000

Conversion of CHARACTERS to FEET:

Characters are the units of input to be converted into equivalent number of *feet*:

Feet = Characters / 120

Program files are always stored as pure text files on computers, in standard ASCII format - that means no formatting is involved - no underlines, no italics, no bold characters, etc. ASCII is an acronym that means *American Standard Code for Information Interchange*.

When a pure text file is stored, the computer operating system such as Windows™, indicates the number of bytes the file occupies on the hard drive (or disk). Although not exact, this file size can be used as a guide - typically, the number of bytes exceeds the number of characters. For example, a text file that shows 3500 bytes may have only 3000 characters, spaces included. Fanuc controls do not consider spaces as characters - spaces are ignored.

DNC Alternative

Although CNC memory capacity can be expanded by adding more memory chips or boards, an inexpensive alternative is the use a of remote computer that serves as a **DNC** source unit.

DNC stands for *Direct Numerical Control* or *Distributed Numerical Control*. Although there is a certain overlap in using these two terms, direct usually means one Personal Computer to one CNC, whereby distributed usually means one Personal Computer to several CNCs. In both cases, suitable software and cabling is required.

CONTROL PANEL

The illustration below shows a typical control panel for a CNC machining center. Turning center control panel will be almost identical. The letter -**M** in the model identification stands for *Milling* - a CNC turning center (lathe) will have the letter -**T** instead (for *Turning*).

Although the control panel layout and its features will vary from one manufacturer to another and even from one control model to another for the same manufacturer, there are enough similarities to group them and examine each group.

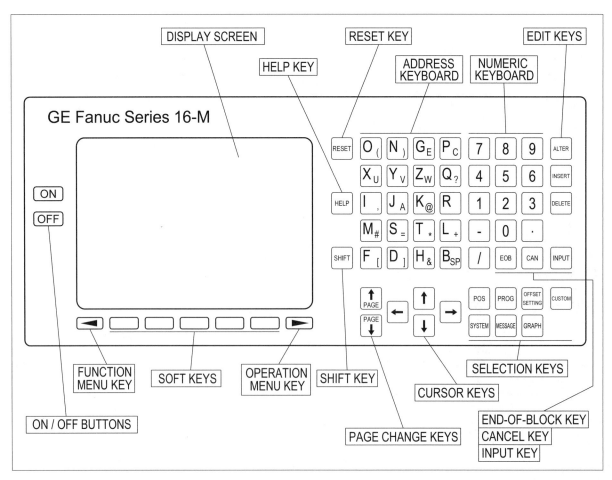

ON-OFF Buttons

The control system has its own independent power source, therefore it has its own ON-OFF buttons. Even if the CNC machine itself is turned on, it does not mean its control system is under power as well. However, the machine power has to be turned on *before* the control system can be turned on via *Control Panel*. The main reason is to separate mechanical, pneumatic and hydraulic elements from the electronic elements.

The **OFF** button of the *Control Panel* is usually the first one to use when the machine is powered down. Usually, the *Power-Off* order is the opposite of the *Power-On* order. As the exact procedures will vary from machine to machine, always consult the manufacturer's recommendations and procedures.

Display Screen and Soft Keys

The largest area of the *Control Panel* is the display screen - the *monitor* of the whole control system.

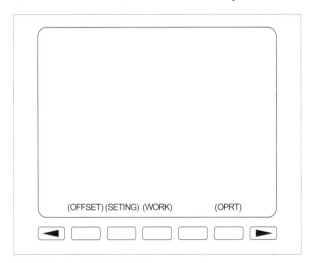

There are three main elements:

- **Display area**
- **Arrow soft keys**
- **Selection soft keys**

When the power is supplied to the control panel, the display shows its default screen. In order to navigate between many screen pages, the control system offers several *soft keys*, located just below the display screen (five keys shown in the illustration + two for navigation).

In order to avoid hundreds of buttons and a very large control panel, modern control systems only display features directly related to some main selection group. For example, if you choose the **POS** selection key, the screen will show only those features that are related to various positions - such as absolute, relative, machine, distance-to-go, etc.

Soft keys always work in conjunction with *Selection Keys* (described next). The illustration shows the meaning of soft keys when the *Offset/Setting* selection key is pressed.

Incidently, the meaning of the word **soft-key** is not related to its physical composition or the feel of one's finger when the key is pressed. The word 'soft' is just a short form of 'software' - as in *'software keys'*. The implication is that it is the software of the control system that determines the meaning of each soft key, based on the selection key. That is the main reason why the soft keys are not identified in any way.

Navigation Keys

Regardless of the number of actual soft keys, there are also two *arrow keys* - one to the left and one to the right of the actual soft keys:

- **LEFT arrow key** ... *Function Menu* selection
- **RIGHT arrow key** ... *Operation Menu* selection

In conjunction with each other, these two arrow keys are used to navigate different pages and sub pages of the screen display.

Selection Keys

As mentioned in the previous section, the actual functionality of each soft-key is always directly dependent on the currently active *Selection Key*. In other words, the selection shown on the screen above each soft key will be different for each selection key. The illustration below shows the most common key selections available:

POS Key

POS indicates settings related to various *position* displays. This is probably the most commonly screen used in part setup. The main modes that can be used are:

- **ABS** ... Absolute display
- **REL** ... Relative display
- **ALL** ... All displays
- **MACH** ... Machine position display

The two main displays - **ABS** and **REL** - are used the most. The *absolute* display is normally used during program processing - it shows the current tool position based on part zero location. The *relative* display is used mainly during setup, for activities such as setting work offset, tool length offset, and others. In this case, at a certain critical position (commonly at machine zero), the CNC operator sets the relative display to zero and makes the required offset measurement, for example, sets **G54** work offset or **G43 H..** tool length offset.

Both **ABS** and **REL** displays are usually shown in larger form than normally, so they can be seen from a distance during part setup.

MACH selection key shows the current tool position always measured from the machine zero (home position). This display is totally independent from any offset setting or part program data.

If you want to see both **ABS** and **REL** position displays, and **MACH** display as well, on the same screen, just press the **ALL** selection key. The display of each part is smaller, but all important data is shown on a single screen. In addition to the four modes, the control system also shows some current activities from the part program, such as current spindle speed, cutting feedrate, active program codes, etc.

PROG Key

PROG indicates the *program mode*. The contents of the screen - what you can actually do - depends on the selection of control *mode*, located on the *operation panel* of the control (described in the next chapter):

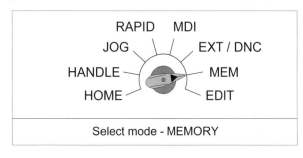

Select mode - MEMORY

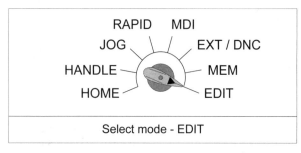

Select mode - EDIT

For the **PROG** key selection to work, either the **MEM** (memory) or the **EDIT** mode has to be selected.

In **MEM** mode, the currently loaded program can be monitored by viewing it on the screen. Current and next block can also be viewed. No program changes are possible in **MEM** mode.

If you want to make changes to the current program, you have to choose the **EDIT** mode from the *Operation Panel*. As these two modes are separate, there is no possibility of making a mistake, for example, by editing part program currently active.

In the **EDIT** mode, program can be edited, viewed on the directory screen, program can also be loaded into memory or deleted from the control memory.

Many controls even offer a feature called *background edit*, which allows programming of the next job, while the current job is running.

OFFSET / SETTING Key

Contrary to some beliefs, this is a dual function key - it does not mean *Offset Settings* - it means *Offset* **and** *Settings* !

Tool length and cutter radius offset can be set by selecting the **OFFSET** soft key. Work offset is set by selecting the **WORK** soft key. In order for either setting to work, the **OFFSET / SETTINGS** key has to be pressed first.

In this mode, various settings may also be done at the control. The most common setting is the selection of units - *metric* or *imperial* (*mm* or *inches*). Another setting may relate to *mirror image* and *macros* - several others are also available from this screen. Although these settings can be done at the machine, it is always important to follow this recommendation:

> Settings that can be done at the control
> AND in the program,
> should always be covered by the program

SYSTEM Key

SYSTEM selection key is the entry into the inner workings of the CNC system. There is no harm looking at this display but making any changes is not part of general operator's duties. Under the **SYSTEM** key selection, you will find *Parameters Setting Screen, Diagnosis Screen, PMC Screen* (PMC stands for *Programmable Machine Control*, which is a version of PLC - *Programmable Logical Control*). PMC/PLC is used by machine tool manufacturers for many different applications. One of the more common ones is connecting various devices together. For example, a transfer system between two CNC machines uses PMC/PLC, The same applies internally for almost all mechanical functions.

> SYSTEM selection key should only be used for viewing various data - changes should always be left to qualified technicians

MESSAGE Key

Typical screens that appear in the *Message* mode relate to various alarms and operation. They may also show alarm history. Message screen will also appear when a macro program includes message to the operator via a part program.

GRAPH Key

An option on many controls - if available, it shows toolpath motions graphically. The main benefit of this optional feature is that a program can be graphically tested prior to actual machining.

CUSTOM Key

In the majority of CNC applications, the *Custom* selection key is used by the machine tool manufacturers for some unique activities of their CNC machine. For example, by pressing the *Custom* key, you may find procedures related to automatic tool and pallet changers,

Address Keyboard

In CNC programming, the word *address* refers to any letter used in the part program. For example, rapid motion **G00** uses the G-address, tool change command **M06** uses the M-address, and spindle speed **S1200** (r/min) uses the S-address. These are just some of many addresses that are used in a typical part program.

The control panel keyboard is not just an exclusive address oriented keyboard, as it also includes digits and special symbols. In reality, the control system keyboard is a multifunctional *alpha-numerical* keyboard, which is actually a description used by some control systems.

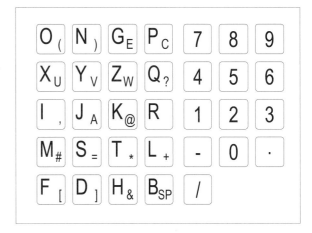

Note that the keys containing letters also contain secondary characters. Grouping of the letters is designed to be convenient for CNC program data entry, and does not follow the common QWERTY keyboard layout.

Secondary characters are accessible via the *Shift* key. Not all secondary characters may be available, even if they appear on the keyboard. A pair of parentheses **( )** is used for inclusion of comments into the program, while square brackets **[ ]** are used only in macros. Macros also use the **#** symbol, equal sign **=**, asterisk ***** for multiplication, plus sign **+** for addition. From the numeric pad, macros also use subtraction sign **-** and forward slash **/** for divisions. Question mark **?**, comma '**,**', ampersand **@**, and the 'at' symbol '**&**' are not normally used in part programs. **SP** is a symbol for 'space' to separate words.

Shift Key

As described already, the *shift* key is used the same way on the control panel as it is used on standard computer keyboard - it selects the secondary character, if available.

Numeric Keyboard

Numeric section of the keyboard contains only primary keys - no shift is required - it covers all digits from 0 to 9 as well as the negative sign and the decimal point. There is also a symbol for the forward slash, used as a block skip symbol in CNC programming.

Edit Keys

There are three *Edit* keys on the control panel. They are used to edit programs stored in the CNC memory:

- **Alter** ... changes existing program entry
- **Insert** ... adds a new entry into an existing program
- **Delete** ... removes program entry

Page and Cursor Keys

Both *Page* and *Cursor* keys are navigational keys. They narrow down the available screen selections:

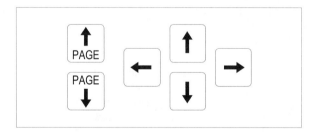

Page refers to a single display of the control screen. For example, a long program may not fit on a single screen and can be viewed only by scrolling from one page to the next, in ascending or descending order. Each time the *Page* key is pressed, another display screen will appear. *Page-Up* and *Page-Down* arrows select the direction of the page scrolling.

Cursor is normally used on the selected page. Its main purpose is to narrow down the selection within a page. Vertical cursor arrows are used to select a particular line displayed on the screen, while horizontal cursor arrows are use to select a particular item with the selected line.

EOB - CAN - INPUT Keys

EOB or **E-O-B** is just a short form for the *End-Of-Block* character. When entering a part program into the control system in *Edit* or *MDI* mode, each program block has to be separated from the next program block. *End-Of-Block* key is used for that purpose.

CAN is the *Cancel* key. As the name suggests, this key is used to cancel an erroneous data entry.

INPUT is a special key used for data settings. Unlike the **INSERT** key, it is not used for program data. Its main use is to set various offsets, settings, parameters, etc.

> **Do not confuse INPUT with INSERT**
>
> **INPUT is used for data settings**
> **INSERT is used to add program data**

Reset Key

Pressing the **RESET** key, many default conditions are restored. One of the most common use of this key is to cancel alarm or error condition.

Keep in mind that the actual cause of any alarm has to be addressed first, and the problem resolved before the *Reset* key can become functional.

Help Key

The **HELP** key is quite self-explanatory - it provides details relating to various commands - commonly used in MDI mode. Various alarm details, as well as details relating to operation methods and parameter table can be found here.

The control system and its display/keyboard portion is closely connected to the operational features of the CNC machine, the *Operation Panel*. The next chapter will cover this subject in detail.

5 OPERATION PANEL

The part of the control system that is designed and manufactured by the machine tool builder (*not* the control system builder) is called the *Operation Panel*. This panel is much richer in visible features than the *Control Panel* and consists of the following major groups:

- **Main buttons and switches**
- **Status indicator lights**
- **ON/OFF switches (toggle or button type)**
- **Rotary switches**
- **Edit key**
- **Emergency Stop switch**
- **Setup handle**

In general, the main purpose of the *Control Panel* is to provide the software aspects of operating a CNC machine, while the *Operation Panel* provides the hardware aspects. Both of these two portions of a CNC system are related and interfaced at the manufacturing stage. They always work together.

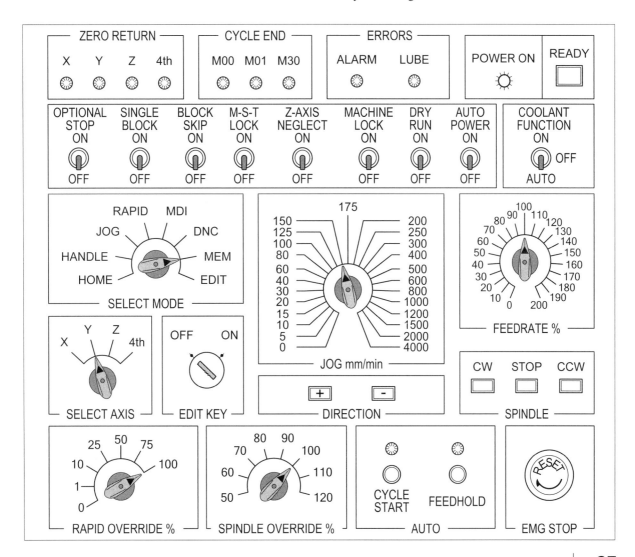

OPERATION PANEL LAYOUT

There are many different versions of machine operation panels. They vary in features and layout, some use toggle switches, others use push-buttons. Some controls use symbolic descriptions, others use literal descriptions. Regardless of these differences, overwhelming majority of features is common to all control systems. Features that are less common often reflect a special option, unique to a particular machine tool builder.

Modern CNC systems use pictorial symbols rather than words (literal descriptions). The most common pictorial symbols will be shown later in this chapter. On the previous page is a schematic layout of a typical operation panel using literal descriptions for easier orientation. Although the shown layout is only a composite based on several operation panels, it does represent all major features found on any modern CNC control system. Not all features can be shown here, for example those relating to ATC *(Automatic Tool Changer)* or APC *(Automatic Pallet Changer)*. Also, features that are particular to a specific CNC machine cannot be shown either. As always, it is important to consult all manuals that are supplied with the CNC machine. They are always based on the actual machine model and its control system. Various manuals provided to the user list in detail all functions available and describe how they work.

When viewing a machine *Operation Panel*, its layout is composed of light indicators and various switches and buttons. Needless to say, all CNC operators should be thoroughly familiar will all features of both the *Control Panel* and the *Operation Panel*. Many operator related features are also directly related to part programming, so it is equally important that CNC programmers have a good (or even better) knowledge of the machine/control operating features.

MAIN BUTTONS AND SWITCHES

From all different buttons located on the CNC machine *Operation Panel*, there are two main buttons that are somewhat more important than others, although all are equally important.

Power ON / Ready Light Indicator

As the button label *Power ON* suggests, this is the main power switch (button) for the machine as well as control system. When the machine power is supplied, the control does not get power automatically right away, the power has to be turned on in subsequent steps.

It is not unusual to perform three or even four steps to get all power needed before using the CNC machine - for example:

- Main switch (breaker) on the wall - ON
- Machine main power switch - ON
- Control ON - first time to power electronics
- Control ON - second time to power hydraulics

Power-Off procedure is usually the reverse order of the *Power-On* procedure. As various CNC machines vary quite a bit in the way they are powered, the actual procedure may also vary quite a bit, so always consult machine tool manual first.

> **Check CNC machine manual for exact *Power ON/OFF* procedure**

When all power has been supplied, the *Ready* light turns on, to indicate machine is ready for initial setup, including power supply to the control unit, as described in the previous chapter.

Cycle Start / Feedhold - Auto Operation

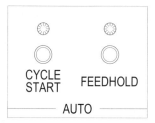

Another push button that belongs to the main category *Auto*, contains the key *Cycle Start* button - typically a push-button with a light that is built-into the switch or provided above it. The words 'cycle start' simply mean that a series of events will take place when this button is pressed. Typically, there are three main events activated by *Cycle Start* (others are also available):

- Part program processing in continuous mode
- Part program processing in single block mode
- MDI commands operation

The other button in the same *Auto* category is the *Feedhold* button, usually located next to the *Cycle Start* button. It can only be used when the part program is running (being processed or executed). Its sole purpose is to stop all axis motion. Feedhold is particularly useful during part setup, often used to determine critical clearances in tight spots. Even during actual machining - and if used properly - feedhold button can be very handy for removing chips, adjusting coolant nozzles, and other last-minute adjustments.

STATUS INDICATOR LIGHTS

The CNC system along with the CNC machine can perform many operations, often in a very short time. Many activities come and go and it would be very difficult to see what state the control system is at any given moment without special *indicator lights*.

Even the *Power On, Ready, Cycle Start* and *Feedhold* buttons are lighted when in effect. However, many additional indicator lights are used during program processing - they indicate the **current control status**.

Indicator Lights - Three Groups

Most control systems include a series of small light indicators that will often vary in color (green or amber), but will share one common feature - **they cannot be changed**. These lights are called the status *indicators* (or just *status lights*). Their purpose is to convey a simple and quick visual message to the CNC operator. For example, a red indicator light will relate to a trouble of some kind, such as low level of machine slide lubricator or a general system or program fault (alarm). An amber light (amber is a shade of orange) will typically identify a warning, but it could also identify a certain condition. A green indicator light identifies a 'go' or 'ready' condition - when everything is in order.

Every *Operation Panel* contains a number of indicator lights - those are lights that indicate a particular control status at any given time. These lights use colors to suggest their importance - typical colors are:

RED	Indicates a severe condition, such as alarm or fault - always pay attention to this problem
AMBER	Indicates a standard condition; a certain status confirmation - usually presents no emergency
GREEN	Indicates a favorable condition, such as 'good', 'setting OK' or 'ready to go'

The three common control status indicator lights can be grouped by their actual purpose:

- Zero return position
- Cycle end
- Errors / Alarms / Faults

Additional lights may indicate status that is unique to a particular CNC machine tool.

Zero Return

When one or more machine axis reaches the machine zero reference position (home position), a status indicator light on the *Operation Panel* will turn on, as a confirmation. Each axis has its own light, so it is easy to check if all axes have reached the machine zero.

These lights are usually green, but can be amber as well - they are assigned to each available axis the machine tool has. On the most common 3-axis CNC vertical machines, there will be three lights identified by the machine axis letters X-Y-Z.

If the fourth axis is also present on the machine, the corresponding light can be identified by the digit 4 (as for 'fourth axis'), as '4th', or it can be the exact name of the fourth axis, such as *A* or *B*. Typically it is the *A-axis* for CNC vertical machining centers and the *B-axis* for CNC horizontal machining centers.

> Most CNC machines have to be zeroed (homed) in *ALL* axes before running a part program

Cycle End

The end of a cycle can be only temporary (using **M00** and **M01**) or permanent (**M02** or **M30**). Fanuc and similar control systems have at least three confirmation lights that relate to *Cycle End* modes:

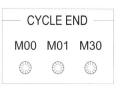

M00	Indicator light turns ON, if *Program Stop* **M00** has been detected in the CNC program
M01	Indicator light turns ON, if *Optional ProgramStop* **M01** has been detected in the CNCprogram
M02	Indicator light turn ON, if *Program End* **M02** has been detected in the CNC program (obsolete)
M30	Indicator light turns ON, if *Program End* **M30** has been detected in the CNC program

> *Cycle End* indicator lights will turn ON only if M00-M01-M02-M30 is detected in the part program

Note that the above illustration does not contain miscellaneous function **M02**. This function is still supported for backward compatibility but is not used anymore.

As all four indicator lights are directly related to the contents of a part program, it is very important to understand how they affect CNC machine operation.

◆◇ M00 - Program Stop

Program Stop **M00** is typically included in the program if the CNC operator has to perform some task for each individual part of the batch, without exceptions. For example, applying special lubricant for a tapping tool, flipping or repositioning the part, readjusting a clamp, doing a mandatory dimensional check, etc., are all good reasons to include **M00** in the part program. Program processing will always be temporarily stopped with **M00**, without any interference from the CNC operator. The **M00** indicator light will only turn *ON* when the miscellaneous function **M00** is detected during program processing.

Typical program application will be at the end of the tool, for example:

```
...
N31 G00 Z2.0 M09
N32 G28 Z2.0 M05
N33 M00 (PLACE PART IN FIXTURE 2)
```

It is a good programming practice to include a descriptive comment (message) with the **M00** function.

◆◇ M01 - Optional Program Stop

Optional Program Stop **M01** is very similar to **M00**, but the CNC operator does have a choice to use it or not to use it. By turning the *Optional Stop* switch *ON* (located on the operation panel), any encounter of **M01** in the part program will force the program processing (execution) to stop until the *Cycle Start* button is pressed.

Both illustrations below show the indicator light status when program processing encounters the **M01** function:

Similar to **M00**, typical program application of **M01** will be at the end of the tool, for example:

```
...
N31 G00 Z2.0 M09
N32 G28 Z2.0 M05
N33 M01
```

Usually, there is no comment (message) attached to the **M01** function.

◆◇ What Happens at the Machine ?

In either case, when **M00** or **M01** program function is detected, many machine activities will be stopped:

- All axis motion will stop
- Spindle rotation will stop   ... M05 state
- Coolant will be turned OFF   ... M09 state

If the program is written correctly, each toolpath that follows **M00** or **M01** function will include spindle speed and coolant functions, even if no **M05** or **M09** had been programmed.

> Look for missing spindle speed and coolant in the blocks following M00 or M01

◆◇ M02/M30 Program End

Most Fanuc controls support two *Program End* functions. One of the functions is the old **M02**, the other is the much more common **M30**. In the old days of paper tape, **M02** was used for a loop tape (that is tape spliced at its ends), because when the end of program had been reached, the tape was physically at its beginning because of the loop. No tape rewind was necessary to use in the program. To accommodate tapes that were too long to form a loop, **M30** function replaced the **M02** function in the program, with almost identical results. In case of **M02**, the tape did not rewind to its beginning, in case of **M30**, the tape was forced to rewind.

Today, there is no physical tape used as a program source anymore, but both functions are still available in virtually all Fanuc control systems. **M30** function is the normal standard, but **M02** function is still supported, mainly for compatibility with old programs. The word *'rewind'* takes on a new meaning, and is now interpreted as *'return to the top of program'*.

> Many modern controls treat M02 the same as M30 and will rewind (return) to the program beginning (top)

Errors / Alarms / Faults

> **When checking the status indicator lights, make sure all light bulbs are functioning properly**

This comment applies to all indicator lights but it is even more important for the group that shows *red* lights. Seeing red - that is, a red status indicator light on the *Operation Panel* - is generally not a good situation. It indicates that something has gone wrong either in the part program itself or during actual machining.

This is the third common group of lights that indicate a certain condition or situation. Usually, there are at least two indicator lights on every control system:

- Alarm ... general alarm or error condition
- Lube ... machine slides lubrication alarm

The first light is often called just *'Alarm'* - this is a non-descriptive name that covers all kind of alarms. Also worth note is that some controls use other words, like *'Error'* or *'Fault'* with the same meaning. Fanuc controls generally use the word *'Alarm'*. Regardless of the definition, the results are the same. Typical alarms that will cause an indicator light to turn ON (usually as a red light) belong to at least three categories:

- Program syntax error ... programming error
- Lubrication error ... low lubrication
- Machine problem ... various causes

Other problems may also be the cause of an alarm.

⇨ Syntax Errors

Syntax error is an error that originates in the part program. This type of error can be easily detected by the control system - syntax error simply means that proper programming format was not followed. For example, the negative sign in -X100.0 is a syntax error, because only a letter can be the first character. On the other hand, X-100 is not a syntax error, because the format itself is correct, meaning 100 microns (metric) or 0.1 of an inch, depending on the selected units! Yes, decimal point *IS* very important.

> **Syntax error is an error caused by an illegal program input.
> This error CAN be detected by the control system**

When you encounter this type of error in the program, the syntax can be wrong in several cases, such as:

- **First character is not a word** (-X10.0 or 10.0X)
- **Letter O was programmed instead of digit 0** (M00) (probably the most common program input error)
- **Digit 2 was programmed instead of letter Z** (2-10.0) (a common error if program written by hand)
- **Space was programmed after the letter** (X 10.0) (also a space between digits)
- **Letter not acceptable by the control was programmed** (Y for a standard 2-axis lathe)
- **Correct but unavailable command or function** (G02 X.. Y.. Z.. I.. J.. is wrong if helical interpolation is not available - same applies to other control options - check manual for the list of commands)
- **... other possibilities**

Syntax error is not exclusive to part programs delivered to the CNC machine. Errors of this kind can also happen at the machine, for example, when using MDI mode *(Manual Data Input)* or even setting various offsets. Syntax has to be accurate and in accordance with the control standards.

⇨ Logical Errors

In the majority of cases, there is no special control feature to 'catch' logical errors. In comparison with syntax errors, logical errors are much more difficult to detect, as they consist of correct program format but wrong input data. For example, the earlier entry of *X100* is as good as any. X100 and X100.0 are far apart. Quite a bit apart, in fact. *X100* is equivalent to X0.1 in metric and to X0.01 in imperial units!

There is, however, one case, where the system alarm *will* work, even for a logical error. The subject is *cutter radius offset*, described in a chapter of its own. For now, if the control system detects that a cutter radius set at the control is larger than any inside radius of the contour, there will be alarm and the indicator light will flash red. In addition, an error message will appear on the screen.

The reason for such alarm is that the cutter radius offset uses a feature called *'look-ahead'* type. During program run, the control will evaluate at least one block that is to be processed (next block) and checks if such cut is possible - if not, hence the alarm.

General alarm condition will also occur if there is a malfunction of the machine itself or some of its main elements. Usually, the error message will point towards the problem cause. One machine related alarm that has its own indicator light is the *LUBE* alarm.

Lubrication Alarm

Lube alarm - *lubrication alarm* - is the result of a situation when the machine slide lubrication supply is detected as being below certain level. This alarm will occur once in a while, for the simple reason that the CNC operator does not normally watch the level of lubricant in the storage container. In fact, there is not need to check how much lubricating oil is available - just wait for the alarm, then fill-up the container with appropriate lubricant. There will be no damage due to insufficient slide lubrication. The control will simply not allow it, as the possible damage could be very substantial.

Many CNC operators are sometimes puzzled why the lube alarm light comes on, when the lube oil container appears to have plenty of oil. Keep in mind that what the lube alarm really means is that the lubrication oil is being taken from a reserve. Although the oil container does seem to be fairly full, the machine has started lubricating its slides from its reserved supply. The natural question is 'why so soon?' - and yes, there is a simple explanation.

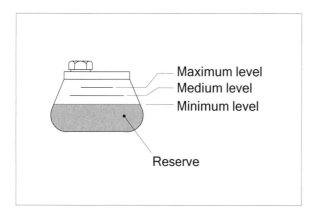

Most common machining operations typically take a very short time, usually only a few minutes to complete. Machining operations taking an hour or even several hours may not be common to small shops, but in certain industries they may be the norm. For example, three dimensional machining in mold work may use a single tool and let it cut for much longer than one hour. For example, what happens if a three hour cutting time experiences a lube alarm after two hours of machining? This is where the reserve comes into the picture.

Stopping machining before all cutting is completed, when the tool is just fine, is not only inefficient, it can cause various flaws to the machined part. In such cases, machining will not stop, when the lube alarm comes on. In fact, the machining will continue normally for the remaining time, until machining with this tool is fully completed. From the time of the first lube alarm appearance to the end of the cutting, the lubricating oil will be taken from the reserve in the container. That is the basic design about having a rather large reserve.

In summary, until it is interrupted by the program or by manual interference, the slide lubricant will be taken from the reserve the moment the lube alarm has occurred. Practically, it means that the lube alarm light will turn *ON* when oil level indicator reaches 'low level', and it will remain *ON* until the oil level in the reservoir is increased. If any machining takes place when the light comes on, the machining will be completed, as long as there is sufficient supply of lube oil. Having a rather large oil reserve guarantees that even a long machining sequence will be completed.

While the lube alarm is active (red light is *ON*), no further machining will be possible after the control system detects program tool change function **M06**, program stop function **M00**, optional program stop **M01**, or program end **M30**. In order to continue machining after these functions, the lube alarm must be eliminated by adding more lube oil to the reservoir.

> When the LUBE alarm is ON, machining cycle will be completed, but program will not continue until the lubrication oil level is increased

ON/OFF SWITCHES

The CNC operator can turn a number of switches located on the operation panel to **ON** state or to **OFF** state, as necessary to achieve certain operating conditions. These are important switches, used daily, and there are many of them located within the operation panel.

Depending on the controls system (manufacturer's preferences), these switches can be of two types:

- **Toggle switches**   ... **UP and DOWN**
- **Button switches**   ... **IN and OUT**

Both types work on the same principle - *they both select one of two modes available*. For example, you flip the switch up or down (for toggle switches), or you push the button once and push it again (for button switches).

Toggle and Button Switches

Regardless of the switch type, either of the two selections will make the particular switch inactive (OFF) or active (ON) and its function disabled or enabled.

The English word 'toggle' means a 'selection from two opposite options'. A typical household light switch is a good example - it can be either ON or OFF - other options are not possible. In CNC, the operation panel has quite a number of toggle switches, mostly offering a selection from two possible options. Operation panel toggle switches can be similar to a common light switch, at least in the sense that they can be operated by flipping the switch up or down. Toggle switches can also be of the push-button type - push once for ON selection, push once more for OFF selection.

Normally, all toggle/button switches are OFF and activated only as needed. The one exception is the *Manual Absolute* switch (if available), which should be normally ON. This is a special purpose switch and will be described later in this chapter.

Several toggle switches can be used in combination to achieve a certain desired result, but understanding each switch individually is most important.

In this chapter, the UP/DOWN type of switches will be shown, as they better show their state visually.

Optional Stop

Optional Stop switch is only one of several switches that require a part program input in order to be fully functional. When the miscellaneous function **M01** is contained in the part program, the CNC operator may choose to use the *Optional Stop* switch by setting it to ON position or leave in the normal OFF position.

When the **M01** is processed, the program processing will stop. To continue with the program processing, it is necessary to press the *Cycle Start* button.

Programmers often place **M01** function at the end of each tool, with the possible exception of the last tool. It is up to the CNC operator to decide whether to use it or not. There are a number of situations when the *Optional Stop* switch is useful:

- Checking tool for wear and other flaws
- Part dimension check
- Temporary absence from the machine *(see note)*
- ... any other activity not required for each part

> Leaving machine with *Optional Stop* set to OFF should be combined with *Feedrate Override* switch set to zero - always observe safety rules

If the program includes mandatory program stop **M00**, program processing will *always* stop - no interference from the operator is necessary.

Single Block

All CNC part programs are always written sequentially, in the order of tasks and instructions needed to complete certain machining operations. Normally, for full production, the *Single Block* switch is set OFF, so the program processing is continuous. Setting the switch to ON position is useful particularly during setup, as it allows the CNC operator to process one block at a time.

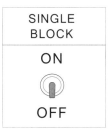

During part setup, using the *Single Block* mode enables the operator to test not only the program itself, but also settings of tools and offsets, check clearances, depth and width of cuts, and other cutting oriented activities.

Well structured part program always helps during part setup. For example, a common program entry for a tool number three (T03) is:

```
T03 M06    (NEXT TOOL SEARCH AND TOOL CHANGE)
```

During single block mode, the next tool search command (**T03**) will be done first but immediately followed by the actual tool change (**M06**) - in the *same block*. There is no pause between the two functions - they will be processed simultaneously as a single block.

There is nothing wrong with the block or program for efficient production, which is always the main objective. On the other hand, the format may slow down the setup in certain cases. Consider a possible scenario that may benefit changing the program structure to two blocks.

If those two functions are programmed in separate blocks, there will be no delay during production, but during setup the two blocks will bring some benefits:

```
T03                     (NEXT TOOL READY)
M06                     (TOOL CHANGE)
```

As the tool search is independent of the actual tool change, it takes place first. Working with *Single Block* mode set to ON, the CNC operator has the option to walk around the machine, check if the right tool is ready for tool change and - if necessary - make corrections *before* the 'wrong' tool is actually brought into the spindle.

This is only a small example of how program structure may affect setup. Programmer's cooperation should not be difficult to secure.

> *Single Block* mode is NOT functional for tapping or threading operations

Block Skip

Block Skip switch is another operation panel feature that requires support in the program. Part program that contains a forward slash symbol (**/**), can be processed in its entirety (*Block Skip* switch setting OFF) or bypassing all blocks containing the slash symbol (*Block Skip* switch setting ON).

Some controls use description *'Block Delete'* instead of the more accurate *'Block Skip'*. Keep in mind that no blocks in the program will actually be deleted - they will be only *skipped* - they will be bypassed and ignored during program processing when *Block Skip* switch is ON.

The most common (and often the only) use of block skip function in the part program is as the *first* character in the block, for example,

```
/ N76 G91 G28 X0 Y0
```

... where the XY machine zero return motion can be skipped, if desired.

Some controls allow block skip function in the middle of a block, for example,

```
N132 S800 M03 / S1200
```

Be extra careful in such situations and think twice.

In this case, the spindle will rotate at 800 r/min if the block skip switch is ON and 1200 r/min if the switch is OFF. If the control system supports this feature, the actual order of commands in the program is *very important* - with two conflicting commands - S800 and S1200, it is the latter one that will be processed. Make sure to check if your control system supports this feature.

Note that the placement of **M03** function is in front of the slash - the spindle rotation clockwise is common to both situations.

◆ Common Program errors with Block Skip

The problems encountered with block skip function can happen because of the operator's error:

- Not activating the *Block Skip* switch at all

- Activating the *Block Skip* switch too late, after the program started processing block with slash

- Switching to Block Skip switch OFF before all skipped program blocks have been processed

Problems can also be caused by two program errors:

- Incorrect block skip symbol (/) application

- Missing modal settings

Of the two problems, missing modal settings are the most common. This is definitely a program error, but a keen operator should be able to detect it. Knowledge of what modal commands are is important, as is the knowledge of what is the effect of various M-functions, particularly **M00** and **M01** functions.

In part programming, modal commands and functions are those that can be programmed only once and remain active without being repeated in each block. Most - *but not all* - G-codes are modal. Certain machine functions are also modal. For example, once the spindle speed rotation **M03** or **M04** is established, it remains in effect.

For both modal G-codes and M-functions, each remains active until either changed (cancelled) by another command or function. As an example, take **M03** - spindle rotation CW function. Spindle rotation CW can be cancelled by **M04** (rotation CCW function) or **M05** (spindle stop function). For G-codes, it is similar. For example, as only one motion command can be active at any given time, it means that commands **G00**, **G01**, **G02**, **G03** as well as any other motion command (such as fixed cycles) will cancel each other.

The following example shows the error caused by incorrect use of modal commands in the part program:

```
N1 G21
N2 G17 G40 G80 G90 T01
N3 M06                              (T01 WORKING)
N4 G54 G00 X75.0 Y125.0 S1000 M03
N5 G43 Z25.0 H01 M08
N6 G99 G81 R2.0 Z-4.6 P200 F150.0
...
...
N23 G80 Z25.0 M09
N24 G28 Z25.0 M05
/ N25 G91 G28 X0 Y0
N26 M01

N27 T02
N28 M06                             (T02 WORKING)
N29 G54 G00 X75.0 Y125.0 S1200 M03
...
```

Evaluate the example and see where the problem is. The program structure is not the best, but the program will work well *if* the block *N25* is skipped. Being able to do such an evaluation is part of being a skilled CNC machine operator.

Overall, the program looks OK and will perform well with *Block Skip* switch turned ON. In this case, the **G90** absolute mode programmed in block N2 remains in effect through the whole program and block N25 will be skipped. If the *Block Skip* switch is turned OFF, *all blocks* will be processed, including block N25. This block includes optional return to machine zero in both X and Y axes, using **G91** incremental method. As there is no **G90** repeated for the next tool, the control will remain in incremental mode and the program will cause many problems. The correct block N29 should be:

`N29 G90 G54 G00 X75.0 Y125.0 S1200 M03`

Further examples in the chapter *'Trial Cuts'* show the proper usage of block skip to avoid problems related to modal values and other activities.

> During program run, make sure
> the *Block Skip* mode is selected *before*
> the program processes the first block
> preceded by the slash

Regardless of whether the block skip function *is* or *is not* used in the program, the normal setting for the *Block Skip* switch is OFF.

M-S-T Lock

The letters **M-S-T** refer to functions related to machine operation and programmed with three types of corresponding addresses:

- M ... Machine functions
- T ... Tool functions
- S ... Spindle functions

These three function groups represent many physical activities of the machine tool, controlled by the program:

M	Machine	M03 -M04 -M05
		M06
		M08 -M09
		M19
		M60
		... and others (depending on machine)
T	Tool	T01 to T_{max} (all range of tool numbers)
S	Spindle	S1 to S_{max} (all range of spindle speeds)

When the M-S-T switch is turned ON, all machine related functions are locked and not available. This setup method allows testing the part program without any tool changes and spindle rotation, although XYZ motions will still be performed. It is often used in Dry Run mode, when there is no part mounted in the fixture.

Z-Axis Neglect - Z-Axis Lock

Another name for the *Z-Axis Neglect* switch is called the *Z-Axis Lock*. In either case, the function of this switch is the same. When the *Z-Axis Neglect* switch is turned ON, and all other switches are OFF, the program can be checked without any motion along the Z-axis, while the rest of the program is processed normally with full functionality.

When using this switch in the ON mode, make sure your ABS (absolute) position is maintained. The initial ABS position should also be available after the *Z-Axis Neglect* switch has been turned back to OFF, its normal setting.

Machine Lock

This is the ultimate machine lock function. While the *M-S-T Lock* switch disables mainly non-axis functions of machine operations, and the *Z-axis Neglect* switch only disables motions along the Z-axis, *Machine Lock* switch locks *all* functions and *all* motions.

When the *Machine Lock* switch is in the ON position, all machine motions will be disabled, along with any related activities. Locking the CNC machine is often used to verify the part program without actually moving any axes. All other settings should be available for best results of the part verification.

Dry Run

This feature of the *Operation Panel* is used quite frequently. The subject title *'Dry Run'* can also be described as saying *'machining the part dry'*. The word *'dry'* simply means machining without a coolant. As most metal machining will require coolant, the term *'running the part dry'* is reserved to the program testing stage where there is no part mounted in the fixture, therefore it can be run *'dry'*, as not coolant is required.

A common application of *Dry Run* is combined with the *Z-axis Neglect*, when both switches are turned ON. In this case, the Z-axis is locked a short distance above the part, so visually following the toolpath is much easier than if the Z-axis were locked too high, machine zero position included.

Manual Absolute

Older CNC machines had a toggle switch identified as *Manual Absolute*, with its ON and OFF mode, like other toggle switches. Modern machines only show a status light, but do not provide a toggle switch. The setting ON or OFF can still be made, but through a much more involved method, suing system parameters. That is not to make the CNC operator unhappy - it is for overall safety.

> For safety reasons,
> *Manual Absolute* mode should always be ON

Manual Absolute is a switch (in whatever form) that updates (if set to ON) or does not update (if set to OFF) the position of the cutting tool when it is moved manually in the middle of program processing. Consider this situation:

Ten holes have to be machined, using drilling fixed cycle mode **G81** with **G99** command. **G81** will drill all ten holes, **G99** controls the retract of the drill from each hole to the feed plane level programmed as R-address in the cycle - this is called the R-level. Assume that the R-level is set at 2.5 mm (0.1 inches). When the tenth hole is completed - where exactly is the drill tip point?

Yes, it is at whatever XY drawing location of the tenth hole, but it is also 2.5 mm above the part. This rather short distance prevents the operator to look into the hole and check it, because the drill is in the way. *Manual Absolute* mode ON comes to the rescue.

By setting the operation mode from *Auto* to *Manual* and/or *Handle* mode (see later explanations), the drill can be moved manually from its current location to a location that allows visual evaluation of the machined hole. of course, the manual motion has changed the programmed XYZ coordinate location. Continuing program processing from this manually reached coordinate location would present serious problems - unless ...

Unless the *Manual Absolute* mode is turned ON !

> *Manual Absolute* mode ON will *update*
> the current coordinate system by the amount
> of any manual motion
> performed during program processing

From the example, you can see that *Manual Absolute* setting being ON makes sense, and setting it OFF does not. For that reason, most controls do not have *Manual Absolute* function as a toggle switch anymore, just an indicator light.

Auto Power

Not every control system has a special switch that allows a selection between *Auto Power* ON and OFF. The purpose of this switch is to allow the user an automatic shut-down of the CNC machine, when the control reads the end-of-program function **M30**. Wire EDM machines often do include this switch, as machining operations are frequently unattended, and may last several hours.

Coolant Function

Coolant function also uses a toggle switch for coolant control, but this switch deviates from the original definition of 'toggle', as it has *three* selectable positions, rather than the normal two.

In the CNC program, coolant (actually coolant pump) is normally turned on by the miscellaneous function **M08** and turned off by function **M09**.

Although the majority of metal machining will require coolant, there are several cases when coolant is either not recommended or not required:

- For initial program verification (*Dry Run* mode)
- For certain materials, such as wood
- When using ceramic or similar inserts
- For some hard-machining applications
- ... other reasons

The three-way setting of the *Coolant* switch provides the following options:

- ON setting ... Coolant is always enabled (ON), regardless of part program data
- OFF setting ... Coolant is always disabled (OFF), regardless of part program data
- AUTO setting ... Coolant is ON *or* OFF, as per program instructions, using M08 or M09 coolant functions

These three settings provide flexibility at the setup.

Program Testing and Setup Mode

Various toggle switches being ON or OFF provide great amount of flexibility during program testing and setup. The question that remains to be answered is *'what else has to be set to make the testing accurate?'*

Keep in mind that program or setup integrity test requires all settings to be completed first. That includes:

- Program (including subprograms and macros) has to be in CNC memory or available in DNC mode
- Setup fixture has to be in fixed position
- Part zero has to be set (G54 - G59)
- Tools have to be mounted, loaded, registered and ready to use
- Offsets have to be set - that includes work offset, tool length offset and cutter radius offset
- Part itself may or may not be mounted in the setup fixture, depending on the type of test performed

ROTARY SWITCHES

Unlike toggle switches which typically select one of two (or three) modes possible, rotary switches select from several different modes. Typical rotary switches found on CNC systems are:

- MODE selection
- AXIS selection
- RAPID override
- FEEDRATE override
- SPINDLE override
- JOG feedrate selection

Mode Selection

The *Mode Select* switch on the operation panel typically offers eight setting in two groups. Each mode has four settings, and only one setting selection can be active at any given time.

- Group 1 = Manual Mode ... four settings
- Group 2 = Automatic Mode ... four settings

Manual Mode includes selection of the following four modes of operation:

- HOME
- HANDLE
- JOG
- RAPID

AUTO mode includes selection of the following four modes of operation:

- MDI
- DNC / EXT / REMOTE / TAPE
- MEM (MEMORY)
- EDIT

HOME Mode Selection

Most CNC operators use the expression *'Home'* when they refer to the *machine reference position* also known as the *machine zero*. For example, *homing the machine* is a term that means all axes will be located at the machine reference point (machine zero).

This is a very special position of any CNC machine. It is the *true machine origin*. Every number that represents axis dimension in the program or settings is somehow related to this point.

In this mode setting, you can move any axis to machine zero. Some machines allow multi-axis motion, other only one axis at a time. Home position also serves as the starting location from which to start a new program, at least for the first part. Most CNC machines require that all axes are zeroed before any program processing can be done. Zeroing the machine - *sending all axes home* - is one of the first activities after the main power had been turned on for the machine. Several newest CNC machines have a special memory that allows avoiding the manual return, but they are still in minority.

The process of sending each axis **HOME** after *Power ON* is fairly consistent for all machine types:

- Turn power to the machine
- Turn power to the control
- Switch to *HOME* mode selection
- Select axis
- Press the + button
- Repeat for other axes

You may need to hold the + **RAPID** button for a second or two to start and then let go.

> Return to primary machine zero (Home) is always a *positive* direction in all axes

Generally, it is a good idea to set all relative (**REL**) positions to zero. There are two major display positions - **ABS** (absolute) and **REL** (relative):

- ABS ... is used for monitoring current tool position
- REL ... is used for setup

What can go wrong ?

As the saying goes, *anything* can go wrong, but in the **HOME** mode, the most common problem is *over travel*. Over travel happens when the selected axis tries to move outside of the machine travel limits. Every CNC machine has a precision limit switch that prevents over travel. When this switch is tripped, the control system will issue an O/T (over travel) alarm.

The most common cause is that the axis position was too close to the home position. The solution is simple - reset the alarm and move the axis that overtraveled further away from machine zero - minimum of 25 mm or one inch should be sufficient.

Necessary precautions

The best precaution to overtravel when homing the axes is not at the beginning of the day but at the end of the day. When the machine is ready to be shut off, make sure that each axis is at least 25 mm or one inch away from machine zero. There should be no overtravel when you use this method.

MDI Input

If you find moving each axis separately to the **HOME** position time consuming, there is an alternative. Use the **MDI** *(Manual Data Input)* mode and the proper program command. Keep in mind that there is a big difference between sending machine home in absolute and incremental mode. For setup at the machine, the *incremental* mode is the one to use.

Program command **G28** for the primary machine zero (or **G30** for the secondary machine zero) is a command that is not always easy to understand, particularly by CNC operators.

G28 or **G30** is simply defined as *Machine Zero Return*

The real definition is longer and more complex:

> **G28 is the machine zero return command that will move the specified axis to primary machine zero through an intermediate position**

The same applies to **G30**, except the motion is to the *secondary* machine zero position.

The intermediate position is often used in programming to avoid an area that can cause collision - this is not the case when zeroing machine axes, and the intermediate position is not necessary. Here are some examples how to zero machine *(send all axes home)*:

On a CNC machining center:

```
G91 G28 X0            (MACHINE ZERO IN X ONLY)
G91 G28 Y0            (MACHINE ZERO IN Y ONLY)
G91 G28 Z0            (MACHINE ZERO IN Z ONLY)
G91 G28 B0            (MACHINE ZERO IN B ONLY)
G91 G28 X0 Y0 Z0      (MACHINE ZERO IN XYZ)
```

On a CNC lathe:

Incremental mode uses U and W for X and Z axes:

```
G28 U0 W0             (MACHINE ZERO IN XZ)
```

U0 or *W0* can also be used as a single axis motion.

HANDLE Mode Selection

Selecting the **HANDLE** mode, you are transferring all manual motions from the control panel to the device that is an integral part of all CNC machines - the *handle*.

Fanuc calls this manual device *Manual Pulse Generator (MPG)*, but nobody else does - it is a *handle*, plain and simple. Most handles are manual devices, but many new models are digital, at least in the sense that they include a small screen display that can show much more then position only.

Regardless of the design, the handle is a device used for manual axis motion, commonly used in setup. Its design usually includes a coiled cable that can be extended into the working area. During setup, you select the axis you want to move and the amount of increment, then use the wheel to move by a very precise distance. The handle can be set in such a way that the minimum increment (that is the smallest amount of motion) is:

- 0.001 mm

    or

- 0.0001 inches

These amounts are standard on all CNC machining centers and CNC lathes. Because of the many features the handle offers, a separate chapter has been developed for that purpose - see chapter named *Setup Handle*.

JOG Mode Selection

The word *jog* means nothing more than *continuous feed*. This mode (and the following *Rapid* mode) have no relation to any program features. Both are strictly used in manual mode, particularly during part setup.

RAPID Mode Selection

Rapid mode is logically related to the *Jog* mode and located next to it. While **JOG** mode is used for motions that require manual feedrate, the **RAPID** mode is used in manual operations to provide rapid motion of the selected axis.

In conjunction with the axis selection and direction of motion, any single axis can be moved in two directions. Velocity of the rapid motion will also depend on the setting of the *Rapid Override Switch* (described later).

 For safety reasons, the rapid motion will only take place, while the ± switch is being pressed.

> During setup, it is safer to start with medium or slow rapid motion and increase it as necessary

MDI Mode Selection

MDI stands for *Manual Data Input*. This feature enables the CNC operator to enter one or more program blocks as a group and process them together. MDI mode is for temporary use only, for example, during setup or testing. It is not intended to run production programs.

MDI mode can be called *programming on the fly* - if you want to use this mode, you have to have at least rudimentary knowledge of programming. MDI mode uses actual program codes to achieve certain results.

For example, the following MDI command will move the X-axis by negative 271.450 mm, in rapid mode:

```
G91 G00 X-271.450
```

The motion will start at the current tool position and end 271.450 mm away in the negative X direction.

The process is simple:

- Select the *MDI* mode
- Type-in the desired command block or series of blocks
- Press *Cycle Start* button to execute

Each block must be terminated by the EOB key, which is the End-Of-Block key

DNC / EXT / REMOTE / TAPE Mode Selection

Depending on the control unit, this switch selection will be identified by one of the descriptions listed in the title. Regardless of the description, the switch always refers to a certain *external* mode of operation. Normally, there are two modes of operations:

- MEMORY ... identified as MEM
- EXTERNAL ... identified as DNC / EXT / TAPE / REMOTE

Memory type of program processing is controlled by the **MEMORY** mode selection (see details in the next section).

The other external types of operations are all the same - they are just identified by a different label on the operation panel of the control system.

Abbreviation **DNC** means *Direct Numerical Control* or *Distributed Numerical Control*. The first abbreviation - *Direct Numerical Control* - often means connection between one external computer and one CNC system, while *Distributed Numerical Control* identifies connection between one external computer and several CNC systems. In both cases, suitable DNC software has to be installed on the external computer and properly configured. General rule of DNC setup is that settings at CNC and external computer must be the same.

In DNC operation, the program is processed from the external device and is *not* registered in CNC memory. The greatest advantage of running a program in DNC mode is that its length is only limited by the available hard drive space on the host computer.

Capacity of modern hard drives far exceeds even the most demanding CNC applications. Programs of virtually any length can be executed from an external computer in DNC mode. Program editing, if necessary, is also done at the external computer, not at the CNC unit.

Some CNC machines, even some very recent ones may still use the description *TAPE*. Paper or mylar tape had been used in the early years of numerical control as the only external source available, but it's not used anymore. The description *tape* is more nostalgic than practical, and it means *any* external device.

MEM or MEMORY Mode Selection

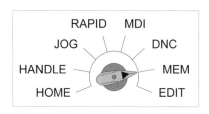

In memory mode, the part program has to be registered in the control memory storage area first. Once registered, it can be processed to run a production job. Keep in mind that memory capacity is limited, depending on the control. CNC memory is not intended for permanent storage of part programs. See *page 21* for more details.

EDIT Mode Selection

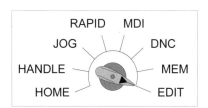

Selecting the **EDIT** mode enables the CNC operator to access and edit any unprotected programs stored in the CNC memory.

The actual editing process depends on the control system, but generally it includes the following steps:

- Select EDIT mode
- Select program to edit
- Set the cursor where the change should take place
- Make the required change
- Exit EDIT mode

Control manual describes each step in more detail.

There are three modes of editing, using the control panel *Edit keys* (see *page 26* for details):

- **ALTER**
- **INSERT**
- **DELETE**

The **ALTER** key will *change* the currently highlighted command or commands.

The **INSERT** key will *add* a new command or commands at the current highlighted command.

The **DELETE** key will *remove* the currently highlighted command or commands.

Axis Selection

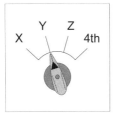

In manual operation, an axis has to be selected before a motion switch is applied. Typically, and axis must be selected before zero return motion, manual rapid motion, jog feed, and incremental jog feed.

Most operation panels show not only the three major axes XYZ, but also the fourth axis, even if it is not installed. On vertical machining centers, if the fourth axis is installed, it will generally be the *A-axis*. Horizontal machining center will always have the fourth axis, specified as the *B-axis*.

For multi axis machine tools, the *Select Axis* switch will always include all present axes of the machine. For example, a CNC boring mill will have the following axis selection:

X - Y - Z - W - B
or
X - Y - Z - W - 4th

OVERRIDES

As the title name suggests, *override* implies a change. In CNC operations, override means a temporary change of machine setting or program data. There are three independent override modes:

- **Rapid override**
- **Feedrate override**
- **Spindle override**

Each override mode has an available rotary switch and can be used independently from the others. The main reason for override switches is the added control at the machine. Override switches are commonly used in setup, running the first part or changing programmed speeds and feeds on a temporary basis.

Rapid Override

In the part program, any rapid tool motion is called by the **G00** command. **G00** has no feedrate associated with it; how fast the tool moves depends on the machine specifications - *actual rapid rate cannot be programmed*. Modern CNC machining centers can rapid at a fairly common rate of 60 000 mm/min (2360 ipm) or even higher. As much as this speed is welcome for production, it is generally too fast for setup and first part run. Using the override switch to slow down the rapid motion is a welcome relief, when needed.

> **Rapid override switch indicates the *percentage* of the maximum amount of rapid motion that the machine is capable of**

Generally, CNC machining centers that do not exceed 30 000 to 38 000 mm/min (1180-1500 ipm) have a selection range different from machines with ultra high rapid motion. Compare the two rotary switches available from the CNC operation panel:

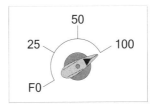 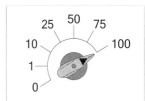

Machine manufacturer may choose a different range. What is important to understand is that machines with very high rapid rate need finer adjustments. A comparative example will follow, but let's look at the F0 selection first.

F0 Selection

On some machines, there is a selection that does not specify the actual percentage of the rapid rate reduction and is identified as **F0**. If available, this will be the lowest percentage, but a percentage that can be changed by system parameter. Other percentages are fixed. For example, if 25% is the lowest fixed setting, you can make the F0 setting to be 20%, 15%, 10%, and so on.

Rapid Override Comparison

Consider the difference between two machines - as an example, one with a maximum rapid rate of 30 000 mm/min (1180 ipm) and another with a fast rapid rate of 60 000 mm/min (2360 ipm):

% Setting	30 000 mm/min	60 000 mm/min	1180 ipm	2360 ipm
100%	30 000	60 000	1180	2360
75%	n/a	45 000	n/a	1770
50%	15 000	30 000	590	1180
25%	7 500	295	295	590
10%	n/a	6 000	n/a	236
F0 = 10%	3 000	n/a	118	n/a
1%	n/a	600	n/a	23.6

Depending on the setting of **F0** (10% used) the table shows how fast the CNC machine will rapid with different override settings.

Rapid Override On Lathes

Most CNC lathes have a rapid override setting in the range of 0-25-50-100%. The main difference on CNC lathes is that the X-axis, because of its shorter travel distance) will have a rapid rate about of only one half of the rapid rate for the Z-axis. For example:

➡ *Older lathe:*

X-rapid rate   =   12 000 mm/min [472 ipm]
Z-rapid rate   =   24 000 mm/min [944 ipm]

➡ *Newer lathe:*

X-rapid rate   =   20 000 mm/min [787 ipm]
Z-rapid rate   =   40 000 mm/min [1575 ipm]

For multi-purpose lathes (mill-turn or turn-mill type), each additional axis will have its own maximum rapid rate assigned by the manufacturer.

Also keep in mind, that some machines use push-button switches rather than rotary switches. In all cases, the functionality of the *Rapid Override Switch* is always the same.

A few suggestions that should come handy for CNC operator at any level of experience:

> Use RAPID OVERRIDE switch for setup and verifying the first part of the run

For running a new part:

> When running a new part, start with RAPID override switch set to between 10 and 50 percent for safety

For production, the fastest cycle time is required. Once the setup has been confirmed as correct and the part program has been optimized, run at 100%:

> After verifying both setup and program, run production at 100% rapid rate for the best cycle time

Modern CNC machines are well built and can withstand high rapid rates repeatedly.

Feedrate Override

Another very useful override switch is the *Feedrate Override* switch. In the part program, a cutting feedrate is programmed using the F-address. Feedrate is associated with cutting motions, using commands **G01**, **G02**, **G03** as well as various cycles and several other motion commands. There are two methods of programming a feedrate, mainly depending on the machine type:

- Feedrate per *minute*   ... typically used for *milling*
- Feedrate per *revolution*   ... typically used for *turning*

Feedrate per minute is normally used to program cutting motion on CNC milling machines and machining centers, while feedrate per revolution is programmed for cutting motion on CNC lathes. Depending on the current units (**G20** for imperial units or **G21** for metric units), cutting feedrate will use the following format:

Feed per minute		Feed per revolution	
Metric	Imperial	Metric	Imperial
mm/min	inches/min	mm/rev	inches/rev

Programming examples - Metric:

F250.0   ... 250 millimeters per minute   250 mm/min
F0.3   ... 0.3 millimeters per revolution   0.3 mm/rev

Programming examples - Imperial:

F10.0   ... 10 inches per minute   10 ipm
F0.012   ... 0.012 inches per revolution   0.012 ipr

> The purpose of the *Feedrate Override* switch is to optimize feedrates provided by the part program

Feedrate Optimization

The programmed feedrate is always 100% and can be decreased or increased at the CNC machine by using the *Feedrate Override* switch.

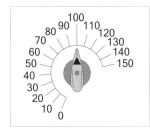 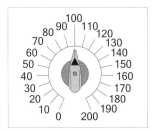

Typical range of a feedrate override switch settings is between 0 and 200 percent, or in some cases only between 0 and 150 percent, in 10 percent increments. When set to 0%, there will not be any axis motion, although there are exceptions on some older machines.

> When running a new part,
> start with FEEDRATE OVERRIDE switch
> set to between 10 and 50 percent for safety

Even experienced CNC programmers often program cutting feedrate a little too low or a little too high. This is not unexpected, as it is always difficult to predict the exact effect of the tooling and setup on speeds and feeds ahead of time. If the CNC operator is qualified and authorized to make changes to the cutting feedrates, it is important to follow a certain procedure for each tool used by the program:

- Start the first part in low override setting (10-50%)
- Adjust the override during dry run (usually up)
- Watch for feedrate deviations of each tool
- Keep track of the 'best' override rate for each tool
- Recalculate the new feedrate for each tool
- Modify actual program settings for the F-address

It is not unusual to adjust spindle speed as well, depending on the actual cutting conditions. Other factors, namely width and depth of cut will also have an effect on speeds and feeds, but that usually involves major program changes, not always quickly done at the machine.

Optimizing Programmed Feedrate

If the cutting feedrate in the program work without adjustments, there is nothing to do but run the program at 100% feedrate override. In many cases, some adjustments will be necessary, even if they are small. The following two examples will show two typical changes - one for a feedrate that to be decreased, and the other for a feedrate that has to be increased.

➡ *Example 1 - Programmed feedrate is too HIGH:*

If the best cutting conditions are achieved when the feedrate override switch is set *below* 100%, it means the programmed feedrate was too *high*.

In the first program example, using *metric* units, the programmed feedrate is set to `F325.0` or 325 mm per minute. This feedrate proves to be too high, and the operator finds the best cutting conditions at 70% setting of the feedrate override switch. What will the change be in the program, so the tool can cut at 100% (no override)?

325.0   = 100%
1%   = 325 / 100   = 3.25
70%   = 3.25 x 70   = 227.5 mm/min

Change program from `F325.0` to `F227.5`.

In a similar example, using *imperial* units, the programmed feedrate is set to `F15.0` or 15 inches per minute. This feedrate proves to be too high, and the operator finds the best cutting conditions at 90% setting of the feedrate override switch. What will the program change be, so the tool can cut at 100% (no override)?

15.0   = 100%
1%   = 15 / 100   = 0.15
90%   = 0.15 x 90   = 13.5 in/min

Change program from `F15.0` to `F13.5`.

➡ *Example 2 - Programmed feedrate is too LOW:*

If the best cutting conditions are achieved when the feedrate override switch is set *above* 100%, it means the programmed feedrate was too *low*.

The calculations are exactly the same as shown in the previous two examples. In fact, a simple formula can be used for all cases:

$$F_n = F_p / 100 \times \% \quad \text{or} \quad F_n = \frac{F_p \times \%}{100}$$

where ...

F_n = **New optimized feedrate**   *New feedrate*
F_p = **Programmed feedrate**   *Programmed feedrate*
% = **Feedrate override setting for best results**

Spindle Override

Both CNC machining centers and CNC lathes have a switch for *Spindle Speed* override. In both cases, the purpose of this switch is to adjust the programmed speed to optimize cutting conditions.

The actual speed of the spindle is programmed in different ways for milling and turning applications, using the address *S*. The machine will always rotate in certain *revolutions per minute*. In milling, the spindle speed is programmed directly:

S2250   ... 2250 r/min spindle speed

In turning, either the cutting speed (CS) or the spindle speed (rpm *or* r/min) can be programmed, depending on the selection of **G96** or **G97**:

G96 S200   ... 200 m/min cutting speed   ... metric
G96 S450   ... 450 ft/min cutting speed   ... imperial
G97 S450   ... 450 r/min spindle speed

The **G96** command in the program interprets the *S* address as cutting speed in feet or meters per minute, whereas **G97** interprets the *S* address as spindle speed in revolutions per minute. In all cases, the rotation of the spindle has to programmed as well. That is done in the program by selecting a miscellaneous function:

- **M03**   ... spindle rotation *clockwise*   CW
- **M04**   ... spindle rotation *counter-clockwise*   CCW

Spindle Override Range

When the spindle speed override switch is used, it will always *increase* or *decrease* the spindle speed in revolutions per minute, regardless of the machine type, units or rotational direction. Typical range of this switch is between 50 percent and 120 percent of the programmed speed, in 10 percent increments. Very few CNC machines have a larger range, for safety reasons.

Optimizing Programmed Spindle

Optimization of the programmed spindle follows the logic and process explained in the previous section related to feedrate optimization. Even the formulas are the same, except spindle speed replaces feedrate:

$$S_n = S_p / 100 \times \% \quad \text{or} \quad S_n = \frac{S_p \times \%}{100}$$

where ...

S_n = New optimized speed   *New speed*
S_p = Programmed speed   *Programmed speed*
% = Spindle override setting for best results

Use the term *spindle speed* in context. It normally refers to actual speed in rpm (r/min or rev/min). Note that the formulas do not specifically define *spindle speed*. The speed input can be either actual spindle speed in revolutions per minute or it can be a cutting speed in meters per minute or feet per minute. The output - *the optimized speed* - will be in the same units.

➪ *For example - 01:*

For milling, the programmed spindle speed is **S2460**, which is 2460 rpm. Optimized spindle speed is at 80%. To run the program at 100%, the programmed spindle speed has to be changed:

2460   = 100%
1%   = 2460 / 100 = 24.6
80%   = 24.6 x 80   = 1968 rpm

Change program from **S2460** to **S1968**.

➪ *For example - 02:*

For turning, the programmed speed is **G96 S350**, which is 350 meters per minute or 350 feet per minute, depending on the current units. Optimization works best at 110%. To run the program at 100%, the programmed speed has to be changed:

350   = 100%
1%   = 350 / 100 = 3.5
110%   = 3.5 x 110 = 385 cutting speed

Change program from **G96 S350** to **G96 S385**.

> **Spindle function must always be an integer**
> **Decimal points are not allowed**

Jog Feedrate Selection

During setup and related activities, we can select not only the *RAPID* mode for moving individual axes manually as rapid motion. We can also control the feedrate, for example, to straighten up stock edge or clean up a face. This is done by following four steps:

- Select *JOG* mode
- Select AXIS to move
- Select jogging RATE in mm/min or in/min *(see next page)*
- Select motion direction ±

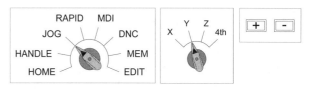

In order to feed at a particular feedrate, most control units offer a rotary switch with a number of selections - this is the **JOG** feedrate setting. Most CNC machines show the selection in millimeters per minute, while some will show the selection in inches per minute. The illustration example shows a range of jogging feedrates between 0 and 4000 mm/min (0 to 157 in/min).

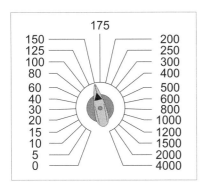

Generally, the middle top setting is a reasonable standard setting, neither too fast nor too slow. *175 mm/min* is just under *7 ipm*. You may want to start at a lower setting and gradually increase.

The jogging feedrate can be set *before* an axis motion starts, which is the preferred method. It can also be increased or decreased *during* motion. This can only be done by using both hands; while holding the directional ± button with one hand, adjust the feedrate setting with the other hand.

Incremental Jog

Most controls have a feature called *Incremental Jog*. This feature allows presetting a specific distance and direction along a selected axis. When applied, the length of motion will correspond to the incremental settings.

EDIT KEY

Access to the program data and many settings can be controlled by several ways. Most CNC machines use the *Edit Key*, where a real key is inserted into the keyhole located on the operation panel. By turning the key, editing of part programs (and some other features, such as certain parameters) can be *enabled* or *disabled*.

Some controls do not use a key but a switch, which is a more convenient method, but less secure.

EMERGENCY STOP SWITCH

One of the most important switches is the *Emergency Stop* (known as the *E-Stop*). This ubiquitous mushroom-like red button switch is always there if something goes wrong. In fact, it is a mandatory feature for every machine. Although it is sometimes called the 'panic button', using overall calm and a clear head are much better ways to approach a problem, even in case of real emergency. Panic distracts from rational thinking and contributes to poor decisions, so keep calm. Note that even when the *E-Stop* switch *is* pressed, machine activity does *not* stop immediately. Physical forces such as gravity and motion deceleration play major role before any machine motion comes to a final full stop.

Also keep in mind that *Emergency Switch* is for real emergencies, particularly when there is a threat to operator's well being. This switch should also be used if the operator detects a potential collision between the tool and the part or fixture, or any other imminent threat.

OTHER FEATURES

New machine features and improvements are in constant state of development. In order to apply these features, they have to be connected (interfaced) between the machine and the control system. In many cases, that necessitates additional switches, dials, and buttons to be implemented on the control panels or the display screen.

For example, a *spindle load meter* is found on many machines, either as a display bar on the monitor or a physical instrument on the operation panel, where a needle shows the load on the spindle during machining. Another examples cover switches for *door open/close*, *chip conveyor*, *pallet selection*, and many others.

CONTROL PANEL SYMBOLS

It used to be that all switches, buttons and other items on the operation panel were identified in English language. Over the years, that has changed. Although localized labeling does exist where market condition warrant, most control and machine manufacturers think globally. Instead of using a text based descriptions, they choose internationally accepted symbols. Many are common, some with only small variations, others are unique, particularly those that control various optional features.

Each machine manufacturer provides a page of two (or more) of the pictorial symbols (icons) used on the front of the operation panel. The illustration shows a fairly large number of symbols. Some symbols serve the same purpose but the icon is different. There is no established standard, so a certain function of one machine will have a different symbol on another machine. The list presented, while quite extensive, is not a complete list by any means. Use it only as a reference to better understanding the icons for CNC machining centers. Lathe controls have additional symbols unique to that type of machine. The same applies to EDM machines, and other types.

Watch for those situations, where the same symbol has a different meaning on two different machines. Also watch for different symbols identifying the same function.

Switches

One set of switches using symbols is the power switch set - the *Power* key may or may not be present, if the *Switch* keys are:

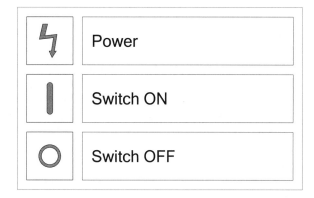

Modes

There are two modes on all CNC machines represented by either a rotary switch or keys with symbols. These modes are the *Manual* mode and the *Auto* mode. Selecting one mode over the other allows selection of only one function out of eight functions available:

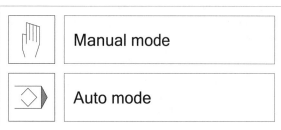

Manual Mode:

In the *Manual* mode, the four selections are:

- Zero return - Home
- Handle
- Jog
- Rapid

Auto Mode:

In the *Auto* mode, the four selections are:

- MDI
- Tape / DNC / Ext
- Memory
- Edit

The main difference between controls is the *Tape / DNC / Ext* mode. You will not see all three describing the same key, but different manufacturers have their own preferences. Although *Tape* is an obsolete input device for a great number of years, the tradition dies hard. *DNC* means *Direct Numerical Control*, and *Ext* means *External*. That is exactly what this key is for - connection with an external device, such as a computer, for example.

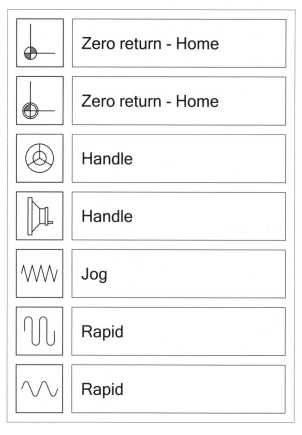

CNC Control Setup for Milling and Turning
OPERATION PANEL

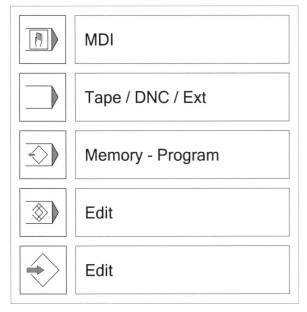

An important part of machining is controlling the three major functions:

- **Rapid motion**
- **Feedrate**
- **Spindle**

Rapid motion is determined by the machine manufacturer, and the maximum allowed is 100%. *Rapid motion override* allows slowing down the rapid motion, if necessary. While the program provides the cutting feedrate and the spindle speed, it is not unusual to adjust either one up or down. For temporary changes at the machine, feedrate can be adjusted from 0% to 150% or 200%. Spindle speed can be adjusted between 50% and 120%. In both cases, 100% is the programmed amount.

Operations

During machining, various operations are controlled by a setting of a particular key. These keys with symbols have the same purpose and function the same way as the rotary switches described earlier in tis chapter. As in other examples, there may be more than one key design.

Several keys can be used to control manual activities of the spindle. The illustrations here show only the keys, but additional dials are located on the operation panel.

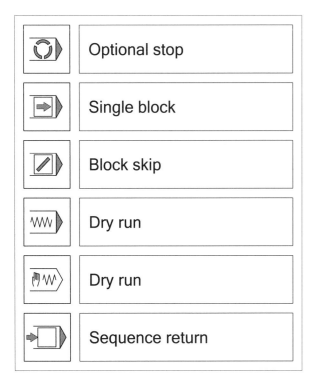

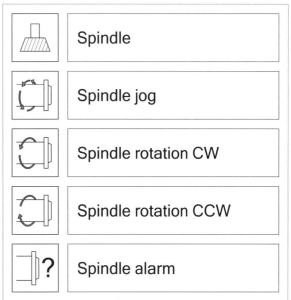

Operations of tools include keys that allow the tool to be placed into the spindle or taken out. *Tool In / Tool Out* key will normally clamp and un-clamp the tool. *Tool clamp / Tool un-clamp* keys are generally variations of the same function.

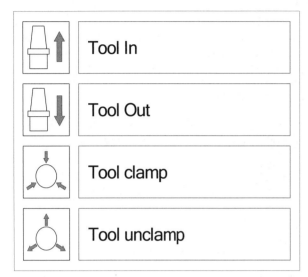

In addition, some miscellaneous keys are also part of the operation panel:

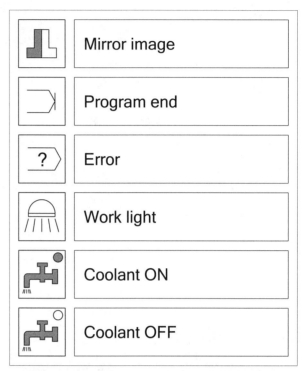

This is just a typical list of symbolic keys used on modern CNC control systems. Check your machine manual for the full listing and meaning of each key.

6 SETUP HANDLE

While the chapter title may be a bit ambiguous for the programmers with limited interaction in actual machining, its meaning should be quite clear to CNC operators at all levels. What is generically described as 'the setup handle', is commonly known under several other names or descriptions. The name of the device defines the main purpose of this ubiquitous hardware, 'officially' called - *manual pulse generator* - also known as the MPG device by its initials. Other names commonly used in machine shop users are *'handle'*, *'dial'*, *'pendant'*, and some other descriptions. For the purposes of this chapter, the all-inclusive term *'handle'* will be used.

PURPOSE

Regardless of how a particular machine handle is designed, it always serves the same purpose during actual machine setup. When setting up a part for CNC machining (milling or turning), the CNC operator needs to use the control and operation panels of a CNC machine to input various offsets, to move the machine slides by a certain distance, or make many other settings. In the category of 'other settings', the most common is setting the work offset using an edge finder or a similar device.

> The single purpose of handle is to allow CNC operator to control axis motion(-s) during setup *away* from the main operation panel

As the definition indicates, the purpose of the handle is to allow various axis motions to take place during setup, when access to the operation panel is too far from the work area or is otherwise restricted.

Whether built into the machine enclosure or used remotely, the handle is normally used in manual mode of operation, selected from the operation panel by a switch set to - *HANDLE*.

The typical handle is a mechanical device sending electronic signals to the control system. Each signal represents the amount of motion to be performed along each axis selected. The use of the handle is generally limited to machine and tool setup.

DESIGN

Although their main purpose does not change from one CNC machine to another, handles used for setup do vary quite a bit in their design. On small CNC machines, the handle is often built in the operation panel or the control panel. For smaller machines, this design hardly presents any problems, as the handle is always within comfortable reach of the CNC operator, during part setup. On the other hand, medium and large CNC machining centers (as well as many CNC lathes) do not provide the same ease of handling, so they provide a handle that is *detached* - such a handle is not a part of the operation or control panel, but a unit that is connected to the control panel via a coiled cord. In this case, the handle pendant also includes switches that allow remote settings. Many modern handles are fully digital and can display much more information than traditional handles.

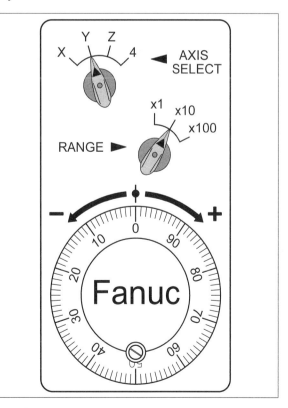

HANDLE FEATURES

As the previous illustration shows, there are three major selections that can be made by a setup handle, regardless of its actual design:

Handle selection	Resulting activity
Axis selection	Selects axis to be set
Range selection	Selects minimum motion amount for one handle division
Dial wheel	Selects the axis direction, based on axis and range selections

It should be understood that *all three features* always work together - there is no such thing as 'set one, ignore the others'.

Handles with digital display have additional features, typically those that would require looking at the main display screen. Good examples are tool position screen, work offsets, even a program listing.

Mode Selection

In order to make the handle functional, proper mode must be selected from the machine operation panel. There are four *MANUAL* mode selections, shown at left, and four *AUTO* selections, shown at right. Some controls use reverse orientations of the two modes.

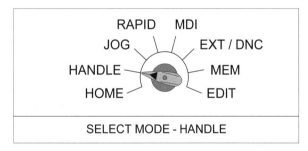

As an example shown in the initial illustration, the rotary switch points out to the Y-axis as the current selection. That means all axis motion generated by the dial will apply to the Y-axis only. Also, the range selection will apply to the Y-axis as well.

Range Selection

Once the working axis has been selected, the next selection relates to the handle *range*. This part of any CNC operation is often most misunderstood. The selection range is commonly identified with certain X-codes. They may look like some 'mysterious' specification, for example, X1, X10, and X100 are typical. Contrary to some interpretations, there is no mystery in these selections at all. The key to understanding such designation is that the letter 'X' does ***not*** refer to a particular axis - the letter X is nothing more than a symbol for the ***multiplying factor***, meaning 'times'. So X1 is actually 'times one', X10 is 'times ten', and X100 is 'times one hundred'.

Of course, the qualifier 'times' in the expression has to have some more detailed additional definition. If X1 means *'times one'*, and X10 means *'times ten'*, then the natural follow-up question is - times *what*? The answer is simple - *'X-times the minimum increment'* of an axis motion.

While the answer is correct in principle, it should be enlarged - after all, what is the 'minimum increment'?

Most control manuals use the term *'minimum increment'* when they really mean *'the smallest axis motion possible'*. Either definition refers to the smallest motion the machine can make within the scope of the selected units (Metric or Imperial). In practice, that means the smallest amount of axis motion is 0.001 mm or 0.0001 of an inch:

> Minimum increment is equivalent to the smallest amount of motion => 0.001 mm or 0.0001 inches

Dial Wheel

Once the desired machine axis had been selected and the range of motion specified, the handle dial becomes the most important instrument during part setup. The handle dial is divided into one hundred equal divisions, generally called *increments* or *resolutions* or *notches* - or - *divisions*.

> One division (notch) always represents the minimum increment within the selected range

Although the handle design will vary from machine to machine, and control to control, its content, purpose, and applications remain the same, regardless of the model.

The following series of tables illustrate the results of various settings and their influence on machine motions. They should answer the most important question a CNC operator might ask is - *what is the length of axis motion when different settings are applied.*

The following tables show the length of motion per dial division (notch) for metric and imperial units respectively, as well as motion amount per turn of the dial.

> Note that when using x100 range,
> even a small turn causes large axis motion

Motion Length per Handle Division - Metric

When a CNC machine is set to *metric* mode of measurement, the units of axis motion are in millimeters (mm). In metric system, the minimum increment (axis motion) is one millimeter divided by 1000, into microns:

One micron = 1 mm / 1000 = 0.001 mm

Increment	x1	x10	x100
1 division	0.001 mm	0.010 mm	0.100 mm
10 divisions	0.010 mm	0.100 mm	1.000 mm
100 divisions	0.100 mm	1.000 mm	10.000 mm

Motion Length per Handle Division - Imperial

When a CNC machine is set to *imperial* mode of measurement, the units of axis motion are in inches (in). In imperial system, the minimum increment (axis motion) is one inch divided by 10000, into one ten thousandths of an inch, commonly known as *'one tenth'*:

One ten thousandth of an inch = 1 inch / 10000 = 0.0001"

Increment	x1	x10	x100
1 division	0.0001 in	0.0010 in	0.0100 in
10 divisions	0.0010 in	0.0100 in	0.100 in
100 divisions	0.0100 in	0.1000 in	1.0000 in

Another way of using the handle is to think of the dial in terms of amount of axis motion in full dial turns.

Motion Length per Handle Turn - Metric

This is a variation on the metric version for motion length of an axis per one division. In the previous metric table, 100 notches represent one full turn of the handle dial in any range setting (360° turn of the dial).

Number of turns	x1	x10	x100
1 full turn	0.100 mm	1.000 mm	10.000 mm
10 full turns	1.000 mm	10.000 mm	100.000 mm
100 full turns	10.000 mm	100.000 mm	1000.000 mm

Motion Length per Handle Turn - Imperial

This is a variation on the imperial version for motion length of an axis per one division. In the previous imperial table, 100 notches represent one full turn of the handle dial in any range setting (360° turn of the dial).

Number of turns	x1	x10	x100
1 full turn	0.0100 in	0.1000 in	1.0000 in
10 full turns	0.1000 in	1.0000 in	10.0000 in
100 full turns	1.0000 in	10.0000 in	100.0000 in

Number of Divisions for a Given Axis Motion

The setup handle can be used for any amount of axis motion, from the smallest increment up. The table below shows formulas to calculate number of handle divisions required to move the selected axis by specified distance:

Range setting	Number of divisions for distance in millimeters	Number of divisions for distance in inches
x1	Motion × 1000	Motion × 10000
x10	Motion × 100	Motion × 1000
x100	Motion × 10	Motion × 100

APPLICATIONS

The setup handle is *not* normally used during actual production (while the part program is being processed). The only time that the handle may be used is to provide manual motion when the program processing is interrupted. This can happen, for example, if you want to inspect a hole, but the cutting tool is in the way. Switching to manual mode and using the handle, the tool can be moved away into a clear position and the hole can be inspected.

> **WARNING:**
>
> Make sure the MANUAL ABSOLUTE switch is ON when moving an axis during program processing

Manual Absolute Switch

The previous warning should be taken very seriously. The majority of modern CNC machines do *not* have the feature called *Manual Absolute* at all, at least not in the form of a switch on the operation panel. This feature can be controlled by system parameters, if necessary.

At the same time, on some older CNC machines (controls), there *is* a physical switch, usually located on the operation panel, and identified as *Manual Absolute*. This switch can be turned *ON* or *OFF*. It is very important to understand the purpose of this switch and its *ON* and *OFF* settings.

Manual Absolute ON

Setting the *Manual Absolute* to *ON* makes the only sense in practical applications. Suppose that you need to inspect a part while the part program is still in progress. This is a common activity between tool changes, usually associated with the M01 (Optional Stop) function.

As is often the case, the part area to be inspected is either too far for comfortable inspection, or the tool itself may be in the way of viewing, or some other obstacle prevents the inspection. In any case, it will be necessary to switch from *Memory* mode to manual mode (such as *HANDLE* mode) and move the tool manually from the programmed position.

Keep in mind that once a certain cutting tool is manually moved during program execution, it no longer represents the required position from the part program. This is where the major benefit of *Manual Absolute* is most apparent.

> When *Manual Absolute* is turned ON, any manual tool motion will UPDATE the work coordinate system

Manual Absolute in its *ON* setting allows a manual motion during program processing. Any manual motion will always update the coordinate settings, which means the control system always 'knows' what is the current tool location from part zero.

There is hardly any practical reason why this switch (if available) or the parameter setting should ever be turned off.

Manual Absolute OFF

Keep in mind that if *Manual Absolute* switch is turned *OFF*, any manual motion performed during part program processing will **NOT** update the coordinate system. The unpleasant result of such a situation is a likely scrap due to wrong positioning of the cutting tool.

> As a general rule,
> *Manual Absolute* setting should always be ON

Safety Issues

Using a handle does not require any special skills, but it does require a certain degree of caution. Study the four tables presented in this chapter and especially focus on the range setting of *x100*.

With such a high range setting, it is very easy to make an axis motion too long and possibly hit the part during setup, for example. Even a small turn of the handle can produce a long tool movement in the *x100* setting range.

Keep in mind that a handle is an electronic device - there is no human feel of touch in the handle. The main purpose of the handle is to transfer electronic signals to the CNC system as manual axis motions.

7   MILLING TOOLS - SETUP

MILLING TOOLS

Tools for CNC milling belong to two basic categories:

- **Tool holders**
- **Cutting tools**

In addition, there are other items related to tool holders, such as collets, set screws, wrenches, and other setup tools. There are many tool manufacturers and a few international standards. Neither CNC programmers nor CNC operators have any influence over the tool holders. Tool holders come in several varieties, all have a conical design that has to match the internal design of the machine spindle. The flange of each tool is the part that allows automatic tool change (ATC). The part that is projected from the spindle face is where the cutting tools are mounted. It can be used for a variety of collets or have a fixed diameter built-in. Collet nuts or set screws secure the cutting tool in the holder. Face mill holders are designed to hold face mills of various diameters.

Tool holders that are not designed for milling include drill holders with a *Morse®* taper built-in (four sizes) or tapping heads.

Tool Holders

The purpose of a tool holder is to safely hold a cutting tool in the spindle. The cutting tool is mounted into the holder and is tightened so it does not move during machining. There are three most common standards for CNC machining centers:

- **CAT standard (also called V-flange)**
- **BT standard**
- **HSK standard**

Other standards exist for small machines, such as tool room mills, knee type mills, etc., many using the old **R8** standard, common to old *Bridgeport®* machines.

CAT standard was developed by an American company *Caterpilar®* and uses unique identifications, such as CAT-30, CAT-40, CAT-45, CAT-50, and CAT-60. These designations refer to the taper of the tool holder, as defined by *AMT* (Association for Manufacturing Technology).

Another common standard is the **BT** standard that originated in Japan. It is similar to CAT design, and it designations are BT-30, BT-35, BT-40, BT-45, and BT-50.

These are the most common holders. They are not interchangeable and the major difference between them is the flange style and its thickness. Also, the retention knobs are different. A simple illustrations show the differences:

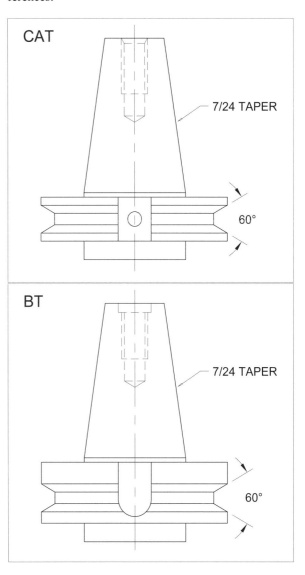

Of the three common tool holders, **HSK** type is a relatively new design. This type represents hollow shank tooling, which has been designed with high speed machining (HSM) in mind. With higher spindle speeds the grip of the tool increases accordingly. These holders do not have a retention knob (pull stud).

Cutting Tools

When it comes to cutting tools, books have been written about their types, designs, and applications. Cutting tools are constantly evolving, new designs come on the markets several times a year. Proper selection of cutting tools starts at the programming stage, but - within reason - some changes can be made at the machine, during part setup.

Cutting tools belong to two basic categories:

- Perishable ... end mills, drills, taps, etc. - mainly HSS
- Indexable ... carbide inserts

All major tooling companies provide an extensive source of information that can be accessed on line. They offer catalogues (in most cases downloadable), and they offer technical tips, do's and don'ts, speeds and feeds recommendations, etc. Using these resources is well worth the time and effort.

Tooling System

Regardless whether you use a CNC machining center with the CAT type of spindle or the BT or HSK type, most tool holders are designed for flexibility. That means, for example, that one face tool holder can accommodate several face mill sizes, within a certain range.

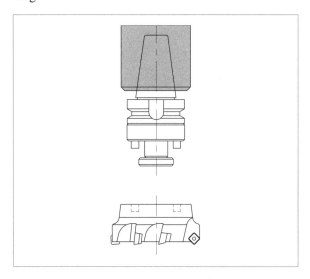

Other tool holders that are part of the tooling system are collet holders, which can be used for end mills, boring bars and other tools. A range of collets is required and collects of all common sizes are available, in metric or imperial sizes.

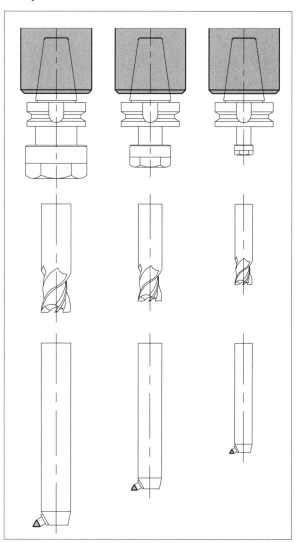

Collet holders have the benefit that they offer concentricity of the cutting tool, as they wrap the tool around the whole circumference. On the other hand, they have to be cleaned from debris after each use, and stored in a safe place.

For some very heavy cuts, the tool can spin inside the collect. If that is a concern, you can used another type of a tool holder that has one or two set screws. For that purpose, the tool has to have a *Weldon*® shank type flat.

ATC - AUTOMATIC TOOL CHANGE

A feature of a CNC machine commonly known as the *'Automatic Tool Changer'* means that one tool can be exchanged with another tool, using CNC program data or MDI input, rather than by any manual interference.

> **The main purpose of ATC is to exchange the tool in the spindle with the tool programmed next**

CNC lathes are not included in this category - any tool change of lathe tools is always automatic. When it comes to CNC mills and machining centers, there are some major differences - differences that will influence the program development, as well as the setup procedures at the control system. Tool changing on milling type machines can be categorized in three groups:

- Manual tool change - no automation
- Fixed type of automatic tool changers
- Random type of automatic tool changers

When there is no tool changer available, does it mean the tool cannot be changed? Not at all. Take an typical example of a tool room oriented shop, perhaps a prototype tool and die shop, or a similar shop where one or two parts are occasionally required. There are many machine shops falling into this category - while they all want to be competitive - and productive - they do not face the same requirements as true production oriented companies. Their use of CNC equipment is not always a high priority; often, it is the desire for precision rather than the best cycle time that determines which work will be assigned to a CNC machine.

For production oriented companies - and they are in majority - an automated tool changer under CNC program control is the best solution.

Manual Tool Change

As the name suggests, there is no provision for storing tools required for a particular job right at the machine. Practically, it means the tools are available in some kind of a portable tool cart, located nearby the machine itself during machining. When a certain tool is required, the CNC operator will temporarily stop the program, remove the existing tool and replace it with the next tool in the machine spindle. When the new tool is in the spindle, pressing the *Cycle Stop* button will continue processing the program with this new tool.

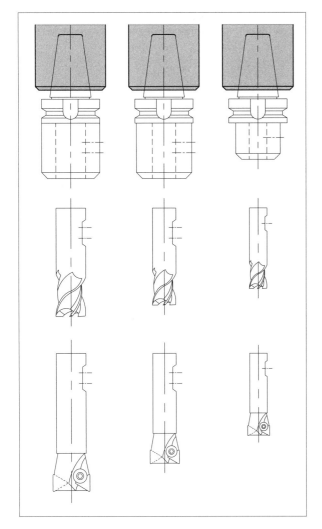

The internal diameter is precision ground, and several sizes are available for a variety of cutting tools.

Similar holder for drills with *Morse®* taper are also available in different sizes.

Regardless which tooling system you use, the tool has to be mounted in the tool magazine, then brought into the spindle. All CNC machining centers have a special feature designed for just this purpose - it is the *Automatic Tool Changer*, or ATC.

Some tool room type CNC mills and even some economy machining centers do not have an automatic tool changer, and all tools have to be changed manually. These machines are not generally used for production, more for prototyping, repairs, tool room work, etc.

While this manual type of a tool changer may be perfectly suited to certain individual types of CNC work, it would be too costly to implement it for production work. From a practical viewpoint, consider a manual tool change as a specialty type of setup only.

For CNC machines using manual tool changer, the part program should have:

- NO tool T-address
- NO tool change M06 function
- M00 (mandatory program stop) after each tool

If there is no ATC involved, using the tool call T-address and the tool change function **M06** are irrelevant, therefore not required. On the other hand, the program must provide means to do the tool change manually. Depending on the person who writes such a program, the **M00** function - *Program Stop* - can be with or without a comment attached:

```
...
N64 M00               ... or ...
...
N64 M00       (DRILL OUT - USE 15 MM END MILL)
```

Attaching a simple comment always helps the CNC operator and makes a better program documentation. Description of the tool also helps to prevent the possible mistake of using the wrong tool.

Tool Magazine

While a manual tool change may be quite useful in certain industries, having an automatically controlled tool change is necessary for production work. In this case, all tools used by the program must be located at the machine. The temporary machine based storage area for these tools is called *tool magazine* or *tool carousel*. The operative word here is 'temporary' - while you can keep frequently used tools in the magazine at all (or most) times you may choose not to.

> **Tool magazine storage is not a permanent depository of cutting tools**

Modern CNC machines have very fast tool change times, measured in two ways:

- Tool-to-Tool time
- Chip-to-Chip time

Tool-to-Tool time is the amount of time expired between one tool in the spindle and the next tool in the spindle.

Chip-to-Chip time is the amount of time expired between making the last cut with one tool and the next cut with the another tool.

Manufactures may sometimes advertise tool-to-tool time, because it is more attractive (meaning shorter) than the chip-to-chip time. As always, compare equivalent specifications.

Tool Magazine Capacity

Only so many tools can be loaded into the tool magazine. Depending on the machine type and size, the number of tools that can be stored in the magazine varies from low 20-30 tools to the high of a few hundred tools.

Control system of the CNC machine will always support all available tools, which means the number of offsets available for tool length and cutter radius will always exceed the number of tools that can be stored in the magazine.

One important part of any tool magazine design is the location where the actual tool change will take place. This position is called the *'tool change position'*, *'waiting position'*, or *'stand-by position'*. Other descriptions may also be used, largely depending on the manufacturers and their preferences.

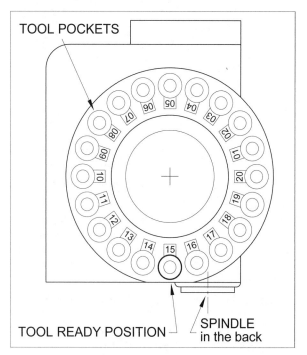

A typical 20-tool magazine is illustrated above. Note the numbered tabs - the are pocket or *'pot'* numbers.

Magazine Labeling

Cutting tools required for a certain job are stored in the tool magazine in individual stations, often called 'tool pots' - short for 'tool pockets'.

Each pot is labeled by a number, from 01 to the number that represents the maximum number of tools the magazine can hold, 20 in the illustration. These labels are fixed, so pot 06, for example, will always be the same pot. When the magazine rotates, the pot labels rotate as well. Rotation is bi-directional, whichever is shorter.

In the illustration, the pot 15 is located at the tool ready position, where the tool will be taken out of the magazine and placed into the spindle.

How these pot labels are used practically during machine setup is covered in the next subject. In order to understand the following setup concepts, you have to understand what the part program is 'telling' you first. Take the two examples as both being correct program entries, typically located at the program start:

- *Version 1 - next tool call T is not programmed:*

```
N1 G21
N2 G17 G40 G80 T01
N3 M06
N4 G90 G54 G00 X.. Y.. S.. M03
...
```

- *Version 2 - next tool call T is programmed:*

```
N1 G21
N2 G17 G40 G80 T01
N3 M06
N4 G90 G54 G00 X.. Y.. S.. M03 T02
...
```

Look carefully - what are the differences? There is only one subtle difference - in the *Version 1* example, there is no *T02* at the end of block N4. This is a significant omission. Providing there is only one T-address per tool programmed, this format identifies a certain type of ATC, called the *Fixed Type*.

Fixed Type ATC

The *Version 1* above represents the fixed type automatic tool changer, if programmed correctly.

Fixed type of automatic tool changer refers to a machine design, where the tool always returns to the same pocket it was located in before the tool change. That means the pot number is - in effect - the tool number.

If three tools are identified in the program as, for example, T01, T06, and T17, the tools have to be placed into pots 01, 06, and 17 respectively.

This format is sometimes used for the random type ATC as well (described next), but represents rather a shabby programming style. For the fixed type ATC:

T.. in the program means the CURRENT tool number

The major disadvantage of the fixed type tool changer is time loss. It takes time to move the magazine from station 01 to station 06 and station 17. Experienced programmers try to number the tools so they are close together in the magazine, but that is not always possible. Even then, there still is a time loss.

A qualified CNC operator may change the tool numbers at the control, if the situation warranties it.

Random Memory Type ATC

The *Version 2* above represents the random type automatic tool changer, if programmed correctly.

Random type of automatic tool changer refers to a CNC machine design, where the tool always returns to the magazine pocket occupied by the previous tool. If programmed properly, the tool magazine will rotate to its new position simultaneously with actual machining activities that are taking place.

For the random memory type ATC:

T.. in the program means the NEXT tool number

The process is the program has three steps:

- Call the tool to be placed into the spindle   N2
- Exchange the tool in the spindle with the new tool   N3
- Call the tool for the next operation   N4

These steps are represented by blocks N2, N3, and N4 of the *Version 2* example.

Keep in mind that tool numbering in the program is not always orderly 01, 02, 03, etc. Some tools are used so often, that programmers use the same number for all jobs. The benefit of this method is that it speeds up the setup process at the machine.

The description 'random memory type' may be a bit confusing - it simply means that the computer - the control system - keeps track of all tools. Of course, some input from the CNC operator is required, in the form of registering the tool numbers in memory, covered shortly.

Tool Waiting

One of the features of modern CNC machining centers is the 'tool waiting'. During machining, the tool that is in the ready position of the magazine is called the 'tool waiting'. The tool is waiting for the ATC, when the next **M06** tool change is programmed. As there is no magazine rotation, the tool in the spindle will be replaced with the tool that is waiting in the 'tool ready position'.

Most programmers understand this concept well:

> **Programming the call for the next tool immediately after the tool change improves cycle time**

The later section on tool repetition *(page 62)* covers this subject in more detail.

REGISTERING TOOLS

Tools used with a CNC system that supports random type ATC have to be registered into the control. The process can be quite quick, but it is still a bit slower than that for the fixed type ATC, where no tool registration is required. Overall, the small delay of the tool registration will be more than compensated by other advantages of the random type ATC.

What steps are involved in the tool registration process? The concept is simple - the CNC operator places all tools to available magazine pots, then 'tells' the control which pot contains the tool number as per program. For example, the program calls for tool number six - *T06* in the program. During tool registration, the operator finds that magazine pot labeled as number 17 is available. In the control, T06 is registered as being located in pot number 17.

The exact procedure may vary from control to control, but all control systems have one common denominator - the settings take place while system parameters can be written to. This is an important requirement - when writing to the parameters is enabled, the control enters into the 'alarm' state and no machining or other machine activity is possible.

In the next illustration, the concept is presented visually. The program calls for four tools - *T01, T06, T13,* and *T17*. The picture shows T01 in pot 05, T06 in pot 17, T13 in pot 02, and T17 in pot 12. This is only an example, but it illustrates that the tools can be placed into any available pot of the tool magazine. Restrictions on tool sizes will be covered in this chapter as well.

TOOL SPECIFICATIONS

Common specifications of cutting tools include not only the cutting tool itself (such as drill, tap, reamer, end mill, etc.), but also the tool holder. Cutting tool holder provides connection between the machine spindle and tool cutting tip. CNC machine manufacturers identify restrictions in three areas of tool specifications:

- **Maximum tool length**
- **Maximum tool diameter**
- **Maximum tool weight**

Maximum Tool Length

In this context, the tool length is *not* the actual length of tool from one end to another. Rather, it is the extended tool length measured from the spindle reference face to the tool tip.

During actual automatic tool change (ATC), the tool is pulled out of the spindle, rotates, and is replaced with another tool from the tool magazine.

Keep in mind that all CNC machine manufacturers always specify maximum tool length without any consideration of any current part setup. The higher the part, the less space will be available for safe tool change. In these cases, the programmer should always select a safe position to make tool change, often away from the part, and not directly above it. Be especially careful with high work setup and long tools.

Maximum Tool Diameter

When not used for machining, at least some tools are located in a tool magazine that can store from a dozen or so tools to a few hundred tools. The tool magazine is usually on the machine side, away from the work area. Stored tools are adjacent to each other, in special pockets (pots), as described earlier. Depending on the size of the tool magazine and the number of tools it can hold, the distance along an arc - called the *pitch* - will vary in size.

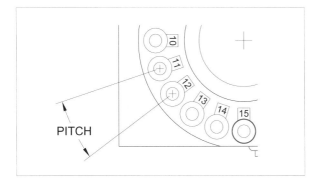

Circular pitch is not the only design of tool magazines. Adjacent pots that accept the cutting holder taper are equally spaced in a square, rectangular, oval, race-track or zig-zag fashion. The pitch between center lines of two adjacent pockets is described in CNC machine manuals.

Also defined in machine tool manuals is the *maximum diameter of the tool*. This maximum diameter specification is based on every magazine pot being filled to capacity. Small tools seldom present a problem, but large tools, particularly face mills, do deserve a special attention. Take the example for a 20-tool magazine used in this chapter. The manufacturer's specification lists the maximum tool diameter as ⌀ 140 mm (⌀ 5.5 inches).

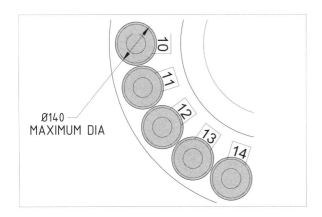

What if you need a face mill that is just slightly larger than the maximum allowed - ⌀ 150 mm (⌀ 6 inches)?

You may be surprised to know that this change is indeed possible, but not without a compromise. First, keep in mind that the difference between the maximum allowed diameter and the larger diameter of the tool used must be relatively small. Second, also keep in mind the increased weight of the tool (see the next section).

The solution? Both pots adjacent to this oversize tool **must be empty**. That's not all - they must be empty, and registered as empty. It's not enough to leave the pots empty, if the controls system does not 'know' it. This is done on the same screen as the other tool numbers, but using a special code, issued by the manufacturer. For example, a negative number or a number like 9999, etc.

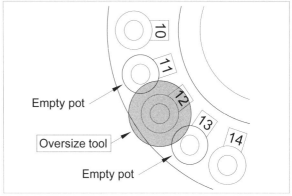

In the illustration, pot 12 contains an oversize tool. Both adjacent pots 11 and 13 *must be* empty and registered in the control system as such.

Maximum Tool Weight

The automatic tool changer has to physically lift the total weight of all cutting tools and items associated with them, such as a tool holder, collet, set screws, etc. Basically a mechanical device, ATC has its limitations not only in the maximum tool length and tool diameter already mentioned, but in overall weight as well.

All CNC machining centers are designed to accept a combined tool weight much higher than the average normally required. The small problem is - what is an average? In order to prevent a failure to the ATC operation, machine manufacturers set limits on the maximum tool weight. For example, a smaller machining centers may have a limit of 6 Kg or about 13 pounds per tool. Larger machines will have the maximum limit proportionally higher. Knowing these limitations may prevent a costly repair job, if tools used are severely overweight.

Does it mean you cannot use a 6.25 Kg tool, for example, on a machine limited to 6 Kg? What about 6.5 Kg or even 7 Kg? Within a specific safety margin, you can - but the part program has to provide for this contingency in the form of a *manual* tool change, rather than the standard ATC - automatically programmed tool change.

ATC PROCESS

The actual process of automatic tool change has been described to some extent already, and more descriptions will follow. When you program or operate a CNC machining center, knowledge of the ATC process is always an advantage. The following program will be used to evaluate details not directly apparent either in the program itself or in the illustration that follows:

```
O1234
N1 G21
N2 G17 G40 G80 T01
N3 M06                (TOOL 01 TO THE SPINDLE)
N4 G90 G54 G00 X.. Y.. S.. M03 T02
N5 G43 Z.. H01 M08
... < machining with T01 > ...
N30 ...
N31 G28 Z.. M05
N32 M01

N33 M06               (TOOL 02 TO THE SPINDLE)
N34 G90 G54 G00 X.. Y.. S.. M03 T03
N35 G43 Z.. H02 M08
... < machining with T02 > ...
N56 ...
N57 G28 Z.. M05
N58 M01

N59 M06               (TOOL 03 TO THE SPINDLE)
N60 G90 G54 G00 X.. Y.. S.. M03 T01
N61 G43 Z.. H03 M08
... < machining with T03 > ...
N88 ...
N89 G28 Z.. M05
N90 M30
%
```

All T-words and tool change function **M06** have been underlined for easier reference. The program can also be presented in a table form, describing individual steps.

One question that may need to be answered first is - *what tool is in the spindle before this new program is ready for machining?* The simple answer is - *it makes no difference*. The last tool from the previous job is one possibility, or a totally empty spindle is another possibility. As long as the tool change in the new program can be done safely, the tool in the spindle really does not matter.

In the following table, the '??? tool' can be an existing tool or a dummy tool, which is an empty spindle.

Block Number	Tool waiting	Tool in spindle
N1 - START	???	???
N2	T01	???
N3	???	T01
N4	T02	T01
... Machining with T01 ...		
N33	T01	T02
N34	T03	T02
... Machining with T02 ...		
N59	T02	T03
N60	T01	T03
... Machining with T03 ...		
N90	T01	T03
N2	T01	T03
N3	T03	T01
...	...	...

As the program shows, all automatic tool changes for CNC vertical machining centers are performed only in one position, and that is machine zero for the Z-axis. For horizontal machining centers the tool change takes place at the machine zero of the Y-axis.

The following set of illustrations represent graphically the same process as shown in the above table:

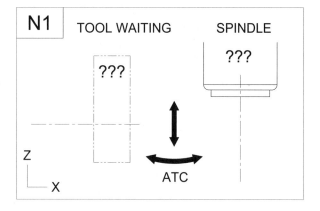

CNC Control Setup for Milling and Turning
MILLING TOOLS - SETUP
61

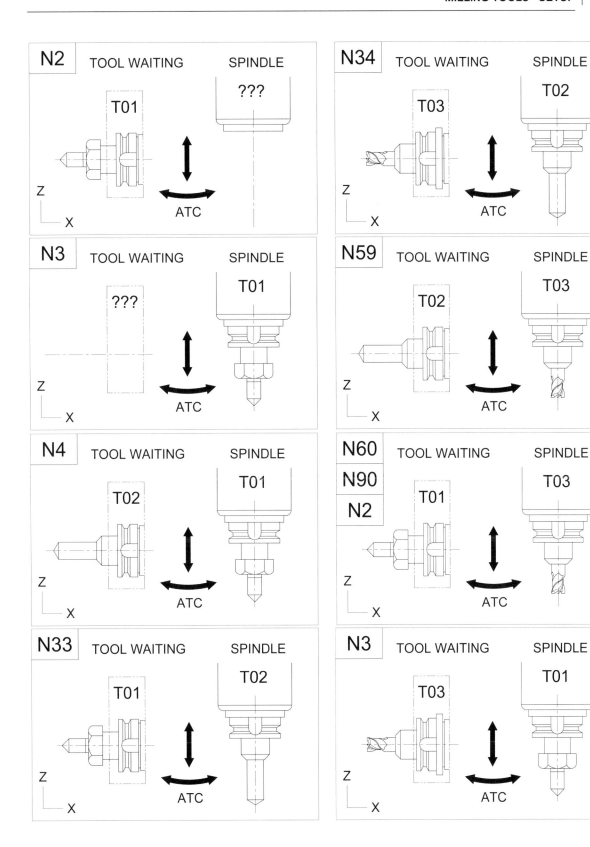

TOOL REPETITION

Where quite a few programs fail the quality test is in the area of *program structure*. In this handbook, the importance of efficient and safe program structure has been mentioned numerous times. The focus of discussion is the practical relationship of the tools in the program. A casual look does not indicate any relationship between tools. After all, one tool does its job, and another tool takes its place. Yet, there is a relationship, albeit very subtle. When this relationship comes to light is in the event of tool repetition.

It is not unusual for the CNC operator to repeat the same tool more than once, typically during the first part run (testing stage). Such situation is common when optimizing speeds and feeds, fine-tuning offsets, making changes to the program, and so on.

If the program structure is 'friendly', the task of repeating any tool from any program block is very simple. Unfortunately, the opposite is also true.

Let's use the previous program structure, but this time we add another two tools, to have a program with five tools. The more tools used by the program, the more important this subject becomes.

Here are the main program blocks for five tools. All tool T-words and **M06** are underlined for easier orientation - these are the critical points of the tool repetition:

```
O1234
N1 G21
N2 G17 G40 G80 T01
N3 M06                 (TOOL 01 TO THE SPINDLE)
N4 G90 G54 G00 X.. Y.. S.. M03 T02
N5 G43 Z.. H01 M08
... < machining with T01 > ...
N30 ...
N31 G28 Z.. M05
N32 M01

N33 M06                (TOOL 02 TO THE SPINDLE)
N34 G90 G54 G00 X.. Y.. S.. M03 T03
N35 G43 Z.. H02 M08
... < machining with T02 > ...
N56 ...
N57 G28 Z.. M05
N58 M01

N59 M06                (TOOL 03 TO THE SPINDLE)
N60 G90 G54 G00 X.. Y.. S.. M03 T04
N61 G43 Z.. H03 M08
... < machining with T03 > ...
N88 ...
N89 G28 Z.. M05
N90 M01

N91 M06                (TOOL 04 TO THE SPINDLE)
N92 G90 G54 G00 X.. Y.. S.. M03 T05
N93 G43 Z.. H04 M08
... < machining with T04 > ...
N117 ...
N118 G28 Z.. M05
N119 M01

N120 M06               (TOOL 05 TO THE SPINDLE)
N121 G90 G54 G00 X.. Y.. S.. M03 T01
N122 G43 Z.. H05 M08
... < machining with T05 > ...
N188 ...
N189 G28 Z.. M05
N190 M30
%
```

The structure seems all right and the program runs fine. Many programmers use this structure and are quite happy with it. CNC operators may be less happy with the structure the first time they have to *repeat* a particular tool. Focus on, for example, tool *T04*, which starts at block *N91*.

Repeating the Same Tool

When the *Optional Stop* **M01** is activated in block N119, tool T04 is completed. You measure the part and find that its size needs to be adjusted by a wear offset. You adjust this offset for T04 as necessary. Now, you have to repeat tool T04 to apply the offset change and make the part to size. The question is - how do you repeat tool T04?

Every control system has a search function that can find a particular address, such as a block number. In the example, block N91 makes the tool change and brings T04 to the spindle. However, when block N119 is the current block, in order to repeat the tool T04 is to search for block N92 - after all, the tool T04 is still in the spindle, so tool change is not necessary.

Repeating the tool while it is still in the spindle is not that much of a problem. You may also do a backward search at the control (from the bottom up) and search for N92, starting from N119.

Repeating a Different Tool

This time, you finish with T04, the current block is N119, but you discover that there was a problem caused by the earlier tool T02, which starts at block N33. You correct the problem caused by this tool, but have to repeat tool T02 for the problem to be solved. Where do you start? What block do you search for?

The immediate answer may be to start from block N33, where the tool T02 was originally placed into the spindle. This is *not* a correct solution.

Consider the current situation carefully. What is the exact status of the program processing? In this example, the tool T04 is in the spindle and all machining with this tool has been completed. The control shows the current block as N119, when the *Optional Stop* has been activated after machining with tool T04.

What will happen if you search for, and start processing from block N33? As the block N33 contains the tool change function **M06**, the control system will do as expected - it will take the tool currently in the spindle (that is tool T04) and replace it with the tool currently located in the magazine 'waiting/ready' position - which is tool T05. However, *this is the wrong tool!*

Solving the Problem

As good as the presented program example looks - and even works in most cases - it has one major flaw - it does *not* allow for easy repetition of *ANY* tool from anywhere in the program. The problem is not the CNC operator's to solve - this is a programming problem. Yet, the solution is quite simple, but it must be done at the programming level, either by the CNC programmer or a qualified CNC operator with solid programming knowledge.

Note the 'next tool ready' entries in blocks N4 (makes T02 ready), N34 (makes T03 ready), N60 (makes T04 ready), N92 (make T05 ready), and N121 (makes T01 ready). All that needs to be done is a simple *repetition* of the block containing the 'waiting tool' call - a block after the **M06** tool change function.

The 'waiting call' blocks N4, N34, N60, N92, and N121 do not change at all, there is no need. In fact, there is no change at all, except a simple *addition*. If the 'waiting tool' call is repeated *before* each **M06**, the problem will be solved. The following example is the modified program *O1234* - Only the additional changes are underlined (block numbering has also changed):

```
O1234
N1 G21
N2 G17 G40 G80 T01
N3 M06                   (TOOL 01 TO THE SPINDLE)
N4 G90 G54 G00 X.. Y.. S.. M03 T02
N5 G43 Z.. H01 M08
... < machining with T01 > ...
N30 ...
N31 G28 Z.. M05
N32 M01

N33 T02
N34 M06                  (TOOL 02 TO THE SPINDLE)
N35 G90 G54 G00 X.. Y.. S.. M03 T03
N36 G43 Z.. H02 M08
... < machining with T02 > ...
N57 ...
N58 G28 Z.. M05
N59 M01

N60 T03
N61 M06                  (TOOL 03 TO THE SPINDLE)
N62 G90 G54 G00 X.. Y.. S.. M03 T04
N63 G43 Z.. H03 M08
... < machining with T03 > ...
N90 ...
N91 G28 Z.. M05
N92 M01

N93 T04
N94 M06                  (TOOL 04 TO THE SPINDLE)
N95 G90 G54 G00 X.. Y.. S.. M03 T05
N96 G43 Z.. H04 M08
... < machining with T04 > ...
N120 ...
N121 G28 Z.. M05
N122 M01

N123 T05
N124 M06                 (TOOL 05 TO THE SPINDLE)
N125 G90 G54 G00 X.. Y.. S.. M03 T01
N126 G43 Z.. H05 M08
... < machining with T05 > ...
N192 ...
N193 G28 Z.. M05
N194 M30
%
```

What has changed? The next tool (the tool waiting) has been repeated **before** any tool change. *What is the actual benefit?*

The real and measurable benefit is even more significant than the change of program structure itself may indicate. Consider the initial example, where the tool T02 had to be repeated after tool T04 was completed.

Within the new program block numbers, tool T04 has been completed in block N122. In order to repeat tool T02, all the CNC operator has to do is to search for block N33. This block now calls the repeated 'waiting tool' number. When the block N33 has been activated, the tool magazine starts rotating and the control system starts searching for the tool T02. From there on, all offsets adjustments will be applied, as required.

This is a very simple - and also very effective - solution to a problem that had caused many delays in processing the first part or two of a batch during setup and the first part run.

Tool Change - One or Two Blocks

Two programming commands are required to make a tool change (ATC):

- **Tool number call**   address T..
- **Tool change function**   address M06

Looking at various programs, you may see the combination programmed in three different ways, for example:

```
T04 M06
```

or

```
M06 T04
```

or

```
T04
M06
```

Incidentally, they are all correct, even if the tool change function is before the T-address. Which one is a better choice?

While the first one (**T04 M06**) is the most common, you may find the last one the most practical. That is also the format used in this handbook.

Here is the reasoning - during setup, the CNC operator generally works in a single block mode, at least for certain activities. When it comes to changing the tool, and the block contains both T04 *and* M06, the control starts searching for the tool, and once the tool is positioned in the waiting station, the tool change will take place. This does not give the operator any time to check if the correct tool is to be loaded.

With the two-block method, the tool to be searched for will be executed, but no tool change takes place. The operator now has time to walk over to the tool magazine and check if the correct tool is waiting. The actual tool change takes place only after the *Cycle Start* button is pressed again. The slight delay in the setup does carry over to the production time, as the control processes the part program exactly the same as it does for the one-block format.

DUMMY TOOL

The term 'dummy' tool is a programming expression that refers to a tool with a given number, but with no actual tool present in the magazine or the spindle. In other words, programming this non-existent tool is a method to make the spindle *empty*, when it becomes necessary.

When would you need the spindle to be free of any tools during production? One such case is to clear the space above the part in order to remove a large part, perhaps with a hoist or a crane. Another use for the 'dummy' tool is when a manual tool change has to be made, for example, if the tool is too long or too heavy for an automatic tool change. Some other special cases may also call for the spindle to be temporarily empty at some point during the program processing.

Although any number can be used for a tool to represent an empty spindle, 'dummy' tools are often programmed using the numbers 00 or 99 - T00 or T99. These are legitimate tool numbers and have to be registered the same way as any other tool.

SUMMARY

When it comes to setup of tools for CNC machining centers, good organization always pays off. Tools should be stored in a pallet-like container, with a hole for each tool, so adjacent tools do not touch each other. Tools should always be well taken care of - cleaned of any chips or other debris before using them. The inside tapered are of the spindle should always be cleaned often as well. Both perishable tools and carbide tools should be stored in such away that they can be easily retrieved, when needed. If possible, tools that are used often should stay in the tool magazine and keep the same numbers in all programs.

One last suggestion - you should never tighten set screws or collet nuts while the tool is in the spindle. Most machines come with a tool setting fixture and suitable wrenches - these are the devices to use for tool setup - always off the machine.

8

SETTING PART ZERO

Setting part zero is closely related to setting work offsets. This chapter will address the part zero in general. For more details, refer to *Work Offsets Settings* chapter, starting on *page 73*.

SYMBOLS

When it comes to the part origin - commonly called *part zero* - it is the CNC programmer who determines its position on the part, but it is the CNC operator who has to make corresponding setup at the CNC machine. In drawing or setup sheets, the part origin - *part zero* - uses a special symbol - here are three variations, similar ones may also be used:

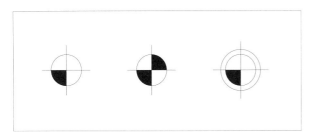

Selecting part zero is one of the most important decisions any CNC programmer has to make when developing a part program. Incorrect selection can lead to many problems - problems that can not be fixed at the machine during setup. One problem caused by an incorrect part zero selection may be a difficulty of physical setup at the CNC machine. Even if this obstacle can be overcome by a inventive CNC operator, there is another, and much more compelling, reason why suitable part zero selection is very important not only for programming, but for subsequent setup as well - *maintaining dimensions*.

In order to maintain accurate dimensional accuracy, the part zero selection plays an important role and the programmer has to address all specific issues in respect of selecting part zero.

Although selecting part zero is the programmer's responsibility, it is very important that the CNC operator understands at least the basic concepts. This chapter covers a good number of such concepts.

WHAT IS PART ZERO ?

Part zero is just one of several descriptions used to identify a critical point applied to both part programming and part setup. Other common descriptions of the same point are *part origin* and *program zero*. Regardless of how this very critical point is described, it has only one meaning:

> Part zero identifies the physical location on the part, from which all other dimensions are based on

In order to understand this very important subject, it is necessary to understand a few statements and definitions, as they are used in general CNC programming and CNC machining. In the summary:

- Part zero
- Part origin
- Program zero
- Program origin

   ... they all refer to the same point !

> Part zero is always determined by the CNC programmer

> Part zero does *not* have be identical to origin of dimensions in a drawing

Part Zero Location

The main focus of selecting part zero location is to consider three outcomes:

- How it affects dimensions and tolerances
- How it effects tooling and machining
- How it effects setup

All three items are important. On prismatic parts, the XY zero is commonly located at one of the part corners, and Z-zero is located at the top of finished face. Some jobs, such certain shapes of castings and forgings, can benefit from the Z0 at the bottom of the part. On parts that are round or close to being round, the XY zero is located at the center, such as the center of a ring, while Z0 can be the top or bottom of the part.

In some jobs, neither the corner nor the center may be convenient, and the part zero is set based on certain features of the fixture. A manufacturing hole or a similar locator can be used for the purpose.

The three common examples of part zero are illustrated below - not all views or dimensions are shown, just the method.

◆ Part corner:

The lower left corner of a prismatic part is very convenient for programming, but *any* corner can be used as part zero. As this is one of the most common part zero settings on vertical machining centers, a different section will evaluate it in much more depth.

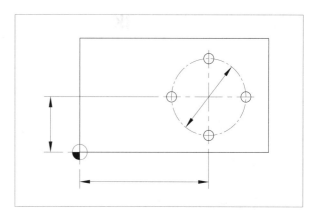

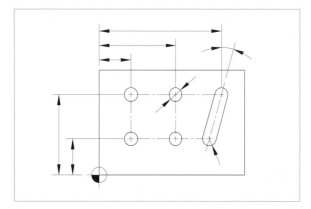

Some programs may actually use the *center* of the round object as part zero. Although such selection should always be justified, it is not wrong - it just makes the setup a bit more difficult. In this case, use an edge finder or a corner finder to establish the corner, than adjust the setting by the X and Y distance to the center of the round object. *Don't forget to compensate for the edge finder radius.* A better method is to incorporate **G52** command (local coordinate system) into the program.

Circular objects can also be located *within* a prismatic part, as the next illustration shows.

◆ Manufacturing hole:

A special hole or a similar opening may be specified on the drawing as the reference point for dimensions. This reference hole, often called manufacturing hole, is also the one that programmers chose as part zero.

◆ Part center:

For round parts, the most logical part zero is the center of the part. Rather than using and edge finder, a test indicator is a better choice.

Round part are usually setup in a three-jaw chuck that is mounted on the machine table in a fixed position, but a V-block or a similar device can also be used to set part that are round.

At the machine during setup, a dial indicator is used to find the hole center, but there is a difference between using the dial indicator for parts that are truly round. As the manufacturing hole is located at an arbitrary place of the part, it is also important to set the edges correctly, so they are parallel to the axes, for example.

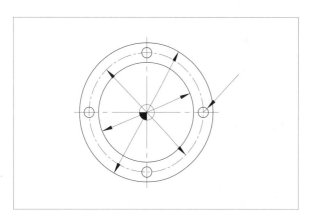

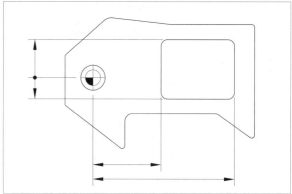

VISES FOR CNC WORK

A machinist's precision vise, preferably a model designed for CNC work, is one of the most common work holding devices for vertical CNC machining centers. One or more vises can be set on the machine table, either in vertical or horizontal orientation. Angular orientation of a vise is theoretically possible, but very impractical. Angles should always be handled by the program.

It is quite common to have more than a single vise mounted on a tombstone for horizontal machining or used in many other ways on any machining center.

CNC Vise - Basic Design

Not all vises are made the same. For CNC work, a vise that has been specially designed for CNC machines should always be used. There are many benefits to CNC vises, such as precision components, high clamping force, jaws that can be quickly reconfigured, ease of handling, quick cleaning, etc. *CNC vises* are sometimes called *machinist vises*.

Some top names in the category of CNC vises are *Kurt®* and *Chick®* in North America, and a number of others from Europe and Asia. The following simplified drawing is a schematic representation showing the main parts of a common CNC vise, as they relate to its setup:

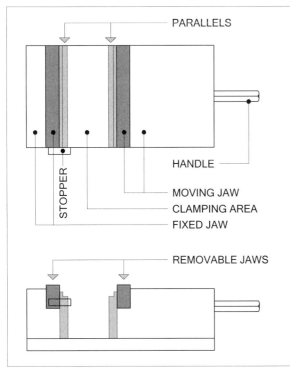

Parallels are optional - their use depends on the job. Their purpose is to provide maximum exposure of the part above the top of vise. Parallels are precision ground flat plates made of hard steel and come in a box as pairs of different sizes. For ultra precision work, the parallels may not be a suitable choice, as they are not physically attached to any part of the vise. Sometimes, parallels have the tendency to move under the part they support. Innovative CNC operators find a number of ways to prevent that, for example, by placing a piece of Styrofoam block between the parallels.

Removable jaws are often made of durable aluminum, and can be machined to any odd shape of the part. Making a step in the removable jaws is often a better choice than using parallels. Removable jaws are also called *interchangeable* jaws.

When selecting a CNC vise, look not only at the features already presented, but also at general parameters. The most important is the *width of jaws*, and *maximum jaw opening,* and *depth*. Vise specifications list all three dimensions.

Part Stopper

A very important part of any CNC vise is a *part stopper*. This is a removable device that can be attached to the side of vise, in order to provide a fixed point for repetitive jobs and production run of multiple parts. Several locations are usually available for greater setup flexibility. Combination of a fixed jaw and the part stopper create a fixed location for all parts using this setup.

CNC Vise - Setup

CNC vise is normally mounted on the machine table. A program comment or a setup sheet may describe or show the orientation of the vise, but do not expect anything else. Correct setting of a vise (or any fixture) is an important part of CNC operator's responsibilities. Here are some of them:

- **Vise position has to allow quick part change**
- **Fixed jaw has to be lined-up with machine axis**
- **Part stopper has to be attached (if required)**
- **Parallels have to be secured (if required)**
- **Contact faces have to be clean**

The concepts shown so far, and those following, can be applied to any part setup, using any type of fixture or a work holding device. The most important is one rule:

> Part zero must be set in a fixed location

APPLICATION EXAMPLE

In order to understand the concept of part zero and its effect on part setup, the following drawing will be used:

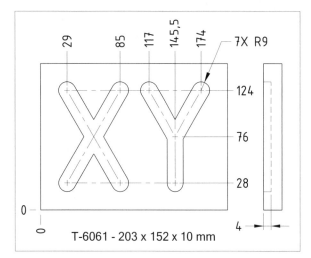

The part itself is not important - it was designed with one objective only - to present a shape that is easily recognizable in *any* orientation of the vise located on the machine table.

The way the drawing is dimensioned, it is clear that all dimensions originate in the lower left corner, which is also selected as part zero. All subsequent setup options are presented for the purpose of *understanding* the subject. Keep one critical fact in mind:

> Part setup orientation is determined by the program

Part Orientation

How the part is oriented on the table is decided at the programming level. Most programmers inform the CNC operator how the part is oriented and where the part zero is located. This is done either through a setup sheet for more complex parts or as a comment at the top of the program for simpler parts. For our example, the comment may be similar to this:

```
(X0 Y0 AT LOWER LEFT OF PART)
(Z0 TOP OF PART)
(SET PART IN VISE HORIZONTALLY ALONG X-AXIS)
```

Comments such as this one are clear enough to give the operator all instructions required to setup the vise with the part. What about having no setup sheet and no instructions at all?

Part Orientation From Program

With no information of part setup, the only other source is the program itself. Here are three program examples of the same part. Can you tell which one is a horizontal orientation and which one is vertical orientation? Can you also tell where the part zero is located?

```
(PART 1)
N1 G21
N2 G17 G40 G80
N3 G90 G54 G00 X28.0 Y-29.0 S1700 M03
N4 G43 Z2.5 H01 M08
N5 G01 Z-4.0 F125.0
N6 X124.0 Y-85.0 F200.0
N7 G00 Z2.5
N8 X124.0 Y-29.0
N9 G01 Z-4.0 F125.0
N10 X28.0 Y-85.0 F200.0
N11 G00 Z2.5
N12 X28.0 Y-145.5
N13 G01 Z-4.0 F125.0
N14 X76.0 Y-145.5 F200.0
N15 X124.0 Y-174.0
N16 G00 Z2.5
N17 X124.0 Y-117.0
N18 G01 Z-4.0 F125.0
N19 X76.0 Y-145.5 F200.0
N20 G00 Z2.5 M09
N21 G28 Z2.5 M05
N22 M30
%
```

```
(PART 2)
N1 G21
N2 G17 G40 G80
N3 G90 G54 G00 X-29.0 Y-28.0 S1700 M03
N4 G43 Z2.5 H01 M08
N5 G01 Z-4.0 F125.0
N6 X-85.0 Y-124.0 F200.0
N7 G00 Z2.5
N8 X-29.0 Y-124.0
N9 G01 Z-4.0 F125.0
N10 X-85.0 Y-28.0 F200.0
N11 G00 Z2.5
N12 X-145.5 Y-28.0
N13 G01 Z-4.0 F125.0
N14 X-145.5 Y-76.0 F200.0
N15 X-174.0 Y-124.0
N16 G00 Z2.5
N17 X-117.0 Y-124.0
N18 G01 Z-4.0 F125.0
N19 X-145.5 Y-76.0 F200.0
N20 G00 Z2.5 M09
N21 G28 Z2.5 M05
N22 M30
%
```

```
(PART 3)
N1 G21
N2 G17 G40 G80
N3 G90 G54 G00 X29.0 Y28.0 S1700 M03
N4 G43 Z2.5 H01 M08
N5 G01 Z-4.0 F125.0
N6 X85.0 Y124.0 F200.0
N7 G00 Z2.5
N8 X29.0
N9 G01 Z-4.0 F125.0
N10 X85.0 Y28.0 F200.0
N11 G00 Z2.5
N12 X145.5
N13 G01 Z-4.0 F125.0
N14 Y76.0 F200.0
N15 X174.0 Y124.0
N16 G00 Z2.5
N17 X117.0
N18 G01 Z-4.0 F125.0
N19 X145.5 Y76.0 F200.0
N20 G00 Z2.5 M09
N21 G28 Z2.5 M05
N22 M30
%
```

While all three programs will machine the part as per drawing, the setup is different for each program:

- **Part 1 - Vertical orientation**
  XY zero at the upper left corner

- **Part 2 - Horizontal orientation**
  XY zero at upper right corner

- **Part 3 - Horizontal orientation**
  XY zero at lower left corner

> The key to interpret part orientation from the program is look for the first few motions and their X and Y signs

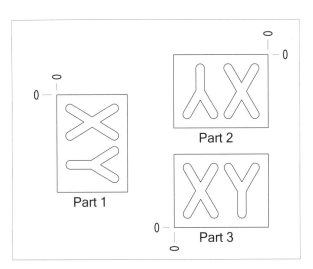

Although the majority of programs start in absolute mode, looking for the **G90** command is beneficial, just in case the program is in incremental mode. Compare these locations with the drawing dimensions. Not providing the part orientation and zero location in the program is sloppy work, but a skilled operator can overcome this by carefully evaluating the program itself.

These are only three mathematical possibilities out of eight, all possibilities are identified in the next section.

Vise Orientation

Note the words *mathematical possibilities* in the last paragraph. These are all possible combinations of vise setup with the part mounted. *Possible* does not always mean *practical*. In fact, there are sixteen possibilities, but these include the vise handle located away from the operator or at left of the vise, above the table. While the first is not practical at all, the latter is practical only marginally. The remaining eight possibilities will include four where the vise is positioned vertically on the table and another four where the vise is positioned horizontally on the table, both in operator's view.

Out of the four possible setups, always reject the ones that will not guarantee positioning accuracy first.

- *Vise => Vertical - Part => Horizontal (A and B)*

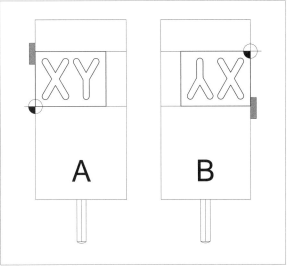

Setup A should be rejected, as its part zero is not located against a fixed point. *Setup B* is correct, but programmers try to avoid it if possible, as it requires all X and Y locations to be programmed in the third quadrant, meaning all XY positions will be negative. There is nothing wrong with that - just a bit inconvenient.

SETTING PART ZERO

→ *Vise => Vertical - Part => Vertical (C and D)*

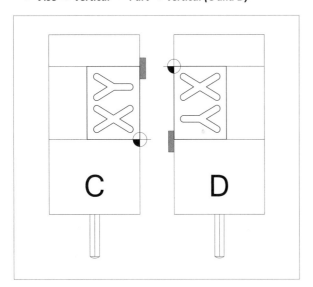

Setup C should be rejected, as the part zero is not against a fixed location. *Setup D* is correct - all X-locations will be positive (no sign), but the Y-locations will all be negative (second quadrant).

→ *Vise => Horizontal - Part => Horizontal (E and F)*

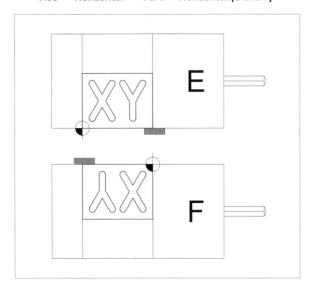

Setup E is correct - both X and Y locations will be positive (first quadrant). This is the preferred setup for parts dimensioned from the lower left corner. Just watch the distortion (bulging effect), if the part is long and thin. *Setup F* should be rejected as the part zero does not lie at a fixed point of the vise.

→ *Vise => Horizontal - Part => Vertical (G and H)*

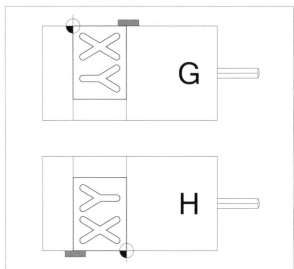

Setup G is correct, with X-positive, Y-negative input (fourth quadrant). *Setup H* should be rejected for the same reason as the other rejections - part zero is not at a fixed point of the vise.

Part Dimensions

An important determination of the part orientation in a vise is not only the X0Y0 at a fixed position, but also the size of the part, namely its *length* and *width*. The part used for the examples is 203 x 152 mm (~8 x 6 inches). CNC vises come in different sizes, from small to medium to large. Assuming the vise design allows it, the part in the example can be located horizontally or vertically in the vise. When the programmer chooses the part orientation, another factor comes into effect, in addition to the factors already mentioned - *distortion*. The part is only 10 mm (~0.4 inches) thick. Needless to say that if the part is located horizontally, the distortion (bulging) will be greater that if the part is positioned vertically. In some cases it may be necessary to add extra support or change the setup altogether.

Also a factor is possible chatter that happens during contouring when a long thin part is not sufficiently supported.

Keep in mind that the programmer selects orientation of the part, which has to be maintained at the machine. From the examples, only *Setup B*, *Setup D*, *Setup E*, and *Setup G* are correct - the choice depends on the part orientation as programmed.

FINDING A CENTER

Sometimes it is necessary to find a center of a circular part or a middle position between two walls. This position could be used for part zero setup or any other purpose. The method shown here is the same method that a probing macro would use for the same purpose.

Using an edge finder, finding the center takes place in two steps:

- **Set the X-axis - do *not* move the Y-axis**
- **Set the Y-axis - do *not* move the X-axis**

Each axis has to be set separately, without moving the other axis. Once the part is set in a fixed position on the machine table, bring the edge finder close to one side and a small depth. Positions XS1, XS2, YS1, YS2 are the starting positions for X-axis and Y-axis. X1, X2, Y1, and Y2 are touch-off positions when the edge finder is centered. There is no need for edge finder diameter adjustment and the direction of measurement makes no difference either.

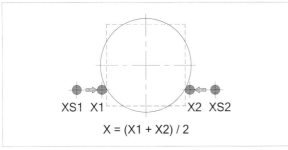

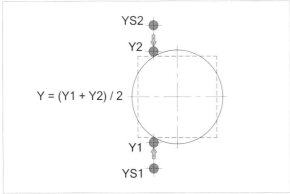

For example: **X1 = -750.0**, **X2 = -483.0**, so the X-center will be at **(-750 + -483) / 2 = -616.500**.

Even if you measure in the opposite direction, the result is the same: **X2 = -483.0**, **X1 = -750.0**, using reversed formula **(-483 + -750) / 2 = -616.500**.

PART ZERO ON CNC LATHES

All CNC machines require part zero and a CNC lathe is no exception. This type of machine is designed to turn, bore, thread, groove, etc., shapes that are primarily cylindrical or conical.

X-axis Zero

In order to maintain the roundness, the axis of the part rotation is the spindle center line. This center line is always part zero in the X-axis.

> Spindle center line is always the X0 for CNC lathe work

Z-axis Zero

When it comes to selecting Z0 of a part, there are several choices. The possibilities to consider by the programmer can be:

A - Front face of stock

B - Front face of finished part

C - Back face of finished part

D - Back face of part stock (face of jaws)

E - Chuck or fixture face

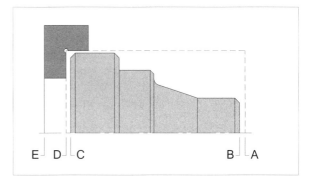

Of course the Z-zero can be in any other position, but these five are at least worth considering. Even without much thinking about the list, you should see that some of the choices are very impractical, for various reasons and should be eliminated right away.

The main criteria of any selection should be dimensional accuracy and ease of setup at the machine. For programmers, the convenience of programming is also important. Not all of the five positions on the list qualify. Here are some comments.

Front Face of Stock - Position A

The face identified as *A* in the illustration is the front face of the stock. During setup using the touch-off method, this is the easiest face to set.

If used as Z0, the programmer would have to adjust each and every Z-coordinate in the program by the amount of stock to be removed from the front face. This is a very inconvenient way, so the face *A* gets rejected. However, it may find some application in the option presented next.

Front Face of Finished Part - Position B

The face identified as *B* in the illustration is the most common selection of part zero on a CNC lathe. As many dimensions originate from the from face, dimensional accuracy is easy to achieve. Equally, this selection is also easy to established during part setup.

One common way to establish the front face of the part as Z0 is to touch-off the front face of the stock, then shift in the Z-negative dimension by the amount of stock to remove from the face. In a way, this is a modified method of the one described previously. The work shift will be described separately, as it can be used in more than one situation.

Back Face of Finished Part - Position C

The face identified as *C* in the illustration is not a convenient one for neither programming nor setup. Even if some dimensions originate at this face, the XZ coordinates in the program would be hard to interpret. This face is not a good choice for Z0.

Back Face of Part Stock (Jaw Face) - Position D

The face identified as *D* in the illustration as the back face of the part (stock) is also the face of the jaws. Providing this face can be touched-off with the tool, many operators prefer it, even if the programmer does not.

It works this way - the programmer will select the position *B* in the illustration, which is the front face of the part. All program coordinates originate from there. During setup, the CNC operator uses this accessible back face to find the required dimensions. Work shift is used in this method, described shortly.

Chuck or Fixture Face - Position E

The face identified as *E* in the illustration is the back face of the chuck or fixture. It is not a very convenient Z0 for any purpose. It is even difficult to touch-off, as the machine limit switch is set further away in the Z-positive direction.

Work Shift

The expression 'work shift' has been used several times in this chapter, as it relates to Z0 setup. Work shift is nothing more than an adjustment between the measured position and another position. In our case, the difference is between the measured jaw face and the front face of the finished part, where the Z0 is located.

Many controls offer a feature called - yes, *Work Shift* - as a separate entry screen on the control display panel. This subject is closely related to work offsets, described in a separate chapter (see *page 73*). Although the display also shows the X-shift option, that does not change, as X0 is always the spindle centerline.

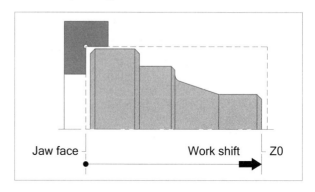

The Z-axis shift is usually set from the measured face to the part zero, but some controls may require the opposite direction of measurement, so always check the machine/control instruction manuals.

9 WORK OFFSET SETTINGS

REFERENCE POINTS

During every part setup, two major locations, called *reference points*, have to be carefully managed. Those are points that establish the relationship between the *known* reference point of the machine (machine zero) and the *unknown* reference point of the part (part zero).

Machine Zero

Reference point of any CNC machine has been selected at a specific fixed point during the initial machine design, by the machine design engineers. It is a fixed point, located within machine travel limits, and its actual position does not normally change. This point (position) is typically called the *machine reference point*, *machine zero*, or simply - the *home position*.

Part Zero

The second reference point that has to be managed during setup is selected by the CNC programmer, during program development for a particular part. This *part* related reference point is selected at a suitable location of the part and is called the *program zero* or - more accurately - *part zero*.

> Part zero is the origin of all point locations as specified in the CNC program

For any CNC machining, during part setup, a connection between these two points has to be established.

Machine Zero to Part Zero Connection

In the early days of numerical control, **G92** preparatory programming command (**G50** on lathes) - called *Position Register* command - had been the primary method of setting actual tool position.

The command definition was quite simple:

> G92 (G50) program command registers the current position of any active tool, as measured *from* part zero (part origin)

Although many CNC machines equipped with such really old control may still be used in many shops, their numbers are dwindling rapidly. All Fanuc controls still support the **G92** and **G50** commands, but only for compatibility with older controls. This support may last for a while, but is certainly not assured in the future. As many programmers will attest, the main difficulty using the *Position Register* command was a requirement that the current tool position is written in the program itself, generally as a separate program block, for example:

```
G92 X250.0 Y200.0
```

The above example 'tells' the control system that the current position of active tool (its command point) is distant 250 mm from part zero along the X-axis and, at the same time, 200 mm distant from part zero in the Y-axis. Sounds simple? Probably, but the simplicity can be misleading. The main problem a CNC programmer faces when using **G92**/**G50** command is plain reality. The problem is that the current position of the active tool is *not always known* when the program is being developed. For many years now, modern control systems use a different system of relationship between various origins. It is called *work coordinate system* - or, in common conversation - *work offset*.

PART SETUP EVALUATION

Before getting deeper into the subject of work offsets, it is important that you know *why* the work offsets are required at all. Consider the very simple part presented on the next page. It contains three holes that have to be spot drilled, drilled and tapped - all very common operations.

When the CNC programmer receives a part drawing and material specifications, all program data will be based on that drawing - *there is no other source*.

Keep in mind that the programmer selects part zero based on the dimensions provided by the design engineer. The objective of the program is to maintain specified drawing dimensions at all times. In some cases, the part zero will correspond to the origin of all - or most - dimensions. In other cases, the programmer selects part zero based on other criteria, such as convenience of the setup at the machine.

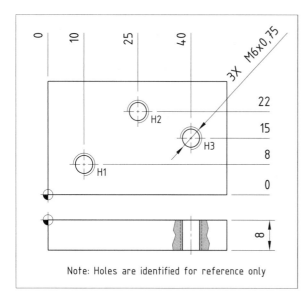

Note: Holes are identified for reference only

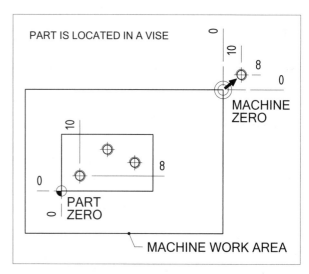

From the example above, the part zero is at the lower corner of the part in X and Y axes, and at the top face for the Z-axis. In this case, part zero matches the dimensioning method. What would be more natural than include the following instruction in the program to rapid to the first hole at lower left?

```
G90 G00 X10.0 Y8.0
```

The question should be - where is the X10.0 and Y8.0 measured from? Yes, from the part zero in absolute mode (**G90**) - but there is a very major problem - the control system has its own zero - it is called the *machine reference point* or *machine zero* - it does not have any information about the part zero location at all. The result of this program statement will be - overtravel. The control system will use the only zero it 'knows' and uses it as the basis for the motion. Since both X and Y values are positive, the motion will try to go into the area outside of the machine limits.

Both origins in the following illustration - program zero and machine zero - are shown. Note that they are both identified as X0Y0 and also note that there is *no* known connection between the two origins, so the cutting tool will try to travel to X10.0 Y8.0 from the known machine zero, rather than from the unknown program zero. The result will be an overtravel condition.

To solve the problem, program must use a *work offset*.

> **The main purpose of work offsets is to provide connection between the two origins (zeros)**

USING WORK OFFSETS

How do work offsets actually work? Let's look at the same three hole example, also used for later topics.

In the introduction to this chapter, **G92** was identified as the current tool position always measured from program zero for each axis. Since this position was not normally known at the time of program development, the **G92** method presented many difficulties. The purpose of work offsets is to establish the relationship between the two origins - machine zero *and* part zero - in a much more practical way. While the **G92** settings were measured from program zero to the current tool position, work offsets are based on a totally different principle:

> **Work offset is always measured**
> *FROM* **machine zero** *TO* **part zero, along each axis**

This simple statement identifies the major difference from the old method - while making the position register work was equally split between the CNC programmer and CNC machine operator, handling work offsets is primarily in the hands of the CNC operator. Programmer still has some responsibility - the program must contain a special preparatory command that selects the work offset required for the part setup.

The programmer has at his or her disposal several standard work offsets, as well as many additional ones, that can can be added as an option to the control system.

Selection of any work offset is done by programming an appropriate preparatory command.

Preparatory Commands G54-G59

Modern controls provide six standard work offsets, using the preparatory commands **G54** to **G59**:

G54 ... Work offset 01
G55 ... Work offset 02
G56 ... Work offset 03
G57 ... Work offset 04
G58 ... Work offset 05
G59 ... Work offset 06

Each work offset can be set for up to six different parts located within the machine work area. As most jobs on a vertical machining center require only one work offset, the following explanations will relate to the first offset in the group - **G54** work offset - although the general logic applies equally to all work offsets.

Is G54 Work Offset the Default ?

If you were following this chapter from its beginning, you may have actually tried the programmed motion at your CNC machine:

```
G90 G00 X10.0 Y8.0
```

What did happen? Did the motion overtravel? Did it work just right? These questions reflect the variety of several possibilities.

G54 work offset has to be set first, by measuring X and Y distances first, then entering the measured data into an offset register. Once the CNC operator sets the work offset as the distance between machine zero and part zero, the tool motion should end at the position that matches the drawing dimension. If the motion is activated *before* the work offset has been set, the tool motion will result in overtravel.

Another question is also important when the operator interprets the program. It is not unusual to find that a particular program does not contain any work offset at all. So, what happens if the work offset command is missing in the program?

Some CNC programmers often rely only on the control system's default settings:

> **Default is a condition that 'assumes' certain settings, unless they are specifically included in the program**

Providing the **G54** work offset is correctly set *and* the control uses *default=G54*, the tool will still move to the correct part location, even if **G54** is omitted in the program itself. This is rather a bad practice in programming, and one advice is equally applicable to both - programming and operation:

> **Never count on default settings !**

As a rule, always make sure the required work offset command is specified in the program.

Establishing the Connection

Just like two excellent solo musicians have to match their individual strengths in a duo performance, two zero locations have to be reconciled for desired results. There is only one simple requirement - making the tool motion to travel to the actual position represented by the program. To achieve this objective, a connection between the machine zero and the part zero must be established. The control system provides at least six work offsets for exactly that purpose and the CNC programmer selects the one or more required for the job.

How Does G54 Work?

Although **G54** is used for the most common applications, any other work offset uses the same principle. If you understand how **G54** works, you should have no problem to understand any other work offset.

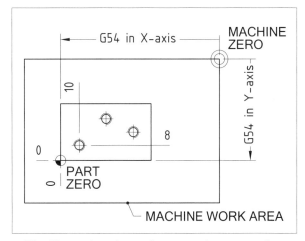

The illustration above shows one important change from the previous illustration - it shows very specific and defined connection between the machine zero and the part zero. As stated previously,

> **Work offsets are always measured from MACHINE ZERO to PART ZERO**

Direction of Measurement

Note that during actual part setup, both distances along X and Y axes will be stored in the control as negative amounts for each axis. The reason is that the measuring distances were in the negative direction from machine zero. Some machines have their machine zero at the upper left corner rather than the more common upper right corner (when viewed from the top). Compare the following illustrations:

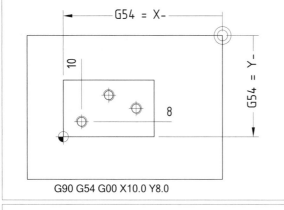

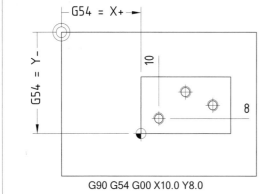

Note that the work offset setting does not change the way tool motion is programmed - in either case, the tool will move to position of the first hole at X10.0Y8.0.

What is different is the internal calculation of the control system, to find the *Distance-To-Go* - also called the *total travel amount*.

Later in this chapter, you will learn how the control calculates the *Distance-To-Go*. This knowledge is not an absolute requirement to setup a job correctly, but it will help in understanding what is a actually happening.

Work Offset Register

Once the offset amounts, measured along each axis, are known to the operator, they have to be registered by the control system. This is done in a special registry of the control, usually selecting two control panel soft keys - *Offset* from the control panel, then *Work* located above one of the soft keys. Although the physical appearance of the screen display will be quite a bit different between individual controls, all principles behind the settings are the same. The table below shows a schematic display for all six work offsets, for a typical three-axis *Vertical Machining Center* (VMC):

00	EXT				
X	0.000				
Y	0.000				
Z	0.000				

01	G54	02	G55	03	G56
X	0.000	X	0.000	X	0.000
Y	0.000	Y	0.000	Y	0.000
Z	0.000	Z	0.000	Z	0.000

04	G57	05	G58	06	G59
X	0.000	X	0.000	X	0.000
Y	0.000	Y	0.000	Y	0.000
Z	0.000	Z	0.000	Z	0.000

If only six work offsets are normally available as standard, why are there seven registry entries shown at the control? The extra registry (00) is set aside for some more advanced work. It will also be explained, later in this chapter. For now, let's stay with the basic six work offsets, particularly with the first one - **G54**.

As you see from the illustration, all offsets have zero values - that means there are no settings applied. Also keep in mind that the actual screen may only show one or two offsets at a time, not necessarily all of them.

Actual Measurements

During setup, the CNC operator has to find the exact distance between machine zero and part zero. This can be accomplished by several methods, typically using special devices and processes:

- **Mechanical edge finder (shaft type)**
- **Electronic edge finder**
- **Magnetic edge (with slot) or corner finders (with hole)**
- **Probe**

All four device groups are generally used to establish only the X0Y0 part origin (part zero point). Zero setting in the Z-axis uses a different method, described in the next chapter. Establishing the XY program zero using a probing device is not a common practice and requires a special macro program - this method is outside the topics covered in this book. The first three methods are used quite frequently, mainly by preference.

> **Ball type wigglers used in manual machining are not recommended for precision work**

USING EDGE FINDERS

For most jobs, a revolving mechanical edge finder is the most suitable device. It is inexpensive even for high end brands and very accurate for most work. The fixed main body is mounted in a collet, and the shaft end is internally connected, using a strong spring. Electronic edge finders are even more accurate but cost a lot more. Also, they can only be used on conductive materials only. Although only marginally similar, these two devices work on the principle of physically touching the part edge that happens to be a zero location.

Modern technology also provides laser edge for virtually any CNC machine. This special method is also worth further study. The last option, using magnetic edge finders or corner finders, offers an excellent alternative in many situations.

Two most common edge finders are single end with the edge finder only and a double end type (described later) that also includes a conical center finder.

Edge finders are used to locate an edge of the part in relation to the spindle centerline. They are handy for parts that have straight edges, such as shoulders, slots and grooves.

Magnetic Single Edge Finder

A very common type of an edge finder is called a *magnetic single edge finder*. As its name suggests, this device is attached to the metal part with built-in magnet - see illustration in the next column.

In order to find an edge, a dial test indicator is used. While it is rotated by hand in the free moving spindle, the axis of measurement is adjusted by the handle, until the reading at both inner walls is identical. Each edge must be measured separately. The main advantage of this type of edge finder is that there is no need to adjust the distance displayed on the control screen. The display is the work offset for the measured axis.

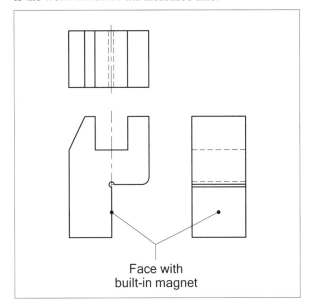

Face with built-in magnet

Magnetic Center Finder

Another type of a similar device also requires the use of a dial test indicator. It is called *magnetic center finder*. As the name suggests, this device will find the corner of a part that will become the part zero. Both axes are measured simultaneously and no adjustments are required.

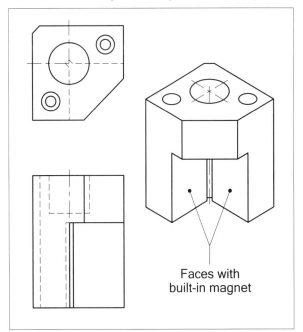

Faces with built-in magnet

Their purpose of either device is to align the spindle centerline with an part edge or a corner. In small fine adjustments, the operator moves the XY position of the indicator, until it runs true.

Whether using the single edge finder of the corner finder, both are precision ground devices and the part edge can be indicated with high degree of accuracy, without any adjustments. Keep in mind that in order to get correct distance measured, you have to start from machine zero and axis position set to zero as well.

> **Magnetic edge or corner finders do not require additional calculations during setup**

In addition, using a dial test indicator with these two magnetic finders, you can also check the squareness of the setup. Both types work well on metal surfaces only and should not be used for plastic or other non-magnetic materials.

DISTANCE-TO-GO

Before discussing practical use of an edge finder (regardless of its type), it is important to understand the concept of what Fanuc controls call **Distance-To-Go**.

Many CNC systems use *distance-to-go* as a description for an axis motion in progress (some controls call this motion *increment* or *total travel distance*). When an axis motion is active, the control shows the exact length of the remaining motion needed to reach programmed target location.

Follow the following two blocks, based on the motion between the first and second hole of the drawing provided at the chapter beginning:

```
G90 G00 X10.0 Y8.0                   (H1)
X25.0 Y22.0                          (H2)
```

In the first block (for H1), the absolute tool position is at X10.0 Y8.0. The following block (for H2) contains a tool motion to the next absolute position of X25.0 Y22.0 (H2). As both positions are in **G90** (absolute mode), they are measured from part zero. The control system has to calculate how much the actual motion between the two holes will be for each axis. To complete such calculation, a suitable mathematical formula has to be built into the CNC software.

Using basic math knowledge, you should see that the actual tool motion length is the difference between two absolute locations - *target* and *current* tool location:

```
Target:            X25.0    (H2)
Current location:  X10.0    (H1)
Difference = Distance-to-Go = 25.0 - 10.0 = 15.0

Target:            Y22.0    (H1)
Current location:  Y8.0     (H2)
Difference = Distance-to-Go = 22.0 - 8.0 = 14.0
```

CNC system will use the same calculation to find the distance between both holes. Some math formulas may accept a lot of data, but the core calculation is never changed.

> ***Distance-To-Go* calculation (total distance) is used for all available offsets**

Understanding the concept of distance-to-go is an important aspect of understanding all types of offsets.

SHAFT TYPE EDGE FINDERS

Using magnetic edge finders requires only a precision dial test indicator attached to the face of a free spinning spindle. By rotating the spindle and moving XY axes positions as required, the objective is to make the indicator run true in one or both axes, depending on which edge finder is used. At this point, the edge location or the corner is known (based on spindle centerline) and can be entered as the work offset for current part setup.

Using much more common *mechanical shaft type edge finder*, the setup procedure is a bit more involved, and some additional calculations will also required.

The simple three-hole part, introduced earlier in this chapter, will be used to illustrate the concept of using shaft type edge finder in detail. Metric edge finder with a 6 mm shaft diameter will be used for the example.

Prerequisites

Regardless of the actual fixture used, one setup condition is absolutely essential:

> **All parts of the batch must be located at exactly the same fixture position**

A vise has to be mounted securely to the machine table and its jaws have to be parallel with an axis motion.

Individual Steps

Using a mechanical shaft type edge finder is divided into three general steps:

- **Step 1** Eccentric rotation of the edge finder (rotation)
- **Step 2** Concentricity at the part edge (centering)
- **Step 3** Shift from the center (often called a 'kick')

In *Step 1*, the edge finder is positioned close to the measured edge, but far enough so its bottom tip can be shifted off its center and rotate eccentrically, without actually touching the part edge. A suitable depth from edge top is also required.

In *Step 2*, the operator uses the setup handle and gently moves the rotating edge finder closer and closer to the measured edge. At a certain point, the diameter of the edge finder tip will touch the part edge. As the operator continues the axis movement, the eccentricity of the tip will become smaller and smaller, until it disappears completely and the $\varnothing\,6$ mm tip is concentric with the spindle center line.

The actual measurement takes place between *Step 2* and *Step 3*. By having the handle set to the smallest increment and moving one division, there will be a 'kick' - that is a common description when the edge finder tip will become eccentric again. By moving the handle back by the same amount, the 'kick' is removed and tip is concentric again. At this point, the measurement is within about 12 microns (0.0005").

As every CNC operator approaches part setup in different ways, the following steps reflect the general suggestions present so far. Once the fixture (such as a machinist's vise) is properly set and the edge finder is mounted in the spindle, the procedure begins from machine zero (spindle is stationary at that time):

01 Make sure X and Y axes show 0.000 (0.0000) position
02 Work with one axis at a time
03 Manually throw the end tip off center
04 Move the edge finder close to the edge to measure
05 Start spindle rotation - 800 r/min (or your preference)
06 Move 3 to 5 mm below the edge top face
07 Using setup handle, gently move towards the edge
08 When contact is established, eccentricity gets smaller
09 Move handle until the shaft runs continuously true
10 Wait for the 'kick' *(see explanation above and below)*

At this point (item 10), the shaft diameter is in contact with the part edge at all times, and the $\varnothing\,6$ mm shaft is exactly 3 mm (its radius) away from the measured edge. Because of the natural resistance between the two surfaces, the shaft will be thrown off its center - this 'kick' means the precise edge location has been established. Write down the position measured for the selected axis, and repeat the procedure for the other axis. You should have two known locations, exactly measured. This procedure takes a bit of practice, so it is best to try the first attempts with an experienced person to assist and guide.

Work Offset Calculations

> One of the most common errors in setup
> is to forget adjusting edge finder diameter to radius

Keep in mind that all measured dimensions are from machine zero to the spindle centerline, which means the actual edges (required for work offset settings) are still off by the actual tip radius (3 mm in the example). This radius amount must always be taken into consideration.

The illustration below shows into which direction the measured dimension should be adjusted, based on the part zero and the axis of measurement. Examples follow.

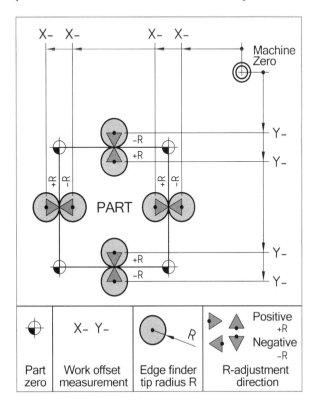

Setup errors also happen even if the tip radius is accounted for. Common reason is that the tip radius was *added* rather than *subtracted*, or vice versa.

The last illustration identifies common setups for *four different* part zeros on vertical machining centers. Typical X and Y work offsets are both measured from machine zero into the negative direction, which requires both work offsets to be stored as negative amounts.

Think of the edge finder center as being either too far (long travel), or not far enough (short travel) from the edge measured. All eight possibilities are listed here:

Part zero location	Axis	Edge finder distance as measured			
		External		Internal	
Lower Left	X	LONG	+R	SHORT	-R
	Y	LONG	+R	SHORT	-R
Lower Right	X	SHORT	-R	LONG	+R
	Y	LONG	+R	SHORT	-R
Upper Right	X	SHORT	-R	LONG	+R
	Y	SHORT	-R	LONG	+R
Upper Left	X	LONG	+R	SHORT	-R
	Y	SHORT	-R	LONG	+R

For the final work offset, you either add or subtract the edge finder radius to the measured negative dimension. The calculation is simple and always the same:

> **NEGATIVE READOUT + RADIUS ±R**

The most important part of the formula is to remember that we are always *adding* the radius, whether it is positive or negative, to a readout that is always negative for the machine zero position as shown.

Part Zero at Lower Left - Example 1:

Setting the part zero at the lower left corner is very common - it makes all XY part locations to be positive. For this example, sample X and Y measurements will be used - keep in mind that actual numbers will be different for each setup:

- Edge finder radius = 3 mm
- Edge finder center measured in X = -618.385
- Edge finder center measured in Y = -307.540

Based on these dimensions, an example can be used. The first one is for the part zero at lower left corner.

Example 1a - If these are EXTERNAL measurements ...

The table shows both measurements as long, meaning the edge finder center is too far and a *positive* radius will be added to both measurements:

X-axis edge = -618.385 + +3 = -618.385 + 3 = -615.385
= Work offset X is -615.385

Y-axis edge = -307.540 + +3 = -307.540 + 3 = -304.540
= Work offset Y is -304.540

Work offset will be set to **X-615.385 Y-304.540**

This example may represent the settings for the three holes as per initial drawing:

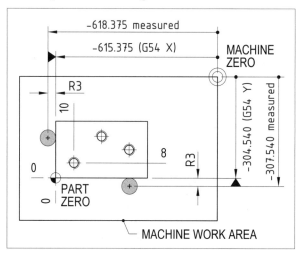

Work offset registry has to be set accordingly:

00	EXT				
X	0.000				
Y	0.000				
Z	0.000				
01	G54	02	G55	03	G56
X	-615.375	X	0.000	X	0.000
Y	-304.540	Y	0.000	Y	0.000
Z	0.000	Z	0.000	Z	0.000
04	G57	05	G58	06	G59
X	0.000	X	0.000	X	0.000
Y	0.000	Y	0.000	Y	0.000
Z	0.000	Z	0.000	Z	0.000

The next few setup examples do not refer to a particular drawing, but show calculations for other possible settings that may exist.

Example 1b - If these are INTERNAL measurements ...

For internal measurements and part zero at the lower left corner, the edge finder center is not far enough and a *negative* radius must be added to both measurements ('friendly' numbers used for convenience only):

X-axis edge = -600.0 + -3 = -600.0 - 3 = Work offset is -603.0
Y-axis edge = -300.0 + -3 = -300.0 - 3 = Work offset is -303.0

Work offset will be set to **X-603.000 Y-303.000**

Part Zero at Lower Right - Example 2:

For this second example, the part zero is at the lower right of the part, using the following measurements for edge finder center, again with 'friendly' numbers:

- Edge finder radius = 3 mm
- Edge finder center measured in X = -600.000
- Edge finder center measured in Y = -300.000

Example 2a - If these are EXTERNAL measurements ...

The table on the previous page shows a short X measurement - it means a negative radius has to be added to the measurement:

X-axis edge = -600.0 + -3   = -600.0 - 3
           = Work offset is ... -603.0

The Y-distance is too long, meaning a positive radius has to be added to the measurement:

Y-axis edge = -300.0 + +3   = -300.0 + 3
           = Work offset is ... -297.000

Work offset will be set to **X-603.000 Y-297.000**

Example 2b - If these are INTERNAL measurements ...

For internal measurement at the same lower right part zero, the calculations must be reversed - positive radius will be added to the X-measurement, and a negative radius will be added to the Y-measurement:

X-axis edge = -600.0 + +3   = -600.0 + 3
           = Work offset is ... -597.0
Y-axis edge = -300.0 + -3   = -300.0 - 3
           = Work offset is ... -303.0

Work offset will be set to **X-597.000 Y-303.000**

Part zero for the other two corners can be calculated in a similar way. Hopefully, the explanation and some examples have made this important subject clear - always think twice before committing a particular calculation to any control system setting.

For reference, when adding or subtracting positive or negative numbers, the following table may be of help:

Calculation	Result
PLUS and POSITIVE (+ +)	A + (+B) = A + B
PLUS and NEGATIVE (+ -)	A + (-B) = A - B
MINUS and POSITIVE (- +)	A - (+B) = A - B
MINUS and NEGATIVE (- -)	A - (-B) = A + B

Original Example Revisited

The simple three hole part introduced at the beginning of this chapter can now be set and work offsets established by measurement, using the described methods.

Once the 3 mm adjustment is made, the settings can be input as the work offset G54 at the control.

The original program block

G90 G00 X10.0 Y8.0   (H1)

... should be changed to reflect the fact that the setting is for the **G54** offset and none other. Including the work offset results in a program that does *not* take chances:

G90 G54 G00 X10.0 Y8.0   (H1)

WORK OFFSET Z-SETTING

The following is the G54 work offset setting for the drawing example, as explained already:

01	G54
X	-615.375
Y	-304.540
Z	0.000

Note that the Z-setting is *zero*.

Where is the Z-setting?

One very natural question can be asked at this point. What about the work offset for Z-axis? Having it set to zero does not seem to make much sense, you may ask.

Typically, Z-axis is the axis that controls the depth of cut. As such, it is measured differently than the XY settings and requires a different offset, called *tool length offset*. This important subject requires thorough knowledge, and will be discussed in a separate chapter.

In some special cases, the Z-setting does *not* have to be zero. This will be explained briefly in this chapter as well as in the chapter covering the tool length offset.

Setting the Z-axis

The setting of the Z-axis is done either at the machine or off-machine. In either case, the amount measured is entered into the tool length offset. The frequently asked question is - *why*? Let's look at the following example. It uses a setup method common to small shops and job shops, so called *touch-off* method. Once the particular tool is placed to the spindle and the Z-axis is at home position, the tool length measurement can begin.

In the following illustration, the tool measurement is *275.627 mm* in negative direction:

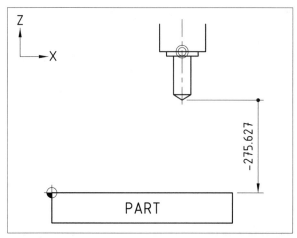

Can this Z-setting be entered into the **G54** work offset?

Consider this updated **G54** setting:

01	G54
X	-615.375
Y	-304.540
Z	-275.627

The answer is yes, as shown, the setting is correct. In the program, there will be an additional command, moving the tool along Z-axis to a clearance position of 2 mm above the part top face (Z2.0):

```
G90 G54 G00 X10.0 Y8.0           (H1)
Z2.0                       (CLEARANCE)
```

Based on the previous distance-to-go calculations for the X and Y axis motion, the same formula will be used for the Z-axis motion:

Target:	Z2.0	(all holes)
Current location:	Z-275.627	(from machine zero)
Difference = Distance-to-Go	= 2.0 + -275.625	
	= 2.0 - 275.627	
	= -273.627	

The setting is correct - yet, there is a problem.

The problem can be summed up in this observation:

> **Work offset always applies to ALL tools**

From this statement, it should be obvious that not all tools used by the program will have the same length. In other words, the work offset setting of **Z-273.627** is applicable to *all tools* using **G54** work offset. As this is not an acceptable situation, the control system offers a special offset, called the *tool length offset* (**G43 H..**). In this registry, each measured tool length is entered individually. This and other details relating to tool length offset are described in a separate chapter.

For the following subjects and various examples, the setting used as the starting point will be the one without any Z-setting (Z0.000).

01	G54
X	-615.375
Y	-304.540
Z	0.000

MULTIPLE WORK OFFSETS

As the heading suggests, multiple work offsets will require respective settings not only in the **G54** registry, but also in one or more of the other registries. There are many uses for more than one work offset, but for machining purposes, the most common use is when multiple parts are setup on the machine table, each with its own part zero, measured in and registered to any of the six work offsets available. Take an example for two vises set on the machine table (not to scale):

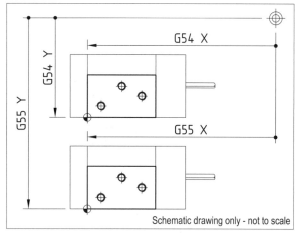

Schematic drawing only - not to scale

Although both parts appear to have the X-setting identical for **G54** and **G55**, that does not mean only one setting is required for the X. The X-origin for one vise will not likely be the same as the X-origin for the other vise.

The difference may be small, but for precision work it is absolutely necessary to have two offsets available.

Settings for the **G54** and **G55** work offsets may be similar to this:

01	G54	02	G55
X	-615.375	X	-614.859
Y	-304.540	Y	-441.363
Z	0.000	Z	0.000

Program Entry

In the program, both offsets have to be used in order to reach the proper hole:

```
G90 G54 G00 X10.0 Y8.0    (H1 - UPPER VISE)
Z2.0                       (CLEARANCE)
<... drill H1 ...>
Z2.0                       (CLEARANCE)
G55 G00 X10.0 Y8.0         (H1 - LOWER VISE)
<... drill H1 ...>
```

Note that the programmer calls for the same hole H1 location, but within two different work offsets.

In practice, the program will use a suitable fixed cycle. The following example is a complete program for spot drilling the three holes in two vises:

```
N1 G21
N2 G17 G40 G80 T01
N3 M06
(=== UPPER VISE ===)
N4 G90 G54 G00 X10.0 Y8.0 S1100 M03 T02
N5 G43 Z2.0 H01 M08
N6 G99 G82 R2.0 Z-3.3 P200 F175.0    (H1)
N7 X25.0 Y22.0                        (H2)
N8 X40.0 Y15.0                        (H3)
(=== LOWER VISE ===)
N9 G55 X10.0 Y8.0                     (H1)
N10 X25.0 Y22.0                       (H2)
N11 X40.0 Y15.0                       (H3)
N12 G80 G54 Z25.0 M09
N13 G28 Z25.0 M05
N14 M01
```

Note that the program does *not* change to **G00** in block *N9*. rapid motion command **G00** would cancel the fixed cycle, which is not the intended outcome. Rapid motion between holes is the integral part of all fixed cycles. Also note block *N12* - the initial **G54** work offset has been reinstated. The following two tools for drilling and tapping (not shown) will have virtually the same format, with the exception of the fixed cycle used and related machining conditions. For more details, see *page 127*.

Programming method using a subprogram could also be used, but it would have no influence on the subject of two work offsets discussed here.

WORK OFFSET ADJUSTMENTS

If one or more work offsets have been set properly at the beginning, there will not be any need to make adjustments once the machining has started. For various reasons, this ideal situation may not always be the case. One common example is to use an imperial example on a metric job or vice versa.

Mixing Units of Measurement

In North America, use of the metric system in manufacturing has progressed very significantly in last few years, but the imperial units are still the norm in many companies. In such situations, it is not unusual to program a metric job and use imperial tools for machining, including the shaft type edge finder. For example, a $\varnothing$ 0.200 inch edge finder tip is used to set metric work offset. In such cases, the edge finder radius has to be converted to metric units, base on 1.0" = 25.4 mm:

$\varnothing$ 0.2 mm / 2	= R0.1 inch
0.1 inch x 25.4	= R2.54

When you consider the very low cost of edge finders, it makes sense to purchase a metric edge finder and a suitable collet and use them for all metric jobs. Any conversion creates a possibility of a error. Even a small error will have serious consequences.

As an example used for comparison, first consider the initial XY measurements to the part edge using metric edge finder with $\varnothing$ 6 mm tip:

01	G54
X	-615.375
Y	-304.540
Z	0.000

How would this setting change if an imperial edge finder with $\varnothing$ 0.2 inch tip were used? It may come as a surprise, but the offset will *not* change at all. The reason?

If the setup is identical to the one described and only the edge finder is different, it is the initial XY position of the spindle center line that will be different, *not* the resulting work offset - after all, the part has not moved:

- Edge finder radius = 3 mm
- Edge finder center measured in X = -618.385
- Edge finder center measured in Y = -307.540

Now for the imperial edge finder:

- Edge finder radius = 0.1 inches = 2.54 mm
- Edge finder center measured in X = -617.915
- Edge finder center measured in Y = -307.080

There is a difference of 0.46 mm between the radius of the metric edge finder and the imperial edge finder.

Height of Part

Using the example of a two-vise setup (G54 and G55) for the three holes will illustrate another common problem that has to be identified and corrected. The problem is the difference between the Z-height of the upper vise and Z-height of the lower vise.

Ideally, using the same type of vise for both setups, the top face of both parts should be identical. That may be the case in some cases - but what about the cases where there is a difference? Consider the following setup as it relates to the part heights of the upper and lower vise, as shown already for the three hole example:

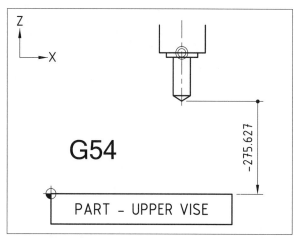

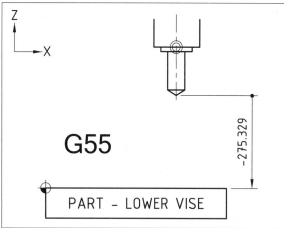

There is only a very small difference between the heights of the parts located in two vises:

- Upper vise: **-275.627**
- Lower vise: **-275.329**
- Difference: 275.627 - 275.329   =   0.298 mm

The dimensions show the part in the lower vise is 0.298 mm *higher* than the part located in the upper vise. How can we tell whether the part is higher or lower? The answer is in the setting amount itself.

During touch-off, the tool tip has to travel a certain distance required to touch the Z0 face. For long tools, this distance is short, while for short tools the distance will be longer. In the example, the travel of 275.329 is shorter than the travel of 275.627, therefore the part located in the lower vise is slightly higher.

This explanation may be interesting by itself, but it is critical for understanding the necessary offset adjustment. Let's evaluate the situation and possibilities.

- **Both vises contain the same part**
- **The same tool is used to move between vises**
- **Upper vise uses G54 work offset**
- **Lower vise uses G55 work offset**
- **Both work offsets in X and Y are set correctly**
- **Both work offsets have Z-setting as Z0.000**

Based on this evaluation, what available options are there? The most obvious answer may be to have a different tool length offset for each part (vise). This solution would work well, except for a major problem in many cases. The more vises used for the setup and the more tools programmed for each part mean many more tool length offsets will be required in the program. Not only the setup time will be greatly increased in such a situation, but - for some complex jobs - you may even run out of the available tool length offsets.

The solution is much simpler than the option just presented - change the Z-setting of the appropriate work offset. In our case, the **G55** Z-setting will not be zero but will contain the amount of the calculated difference, which is 0.298 mm.

For the operator, the difference itself is not enough, as the current Z-setting of **G55** has to be adjusted UP or DOWN - in other words do we add 0.298 mm or take that amount away from the current setting?

In the example, the **G55** lower vise part is higher than the **G54** part, therefore the work offset has to be increased (positive).

01	G54	02	G55
X	-615.375	X	-614.859
Y	-304.540	Y	-441.363
Z	0.000	Z	0.298

The *distance-to-go* will be adjusted - that is *shortened* - by the amount of 0.298 mm for the part within the **G55** work offset (lower vise).

EXT WORK OFFSET

Just by looking at the work offset screen, you see there are six standard work offset registries and an additional one called **EXT**. So far, we have used only the standard registry **G54** and **G55**. Others in the standard series of six (**G54**-**G59**) are used the same way. On some controls, particularly older models, you may also see an additional offset, but this time marked as **COM**. What is the purpose of these additional offsets?

EXT or COM ?

First, the 'mystery' of the two abbreviations:

- **EXT**   ... stands for *EXTernal*
- **COM**   ... stands for *COMmon*

These are not two different offsets - they are the same *external* offset under two names:

> Some older CNC units identify this external offset as COM - for COMmon offset. COM identification has been dropped sometime ago, to avoid a possible confusion with COM standing for COMmunications. Modern control systems use the abbreviation EXT - EXTernal.

This in not a programmable work offset, so there is no G-code associated with this setting. The purpose of external work offset is to allow global changes - changes that apply to all other work offsets used in a program.

Take, for example, a program that uses three fixtures on the machine table, each using a separate work offset. **Fixture 1** uses **G54**, **Fixture 2** uses **G55**, and **Fixture 3** uses **G56** work offset. If any individual fixture needs an adjustment unique to that fixture, one or more settings can be changed for that fixture only. Other two fixtures are unaffected. What about a change that affects *all three fixtures*? For example:

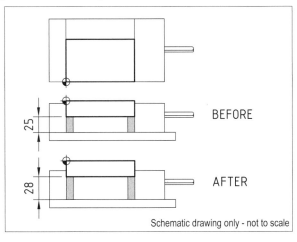

Schematic drawing only - not to scale

The illustration shows a change in the heights of parallels that support the part in the vise. In the original setup, the parallels were 25 mm high, but a change in setup calls for parallels of 28 mm high. There is a difference of 3 mm. How can such change be handled?

Offset Adjustment Options

As only the Z-axis is affected, the XY settings will not change, regardless of the adjustment method used. There are several options:

- Tool length offset change
- Standard work offsets change
- External offset change

Tool Length Offset Change

The change in the height of parallels was from 25 mm to 28 mm, which means the part is now at a higher position than it was in the original setup. The touch-off distance has shortened, so each tool length offset has to be 3 mm smaller that its current setting. For example, if tool length offset 02 (H02 in the program) was originally set to -264.013, it has to be changed to -261.013. The problem with this method is that every tool length offset programmed will have to be changed. With many tools, this is rather impractical and subject to errors that can have serious consequences.

Standard Work Offsets Change

A better method is to change the Z-setting of each work offset used by the program. In the example, there are only three. The illustration shows the settings for all three work offsets used:

01	G54	02	G55	03	G56
X	...	X	...	X	...
Y	...	Y	...	Y	...
Z	3.000	Z	3.000	Z	3.000

What about the **EXT** offset? There is nothing to do. Whatever setting was there before, that setting does not change - remember, the change was made individually for each work offset. In this case, the external offset has a zero value for the Z-setting:

00	EXT
X	0.000
Y	0.000
Z	0.000

WORK OFFSET SETTINGS

External Offset Change

The last version is the most convenient of the three - it requires very little effort and making an error is severely minimized. This method uses the advantage that *all work offsets have to be changed by the same amount*. Instead of placing the 3 mm adjustment into each offset individually, just place it once into the external offset and leave the others as per original setting:

01	G54	02	G55	03	G56
X	...	X	...	X	...
Y	...	Y	...	Y	...
Z	0.000	Z	0.000	Z	0.000

As there was no change in individual offsets, the **EXT** offset has to be changed:

00	EXT
X	0.000
Y	0.000
Z	3.000

External offset setting for any axis will update any and all settings in individual work offsets, for that axis. Think of this feature as a global update - *or* - as the old **COM** abbreviation meant - as an adjustment that is *common* to all other offsets.

> **External offset will update settings for ALL work offsets used by the program**

WORK OFFSETS - DATA INPUT

This chapter has covered setting and changing work offsets. How exactly is the offset entered or changed at the control?

The method of entering or changing work offsets is the same method used for entering or changing other offsets and even some parameters. It uses two buttons or two soft keys with similar names:

- INPUT   ... ABSOLUTE data input
- +INPUT   ... INCREMENTAL data input

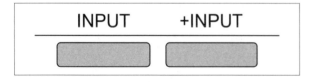

Typically, these two selections appear above soft keys on the control monitor:

INPUT is described as an ***absolute*** data input. This means the current data will be ***replaced*** with the new data.

+INPUT is described as an ***incremental*** data input. This means the current data will be ***updated*** by the new data.

Data Change Example

Study this example - the current work offset setting is -615.375. This setting proves to be too far from machine zero and has to be changed. The corrected offset measurement is -614.893. The difference is:

615.375 - 614.893 = 0.482 mm

One method is to enter the new setting as absolute data input using the **INPUT** button. In this case, this is the best way to change the offset by replacing it.

In other situations, the amount of adjustment has to be updated rather than replaced. Using the same example, the difference amount of 0.482 has to be entered with the **+INPUT** key. The question is will the current offset be incremented by a *positive* amount or a *negative* amount? The correct answer is critical - many errors are caused by the wrong calculation. Lets look at the current setting:

-615.375

It is a *negative* setting. **+INPUT** will always ***ADD*** the new amount:

```
-615.375 + +0.482    = -615.375 + 0.482
                     = -614.893

-615.375 + -0.482    = -615.375 - 0.482
                     = -615.857
```

As you see, the first example is correct - you will be adding a positive amount.

> **+INPUT will always ADD the amount of adjustment**

10

TOOL LENGTH OFFSET

The 'missing' Z-axis setting mentioned in the previous chapter gets another look in this chapter that covers the subject of *tool length offset* principles and settings.

TOOL LENGTH

First, what *is* a tool length? This simple question does not have a simple answer. For programming and setup purposes, tool length is *not* the actual length of the tool from one end to another. The required tool length is the length of the tool measured between the spindle gage line and the tool tip (end point).

Typical CNC program calls several tools used in the machining process. These tools often belong to many categories, they have different diameters, and they also have *different lengths*. They are the focus of this chapter.

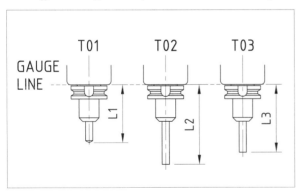

There are at least three common methods of measuring tool length:

- Presetting method   ... off-machine
- Touch-off method   ... on-machine
- Reference tool method   ... on machine

Each method will be described separately.

> The main purpose of measuring tool length
> is to establish relationship between
> the tool tip *(cutting point)*
> and the part zero in Z-axis *(Z0)*

Once the tool length measurement is completed, it has to be entered into the control system, using a specific registry called the *tool length offset*.

Tool Length Offset Register

Typical capacity of the tool length offset register is at least 32 offsets for very small machines (usually more), and up to 999 offset entries for larger machines.

In the part program, it is typical to assign the same tool length offset number as the tool number, whenever possible. For example, the tool selected as T01 will use H01 as the tool length offset number, the tool selected as T04 will use H04 as the tool length offset number, etc. The reason is simplicity - after all, each tool number is unique, and each tool is associated only with one tool length offset (for the majority of work), so keeping both numbers the same makes it easier during part setup.

TOOL LENGTH OFFSET COMMANDS

Fanuc and similar controls support four features related to tool length offsets used in programs for milling applications:

- G43   ... Tool length offset positive
- G44   ... Tool length offset negative
- G49   ... Tool length offset cancel
- H..   ... Tool length offset number

Before you understand how the tool length offset actually works, you should be familiar with the concepts of work offset settings (work coordinate system commands **G54-G59** and higher). Work offsets have been described in the previous chapter.

Using a typical 3-axis vertical machining center as an example, the work offset **G54** may be set to the following amounts:

01	G54
X	-615.375
Y	-304.540
Z	0.000

From the chapter describing work offsets, you know that X-615.375 setting indicates the distance of 615.375 mm measured from machine zero to part zero, parallel to the X-axis. The Y-axis setting of Y-304.540, which is 304.540 mm, is also measured from machine zero to part zero, this time parallel to the Y-axis. Note that the work offset setting for the Z-axis is zero (normally).

Limitations of Work Offsets

What about the Z-setting? Where is the offset? If the settings for X and Y axes indicate the distance from machine reference point (spindle centerline), does not the Z-setting indicate the same distance for the tool length reference point? Of course it does. Work offset commands are applicable to *all* axes. The persisting question is still unanswered. If you can set a work offset in X and Y axes, you can also set it in the Z axis. The Z-setting in the work offset would reflect the tool length setting for the tool used. Let's take a tool that has setting of 336.7 mm in the negative direction. That would change the previous **G54** work offset to:

01	G54
X	-615.375
Y	-304.540
Z	-336.700

How do we check if the new settings work? A four block program segment will do the job:

```
G90 G54 G00 S1200 M03    (INITIAL SETTINGS)
X100.0                   (DISTANCE-TO-GO IS -515.375)
Y70.0                    (DISTANCE-TO-GO IS -234.540)
Z2.0                     (DISTANCE-TO-GO IS -334.700)
```

In order to verify that the setting is correct, watch the *'Distance-To-Go'* screen display when each block is being processed. *Distance-To-Go* uses the following calculation for the actual length of tool motion:

X-axis ...   -615.375 + 100.0 = -515.375 actual motion
Y-axis ...   -304.540 + 70.0 = -234.540 actual motion
Z-axis ...   -336.700 + 2.0   = -334.700 actual motion

The calculation is constant for all axes and results in correct distance to travel by the tool (in all three axes), to reach the programmed position.

Feel free to read further if you wish, but you may want to stop for a moment and think about the results.

No doubt, the tool is now located exactly at the part position as programmed - **X100.0 Y70.0 Z2.0**, all measured from program zero. Is there anything else to consider? Is there a clue somewhere to something else? Think about both questions before reading further.

The answer to the two questions above is simple but not quite obvious. While the work offset setting will do exactly what the program requires, there is a severe restriction:

> **Tool length offset specified as Z-axis setting in the work offset, applies to ALL tools**

Here is the real problem - while the **G54** can be set to accommodate all three axes, including the Z-setting is not practical, for one simple reason - *most programs use more than one tool to machine a part.*

Only one Z-axis setting is available in **G54**, meaning only one tool length can be set in the **G54** registry. Logically, other work offsets could be used with identical XY settings, but Z-settings for additional tools. This is still quite limiting and not very practical. Fanuc has solved this problem by delegating the tool length offset to a different registry, called the *Tool Length Offset*.

> **Tool length offsets are based on the same principle as work offsets**

G43 command behaves the same way for the Z-axis, as work offsets **G54**-**G59** behave for all axes.

G43 and G44 Commands

Most control system manuals list two commands that activate tool length offset:

- **G43** ... Tool length offset POSITIVE
- **G44** ... Tool length offset NEGATIVE

Both commands **G43** and **G44** can be quite misleading, if you try to understand them only by their definition. **G43** is describes as tool length offset positive - *positive what?* Even if command descriptions are not meant to give exact and lengthy definitions, this one can be outright misleading. The word positive and negative relates to the basic way the total tool travel - *Distance-To-Go* - is calculated by the control system:

G43:   Distance-To-Go = Length offset   *PLUS*   Z-target
G44:   Distance-To-Go = Length offset   *MINUS*   Z-target

Although not shown in the following example, the calculation of the total travel distance will also include the Z-setting of the current work offset as well as the Z-setting of the **EXT** work offset.

➥ Z-axis with G43 ... -336.700 + 2.0
     = -334.700 actual motion

> Using G44 is extremely rare in a CNC program

Applying Tool Length Offset

In the program, the tool length offset is typically included shortly after a tool change. For safety reasons, programmers like to rapid along XY axes first, then apply the tool length offset for the Z-axis. In order to select the appropriate length offset from the registry, the **G43** command has to be paired with the tool length offset number, and programmed with the H-address.

The following example shows the tool motions to the programmed target position of **X100.0 Y70.0 Z2.0**:

```
N71 T04
N72 M06
N73 G90 G54 G00 X100.0 Y70.0 S1200 M03 T05
N74 G43 Z2.0 H04 M08
N75 ...
```

As tool T04 is used in the example, H04 is the recommended tool length offset number.

Another technique CNC programmers often use is to split the Z-axis motion into two - mainly for safety during setup:

```
N71 T04
N72 M06
N73 G90 G54 G00 X100.0 Y70.0 S1200 M03 T05
N74 G43 Z25.0 H04 M08
N75 Z2.0
N76 ...
```

When used in single block mode during part setup, this method enables the CNC operator to check the tool tip clearance while it is still safely away from Z0 of the part. This method does not influence total cycles time during automatic mode.

Note that although there are different ways to physically measure the tool length offset at the CNC machine, the program itself does not change, only the offset itself.

> Actual measured tool length will be different for each setup method

G49 Command

Command **G49** cancels any active tool length offset. This simple statement, while true, is a subject of some controversy. One group of programmers insists on using it in the program, in fact, they use it for every tool. Other group never programs **G49** at all. Which group is right?

The previous section covered **G43** tool length offset command. In a rather simple definition, consider the following statement regarding the **G49** command - tool length offset cancel:

> G49 command in a program is not necessary if every tool uses G43 tool length offset

This is the point where we should be very clear here - including **G49** at the end of each tool will cause no problems, *if used properly*. **G49** is often used together with the **G28** command - machine zero return. In reality, G49 is a totally redundant entry, as **G28** machine zero return in Z-axis will cancel the tool length offset anyway.

Now, think about what can happen, if the programmer forgot to include **G43 H..** for a particular tool, and the previous tool canceled the length offset with **G49**? In this hopefully rare situation, the control system has no way of 'knowing' the length of the new tool, so it assumes the spindle gage line location as the tool tip reference point - yes, this is wrong and potentially dangerous. The danger is particularly high, if the new tool is longer than the previous tool. For the machine operators, this short lesson statement is quite simple:

> Always check that each tool is assigned G43 H..

OFFSET MEMORY TYPE

Fanuc and similar control systems offer a variety of options and features available, and one such feature relates to tool length offset. In Fanuc terms, the feature is called *offset memory type*.

There are three types:

- Type A   ... shared register   (one column)
- Type B   ... shared register   (two columns)
- Type C   ... dedicated register   (four columns)

Both types *A* and *B* are called *shared* offset registers, because a single range of offsets is used for both tool length and cutter radius offsets. Type *C* allows tool length offsets to be stored independently of the cutter radius offsets (described in separate chapter).

Offset Memory - Type A

This offset memory type is the most common. It is generally available on lower level controls. Lower level applies to features, not quality, as all Fanuc controls are known for their excellent quality and reliability.

Type A offset register has only one column:

Offset No.	Offset
01	0.000
02	0.000
03	0.000
...	...
51	0.000
52	0.000
53	0.000
...	...
99	0.000

Both tool length and radius offsets are stored (shared) within the available range of offsets (99 offsets shown). That means there can be no duplication of offset numbers between *H* number for tool length and *D* number for radius.

Adjustment to a stored offset is done by adding to or subtracting from the current offset amount.

Offset Memory - Type B

Type B offset register has two columns:

Offset No.	Geometry	Wear
01	0.000	0.000
02	0.000	0.000
03	0.000	0.000
...	...	...
51	0.000	0.000
52	0.000	0.000
53	0.000	0.000
...	...	...
99	0.000	0.000

Type B is similar to *Type A* in the sense that they are both shared offsets. Again, both tool length and radius offsets are stored within the available range of offsets. No duplication of offset numbers between *H* number for tool length and *D* number for radius is possible.

Adjustment to a stored offset is done by adding to or subtracting from the current offset amount in the *Wear* column. **GEOMETRY** column should always contain the original setting and **WEAR** column should always contain adjustments, as necessary.

Offset Memory - Type C

On the higher level controls, the *Type C* is standard. It offers the highest level of flexibility. It is commonly used on CNC machine tools with a large number of automated features.

Type C offset register has four columns:

Offset No.	H-offset		D-offset	
	Geometry	Wear	Geometry	Wear
01	0.000	0.000	0.000	0.000
02	0.000	0.000	0.000	0.000
03	0.000	0.000	0.000	0.000
...	...	...	...	...
51	0.000	0.000	0.000	0.000
52	0.000	0.000	0.000	0.000
53	0.000	0.000	0.000	0.000
...	...	...	...	...
99	0.000	0.000	0.000	0.000

Adjustment to a stored offset is done by adding to or subtracting from the current offset amount in the *Wear* column of the H-offset for tool length and the D-offset for cutter radius offset. **GEOMETRY** column should always contain the original setting and **WEAR** column should always contain adjustments, as necessary. This is the same approach as for *Type B*.

INPUT and +INPUT Key Selection

The same hard or soft keys that were used for work offset adjustment are used for tool length offset:

- **INPUT** ... *replaces* the current setting
- **+INPUT** ... *adds to* the current setting

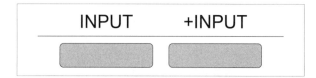

Keep in mind that **+INPUT** *key always* **adds** *the amount entered to the current amount of the offset. For that reason, it is important to enter the adjustment either as a positive amount or as a negative amount.*

Adjustment Examples

For examples of various adjustments, we will use offset number 2 (H02), with the initial setting of **-336.700**, established by the touch-off method (described later). Initial setting used the **INPUT** key, which replaced any old setting. Part program using this offset will be considered correct for the two examples.

✦ Ex. 1 - Measured cut is too deep by 0.28 mm

If the cutting tool makes a cut deeper than required, it means the tool tip was set closer to Z0 (top of part) than necessary and the offset must be adjusted into the *positive (plus) direction*:

-336.700 + +0.280 = -336.700 + 0.280 = -336.420

Keep in mind that the offset setting is stored as a negative amount, so the adjustment amount of 0.280 has to be *added* or stored as *positive*. Initial setting of -336.700 was excessive. Offset adjustment will depend on the memory type:

TYPE A

Offset No.	Offset	Offset No.	Offset
01	0.000	01	0.000
02	-336.700	02	-336.420
03	0.000	03	0.000
…	…	…	…

Initial offset — Adjusted offset

TYPE B

Offset No.	Geometry	Wear
01	0.000	0.000
02	-336.700	0.280
03	0.000	0.000
…	…	…

TYPE C

Offset No.	H-offset Geometry	H-offset Wear	D-offset Geometry	D-offset Wear
01	0.000	0.000	0.000	0.000
02	-336.700	0.280	0.000	0.000
03	0.000	0.000	0.000	0.000
…	…	…	…	…

All three offset types show the same adjustment of the tool length offset.

✦ Ex. 2 - Measured cut is too shallow by 0.17 mm

If the cutting tool makes a cut not deep enough, it means the tool tip was set further from Z0 (top of part) than necessary and the offset must be adjusted into the *negative (minus) direction*:

-336.700 + -0.170 = -336.700 - 0.170 = -336.870

Again, keep in mind that the offset setting is stored as a negative amount, so the amount of 0.170 has to be *subtracted* or stored as *negative*. Initial setting of -336.700 was insufficient. Offset adjustment will depend on the memory type:

TYPE A

Offset No.	Offset	Offset No.	Offset
01	0.000	01	0.000
02	-336.700	02	-336.870
03	0.000	03	0.000
…	…	…	…

Initial offset — Adjusted offset

TYPE B

Offset No.	Geometry	Wear
01	0.000	0.000
02	-336.700	-0.170
03	0.000	0.000
…	…	…

TYPE C

Offset No.	H-offset Geometry	H-offset Wear	D-offset Geometry	D-offset Wear
01	0.000	0.000	0.000	0.000
02	-336.700	-0.170	0.000	0.000
03	0.000	0.000	0.000	0.000
…	…	…	…	…

As in the previous example, all three offset types show the same adjustment of the tool length offset.

The manual touch-off method used for the examples served as an illustration of the offset adjustment. Many modern CNC machining centers have an automatic tool length setting, which simplifies the initial setup. However, manual offset adjustments are part of the daily machining life.

Before discussing various setting methods, including the touch-off method, let's look at machine geometry.

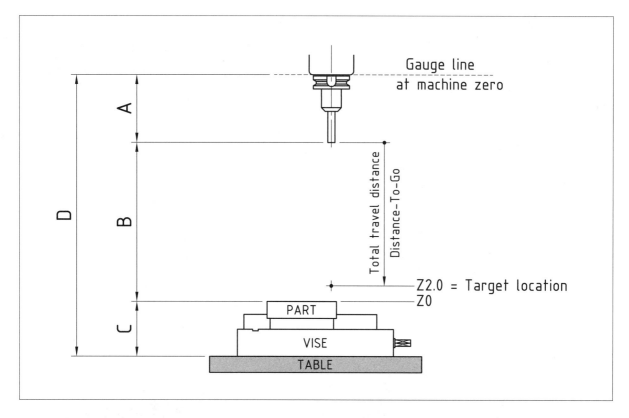

MACHINE GEOMETRY

Looking from the front of a vertical machining center will provide the best way to understand the relationship of various machine features.

The above illustration shows a part, mounted in a vise, with Z0 at the top. There are four dimensions *A* to *D*.

- *'A' dimension*
  ... represents the tool length measured between gauge line a the tool tip
- *'B' dimension*
  ... represents the distance between the tool tip and part Z0
- *'C' dimension*
  ... represents the part Z0 measured from the table surface
- *'D' dimension*
  ... represents the distance between top surface of the table and the gauge line

Only one dimension - *D* - is always known. It is a fixed dimension that is set by the machine manufacturer and is available from machine documentation or one-time measurement.

Which dimensions from the remaining three (A, B, C) have to be known depends on the method of offset setup.

SETTING THE TOOL LENGTH

Once you understand why each tool has to be defined by its tool length separately, it should be easier to understand the way a particular tool is actually used.

Tool length setting can be achieved by at least three common methods:

- **Preset method**   ... using an external setup equipment
- **Touch-off method** ... individual tools measured
- **Touch-off method** ... based on the longest tool

The following sections will describe each method in detail, including advantages and disadvantages of one method over another. Regardless of the actual setup process used at the machine, the program code will always be the same - the setup differences are limited to the machine tool only, *not* to the programming method:

```
G43 Z25.0 H02 (M08)
```

is always common to *all* tool length setup methods.

For comparison, the major benefits - and the inevitable disadvantages - of each tool length setup method are summarized in the following table:

Method	Advantages	Disadvantages
Preset	■ Minimized setup time ■ Off-machine setup	■ Possible high cost of equipment ■ Off-machine setup
Touch-off	■ Easy at the machine	■ Time consuming ■ Setup costs higher
Longest tool	■ Easy at the machine ■ Improved setup time	■ Faster method than Touch-off, but still time consuming

Even the brief descriptions should present good understanding of the pros and cons of each method. In the following sections, you will find additional details, including several practical example.

Tool Assembly Fixture

Before the tool length can be measured (by using any method), it is important to secure the cutting tool into the tool holder, as necessary for machining. For this purpose, most machine tool manufacturers provide a special tool assembly fixture, that can be permanently mounted on a work bench, close to the CNC machine.

> **Never assemble the tool in the machine spindle**

Mounting the tool holder in the machine spindle and then setting the cutting tool into it is a very poor practice that can severely damage the spindle. Always use the off-machine method, such as the tool assembly fixture.

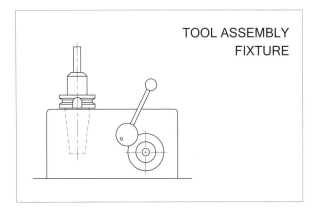

TOOL ASSEMBLY FIXTURE

Preset Tool Method

Preset tool length setting uses the actual tool length measurements found during the setup (*A* dimension - *see page 92*). As the name of the method suggests, 'preset tool' simply means that tool settings take place *before* actual setup at the machine. This method of setup is called *off-machine* tool setup.

A special hardware (even software in some cases) is required to set the tool length off-machine.

Tool Presetter

The device that is used to set tool length off-machine is called the *tool presetter*. There are many designs, mechanical and electronic, in a rather large price range. Some presetters resemble standard tool height gage, others use various indicators. A typical, and relatively inexpensive, tool presetter is shown in the following sketch:

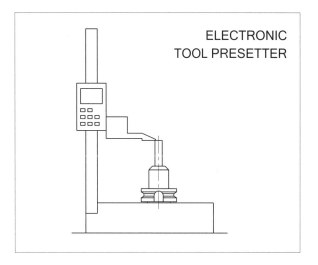

ELECTRONIC TOOL PRESETTER

Note that in the above table, the *Off-machine setup* is listed as both an advantage and a disadvantage. To a large extent, it depends on the overall management of company operations. Setting the tools and their offsets away from the machine significantly reduces lead time at the machine. That is certainly an advantage. On the other hand, it also requires another person to do the setup, plus a potentially high cost of the initial equipment, particularly for a large CNC machine shop. That is an obvious disadvantage. In order to obtain setup advantages only and minimize the disadvantages, a careful justification study may be necessary or even mandatory before any significant investment.

TOOL LENGTH OFFSET

Small and medium machine shops often use a home made solution. They use a relatively inexpensive standard height gage to set the individual tool lengths. In this case the height gage is dedicated to a single purpose of setting tool lengths off-machine. Usually, a presetting fixture is an important part of this method, because it provides the necessary reference point that matches the reference point of the CNC machine.

Preset Method Application

To use this method for tool length offset, it is necessary to know dimensions *A, C,* and *D,* as shown in the last illustration *(see page 92).* In order to understand how this method works at the machine, it is important to understand how the CNC system calculates the total travel distance to reach the target position. This is the *Distance-To-Go* on many controls. For the example, consider the following known and unknown dimensions used in the tool length setup:

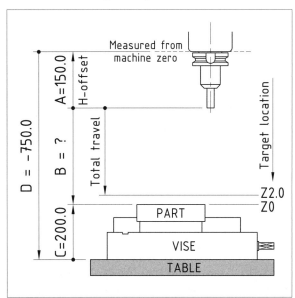

Although easy to interpret, round numbers have been used for the better understanding - keep in mind that in reality, the numbers will most likely have three (metric) or four (imperial) decimal place accuracy.

To calculate the *Total travel*, the CNC system will consider all relative offsets and their Z-axis setting:

- *External* offset (also known as the *Common* offset)
- Current *Work* offset (typically G54 - G59)
- Current *Tool length* offset (Geometry)
- Current *Tool length* offset (Wear - if available)
- The Z-axis *target* location

Total travel is always the **SUM** of all these dimensions. For example, based on the illustration on *page 92*, the setup dimensions are as follows (for **G43** command):

	Description	Example
A	Tool length measured off-machine, using a tool presetter (always positive)	150.000
B	UNKNOWN - represents distance between the tool tip and part Z0 (always negative result)	?
C	Distance measured from the table top to the part Z0 (always positive)	200.000
D	As per machine specification - always known and always negative	-750.000

➡ *Total Z-axis motion - Example for the PRESET method:*

External offset [D]	-750.000
	+
Work offset (Z-axis setting) [C]	200.000
	+
Tool lg. offset Hxx (Geometry) [A]	150.000
	+
Tool lg. offset Hxx (Wear) [A]	0.000
	+
Z-axis target position [program]	2.000 [Z2.0]
TOTAL TRAVEL FOR Z-AXIS	**-398.000 mm**

From all three common methods of setting the tool length offset, this is the only one that requires the dimension *D* - the distance between the spindle gage line and the top of the machine table.

Keep in mind one important statement:

> Regardless of the tool length offset setup used, the internal calculation of the total Z-axis travel motion (Distance-To-Go) will always be the same

Program block that activates the tool length offset, *eg.,*

```
G43 Z2.0 H02 M08    (ABS. POSITION = Z2.000)
```

will require *offset two* setting in tool length offset registry. On the display screen, the various settings used for this example (offset memory *Type B* shown) will look something like this:

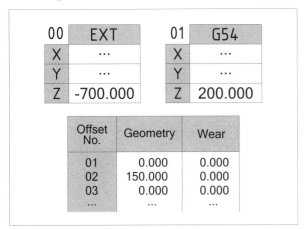

At the machine, some fine-tuning may often be necessary. As the **EXT** and **G54** offsets are fixed, the only adjustment for the actual tool length must be done in the offset registry **02**. For offset memory *Type A*, the adjustment is done directly to the current amount. For offset memory *Type B* and *Type C*, the offset adjustment is done in the *Wear* column.

Offset Adjustment Examples

➼ *Example 1 - Tool is SHORTER than previously set*

After replacement, the tool is 0.5 mm *shorter* than the previous setting. In this case, the preset tool length is not 150.0 but only 149.5 mm. In order to reach the Z0 that has *not* changed, the total travel distance has to *increase*:

Type A offset	Geometry: 149.500	Wear:	N/A
Types B/C offsets	Geometry: 150.000	Wear:	-0.500

In both cases, the total Z-travel will be **-398.500** mm.

➼ *Example 2 - Tool is LONGER than previously set*

After replacement, the tool is 0.5 mm *longer* than the previous setting. In this case, the preset tool length is not 150.0 but only 150.5 mm. In order to reach the Z0 that has *not* changed, the total travel distance has to *decrease*:

Type A offset	Geometry: 150.500	Wear:	N/A
Types B/C offsets	Geometry: 150.000	Wear:	0.500

In both cases, the total Z-travel will be **-397.500** mm.

For the following two methods of tool length offset setup, the same approach will be used for consistency.

Touch-Off Method

While the touch-off method of finding a tool length offset is quite simple to perform at the CNC machine, the procedure itself increases the non-productive time between job setups, sometimes quite significantly. In spite of this main disadvantage, it remains a very common and popular method, particularly in small shops, where the pre-setting alternative is either not practical or economical. The touch-off method is also a common setup method when only a small number of tools is needed for a particular job. Using the touch-off method of tool length setup can also be justified in those machine shops, where there is no large volume of parts to be machined. These include small job shops (custom machine shops), and some tool and die shops where making only one or two parts is the main objective.

The description *touch-off* describes the most significant part of the tool length setup process. To understand how this process works, it is important to understand that the *tool lowest point* - the *tool end tip* - is the major command point for all program coordinates along Z-axis. This is no different for the preset method, but it is more important to understand it for the touch-off method.

From the original illustration on *page 92*, only a single dimension - the *B* dimension - has to be known, and finding this dimension is also the main objective of the touch-off method. The process is quite straightforward:

1. Mount a tool holder with the assembled cutting tool into the spindle
2. Move the spindle to the machine zero position (home) along the Z-axis
3. On the display screen, set the current relative tool position to zero (Z0.000)
4. Move the tool tip close to Z0 position of the part
5. Use a thin shim or even a piece of paper as a feeler
6. Gently, touch the feeler, using the setup handle in its smallest resolution (minimum increment)
7. Upon contact with the feeler, the tool length offset will be established
8. Turn the handle off - do not move it
9. Use automatic registry entry function - *OR* - enter the measured amount manually into the proper tool length offset register
10. For utmost precision, the amount used by the feeler can be further compensated

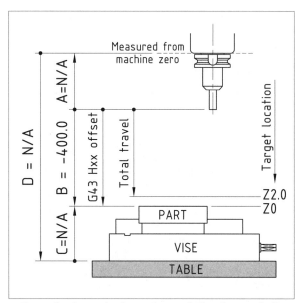

Formula that calculates the total Z-axis travel (Distance-To-Go) is exactly the same as described for the preset method of tool length setup.

Similar illustration and tables can be used for the touch-off method:

	Description	Example
A	Tool length measured off-machine, using a tool presetter	Not required
B	Represents the TOUCH-OFF distance measured between the tool tip and part Z0 (always negative result)	-400.000 stored in Hxx offset
C	Distance measured from the table top to the part Z0	Not required
D	As per machine specification - always known and always negative, but ... not required for the touch-off method	Not required

In this common case of tool length setup, the measured B dimension will be the offset amount measured. The offset will be *negative* and has to be stored in the proper tool length offset register.

In the example, for tool T02, the program will call for:

```
G43 Z2.0 H02 M08
```

Absolute screen position shown will be **Z2.000**

NOTE - There is **no** change in the programming format

→ *Total Z-axis motion*
Example for the TOUCH-OFF method:

External offset [D] & tool lg. [A]	Not required
	+
Work offset (Z-axis setting) [C]	Not required
	+
Tool lg. offset Hxx (Geometry) [B]	-400.000
	+
Tool lg. offset Hxx (Wear) [B]	0.000
	+
Z-axis target position [program]	2.000 [Z2.0]
TOTAL TRAVEL FOR Z-AXIS	**-398.000 mm**

Keep in mind, that just because the work offset dimensions *C* and *D* are not required for the tool length offset calculation, they are still used by the formula that calculates the total travel distance in Z-axis.

On the display screen, the various settings used for this example (offset memory *Type B* shown) will look something like this:

00	EXT		01	G54
X	...		X	...
Y	...		Y	...
Z	...		Z	...

Offset No.	Geometry	Wear
01	0.000	0.000
02	-400.000	0.000
03	0.000	0.000
...	...	...

If the amount measured was 400 mm, it will be entered as -400.0 in the offset register number two.

Offset Adjustment Examples

- **Example 1 - Tool is SHORTER than previously set**

After replacement, the tool is 0.5 mm *shorter* than the previous setting. In this case, the touch-off length is not -400.0 but -400.5 mm. In order to reach the Z0 that has *not* changed, the total travel distance has to *increase:*

| Type A offset | Geometry: -400.500 | Wear: | N/A |
| Types B/C offsets | Geometry: -400.000 | Wear: | -0.500 |

In both cases, the total Z-travel will be **-398.500** mm.

- **Example 2 - Tool is LONGER than previously set**

After replacement, the tool is 0.5 mm *longer* than the previous setting. In this case, the preset tool length is not -400.0 but -399.5 mm. In order to reach the Z0 that has *not* changed, the total travel distance has to *decrease:*

| Type A offset | Geometry: -399.500 | Wear: | N/A |
| Types B/C offsets | Geometry: -400.000 | Wear: | 0.500 |

In both cases, the total Z-travel will be **-397.500** mm.

Reference Tool Method

Tool length offset can also be set by the method called the *reference tool method*. This is also known as the *longest tool* method, although the tool used as reference does *not* have to be the longest tool.

Regardless of the description, the so called *'reference tool'* setup method used for tool length offset measurement is a very attractive alternative to the potentially lengthy *touch-off* method for individual tools, and the often expensive tool *preset* method. Both methods were described earlier in this chapter.

This third method is certainly worth investigating, as it is particularly invaluable for those companies that currently use the common touch-off method, but require a shorter setup time.

> The *reference* tool method of tool length offset measurement can significantly reduce setup time

As attractive as this method is, it may not be the one and only 'best' method in all situations. Setting the tool length on the basis of the 'reference tool' or the 'longest tool' is particularly beneficial for those jobs where cutting tools stored in the tool magazine do not significantly change from one job to another. This is a very significant consideration.

There is a simple principle in this method:

Measure all tool lengths relative to the reference tool (constant) rather than the part Z-zero (changing)

When a new part is setup, only the reference tool has to be touched-off at the new Z0, and all other tools are set at the same time.

What is the Reference Tool?

Using the term *'reference tool'* is far more descriptive than the frequently used term *'the longest tool'*. First, the term *'longest tool'* should not be taken literally, as tool of *any* length can be used for the same purpose. Another term for the same tool length measuring method is the *'master tool method'*.

While the word 'tool' suggests a cutting tool, that is not necessarily so. In fact, the tool does not have to be an actual cutting tool at all. In many cases, a special steel bar (10-12 mm or 0.5 inches in diameter), tapered and rounded at the end, can be permanently set in its own tool holder and used exclusively for tool length offset setup. The additional cost of such a dedicated tool holder is well justified, considering the time saved for hundreds and even thousands of subsequent tool setups.

> **Once this tool is set, it must remain in the holder used for the initial measurement**

This is a very important part of this tool length setup method.

The Longest Tool

As mentioned already, the reference tool used for tool length offset measurement does *not* have to be the longest tool. On the other hand, there are two major benefits of using the setting tool that is *physically longer* than any other tool:

- The first benefit is that all tool settings will be entered as positive amounts into the tool length register

- The second benefit is important for its safety aspect:

  ... with all cutting tools being shorter than the reference tool, tool-to-part collision due to the wrong tool length setting can be virtually eliminated

During actual tool length setup at the CNC machine, there is a certain procedure that will include the *External work offset* **(EXT)** as well as all tool length offsets used by the part program.

CNC Control Setup for Milling and Turning
TOOL LENGTH OFFSET

Setup procedure for the longest tool method is very similar to the touch-off method. Initially, all tools have to be set individually:

1. Mount the reference tool into its holder and place it in the spindle
2. Start at Z-axis machine zero and set the relative position screen to zero in Z-axis
3. Use standard touch-off method for the reference tool, but leave the tool in position after touching the Z0 face !
4. The amount of travel will not be registered to any tool length offset, but to the EXT work offset in the Z-axis setting - the amount will be *negative*
5. While the reference tool is still touching the Z0 location, reset the relative position on the screen for Z-axis to zero
6. Continue the process by measuring all other tools using the standard touch-off method

   NOTE: *The position screen will show distance from the reference tool tip, not from machine zero*

7. Register all measured amounts under the appropriate number in the tool length offset register

   NOTE: *All offsets will be negative, if the reference tool was the longest tool used. Any tool that was longer than the reference tool will have a positive amount*

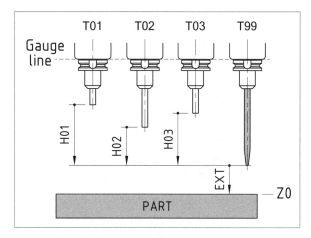

This illustration shows the relationship of three tools to the reference tool. Here, the reference tool is the longest tool, but not a cutting tool. Note it has a number T99 - tool number for the longest tool is only necessary if the tool is controlled by ATC (Automatic Tool Changer). For safety, this tool should be handled only manually.

In a rare case, where the measured tool will have exactly the same length as the reference tool, its H-offset amount will be zero. As the above illustration has shown, the tool length offset amount is the measurement from the tip of any tool to the tip of the reference tool.

In case the longest tool is actual tool rather than a special steel bar, its tool length offset will be zero.

	Description	Example
A	Tool length measured off-machine, using a tool presetter	Not required
B	Represents the TOUCH-OFF distance measured between the reference tool tip and part Z0 (always negative result)	-175.000 stored in EXT offset
C	Distance measured from the table top to the part Z0	Not required
D	As per machine specification - always known and always negative, but ... not required for the touch-off method (including reference tool setup)	Not required

From the general illustration on *page 92*, the *B* dimension example has been updated to represent the reference tool length. Also, note that whatever such dimension actually is, it will be entered to the **EXT** work offset as its negative Z-axis setting. Part program call for offset H02 is the same as before:

`G43 Z2.0 H02 M08`

The following examples will illustrate the total tool travel in the Z-axis (T02 used). For consistent results, the distance between spindle gage line and machine table will remain the same as in previous examples.

➡ *Total Z-axis motion example - LONGEST TOOL*

External offset for the longest tool	-175.000	
	+	
Tool lg. offset H02 (Geometry)	-225.000	
	+	
Tool lg. offset H02 (Wear)	0.000	
	+	
Z-axis target position [program]	2.000	[Z2.0]
TOTAL TRAVEL FOR Z-AXIS	**-398.000 mm**	

On the display screen, the various settings used for this example (offset memory *Type B* shown) will look something like this:

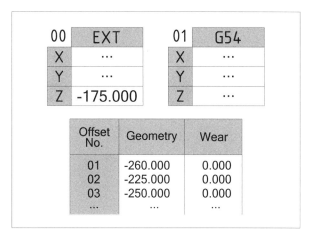

Offsets *01* and *03* contain reasonable settings relative to offset *02*, but are not used for the example.

Offset Adjustment Examples

❖ Example 1 - Tool 02 is SHORTER than previously set

After replacement, the tool T02 is 0.5 mm *shorter* than the previous setting. In this case, the difference from the longest tool is not -225.0 but -225.5 mm. In order to reach the Z0 that has *not* changed, the total travel distance has to *increase*:

| Type A offset | Geometry: -225.500 | Wear: | N/A |
| Types B/C offsets | Geometry: -225.000 | Wear: | -0.500 |

In both cases, the total Z-travel will be **-398.500** mm.

❖ Example 2 - Tool 02 is LONGER than previously set

After replacement, the tool T02 is 0.5 mm *longer* than the previous setting. In this case, the difference from the longest tool is not -225.0 but -224.5 mm. In order to reach the Z0 that has *not* changed, the total travel distance has to *decrease*:

| Type A offset | Geometry: -224.500 | Wear: | N/A |
| Types B/C offsets | Geometry: -225.000 | Wear: | 0.500 |

In both cases, the total Z-travel will be **-397.500** mm.

The second illustration shows a setup of four tools, where the reference tool is an actual cutting tool, which in this example is *not* the longest tool.

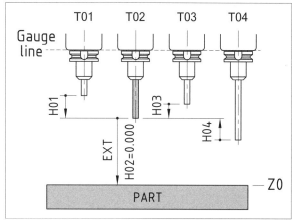

As the reference tool has changed, so has the **EXT** work offset. In the last illustration, offset H01 will be negative, offset H02 will be zero, offset H03 will be negative, and offset H04 will be positive.

	Description	Example
A	Tool length measured off-machine, using a tool presetter	Not required
B	Represents the TOUCH-OFF distance measured between the reference tool tip and part Z0 (always negative result)	-400.000 stored in EXT offset
C	Distance measured from the table top to the part Z0	Not required
D	As per machine specification - always known and always negative, but ... not required for the touch-off method (including reference tool setup)	Not required

The following offset settings reflect the example:

00	EXT		01	G54
X	...		X	...
Y	...		Y	...
Z	-400.000		Z	...

Offset No.	Geometry	Wear
01	-35.000	0.000
02	0.000	0.000
03	-25.000	0.000
04	40.000	0.000
...	...	...

TOOL LENGTH OFFSET

In all examples shown in this chapter, the same machine setup has been used for consistency. The illustrations are not in scale, but they represent reasonable relationships between individual tools.

EXT OFFSET AND WORK OFFSET

In this chapter, the preference was given to placing various measured data into the external **EXT** work offset. What about placing the same setup data into the standard offsets with the range of **G54** to **G59**?

The answer to this question is simple - *yes,* a standard work offset, such as **G54**, can be used for setting instead of the external offset. There is one major difference between them:

- **STANDARD** work offset such as G54 applies to *all* tool motions within this *particular* offset

- **EXTERNAL** offset applies to *all* tool motions within *any* standard offset setting

If used properly, the results will be exactly the same for both types of offsets. The same applies to any extended work offsets, usually in the range of **G54 P1** to **G54 P48**.

WORK AND TOOL LENGTH OFFSET

The illustration below is for reference only - it shows the relationship between work offset **G54** and tool length offset **G43 H01** on simple example:

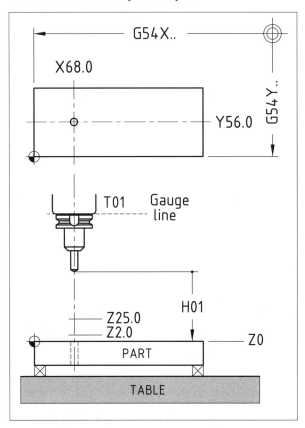

The program beginning will apply both work and tool length offsets (touch-off method shown):

```
N1 G21
N2 G17 G40 G80 T01
N3 M06
N4 G90 G54 G00 X68.0 T56.0 S1200 M03 T02
N5 G43 Z25.0 H01 M08
N6 G99 G81 R2.0 Z-.. F..
N7 ...
```

11

MACHINING A PART

MACHINING OBJECTIVES

The objective of *producing a machined part made to specifications* is and always should be the *main* - and the *ultimate* - objective of the complete CNC process. This is the area where a CNC machine shop can succeed - or fail. This area of manufacturing allows no compromises whatsoever. Either the part is machined to engineering specifications or it is not. There is no room for the *'middle'* in here, and there should never be such descriptions as *'close enough'* or *'good enough'*.

For the main machining objective, there should be only *one* expected outcome - that is to be *'right on'* - to be *exact* - in all aspects of the part development. Part development starts with the initial design, but in practical terms of a machine shop, it starts at the programming stage and ends when all parts have been machined.

PROCESS OF MACHINING

Large, medium and small companies all have their own ways of establishing a particular machining process. A large company may involve several people, each with unique expertise, while a small company may only have one person or two persons, each assigned with multiple responsibilities. Regardless of any applied management methods, the machining process itself shares many steps. In essence, the process of CNC machining a part narrows down to the responsibilities of the CNC operator and/or the setup person. It can be summed up into a series of general steps:

- Drawing evaluation
- Material inspection
- Program evaluation
- Part setup
- Tooling and setup
- Program input
- Program verification
- Machining a part
- Part inspection

This process is typical for a CNC operator who is responsible for both the setup and machining of a batch of parts on either a CNC machining center or a CNC lathe.

DRAWING EVALUATION

The first time an engineering drawing is evaluated for production purposes, it happens at the desk of CNC programmer. The programmer evaluates the drawing, selects the most suitable machine setup, cutting tools and method of machining. The whole part program is built on these decisions established by the programmer.

The second time the drawing is evaluated, is when it gets to the machine shop, together with the program and other information, such as a setup sheet. The purpose of this evaluation is much different from that of the programmer. Operators look for drawing data that directly influence the actual machining, such as:

- Dimensional tolerances
- Surface finish requirements
- Stock left for subsequent operations
- ... and other specifications

Keep in mind CNC operators have no control over the part program, but they do have a large control over the part setup. The program has to include those features that allow machining with certain degree of flexibility.

Tolerances

Machining flexibility can be illustrated by a program example and its influence on dimensional accuracy. Consider this 100 x 70 mm stock and the program that machines the contour (complexity of the part is not important for the example):

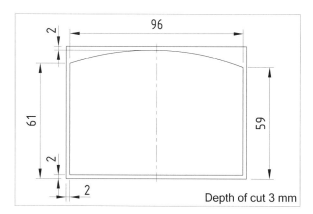

This is program listing for the drawing - one contour:

```
(*** VERSION 1 ***)
N1 G21
N2 G17 G40 G80       (T01 = 12 MM = IN SPINDLE)
N3 G90 G54 G00 X-8.0 Y-8.0 S1200 M03
N4 G43 Z2.0 H01 M08
N5 Z-3.0
N6 G01 G41 X2.0 D51 F200.0
N7 Y61.0
N8 G02 X98.0 Y59.0 I45.082 J-141.063
N9 G01 Y2.0
N10 X-8.0
N11 G40 G00 Y-8.0 M09
N12 G28 Z-3.0 M05
N13 M30
%
```

By including the cutter radius offset in block N6, the programmer provided the necessary function that allows flexibility at the machine. Setting the actual amount of D51 offset is done at the machine. As the program uses drawing dimensions and the tool diameter is 12 mm, the nominal offset amount will be D51 = 6.000.

On the other hand, only a single cutting tool is used for the contour. When only a few parts are to be machined, or when the tolerances are not critical, one tool program is more efficient than program with two tools.

One tool can also be programmed with roughing and finishing motions, contouring the part twice. Again, this is a programming decision. The above program will be changed to reflect the two tool motions:

```
(*** VERSION 2 ***)
N1 G21
N2 G17 G40 G80       (T01 = 12 MM = IN SPINDLE)
N3 M06
N4 G90 G54 G00 X-8.0 Y-8.0 S1200 M03
N5 G43 Z-3.0 H01 M08
(=== ROUGHING WITH D61 ===)
N6 G41 G01 X2.0 D61 F250.0
N7 Y61.0
N8 G02 X98.0 Y59.0 I45.082 J-141.063
N9 G01 Y2.0
N10 X-8.0
N11 G40 G00 Y-8.0 M09
(=== FINISHING WITH D51 ===)
N12 G41 G01 X2.0 D51 F200.0
N13 Y61.0
N14 G02 X98.0 Y59.0 I45.082 J-141.063
N15 G01 Y2.0
N16 X-8.0
N17 G40 G00 Y-8.0 M09
N18 G28 Z-3.0 M05
N19 M30
%
```

A subprogram can be used for complex contours.

Note that the programmer used two different radius offsets for the same tool! This is very important. Also note that the roughing cut offset is D61, while maintaining the finishing cut offset is D51, which follows the common addition of 50 to the tool number for controls with shared offset registry.

> Always watch for multiple offsets for a single tool

The above program could use a subprogram as well as two different tools for more control of dimensions.

Radius from I and J

The drawing for the example has an arc definition based on three points. In this case, the radius is not known and can be calculated during programming. Regardless of whether the radius is or is not known, many programs containing arcs are programmed using arc vectors I and J, rather than the direct radius R. In these cases, the arc radius is also not known, although listed in the drawing. That raises a question - is there a way to find the radius of an arc, if only I and J arc vectors are given?

The answer is YES and the method shown here can be very useful to verify that the IJ vectors in the program are correct. Here is a program section that matters:

```
...
N71 G01 X12.893 Y48.117
N72 G02 X80.508 Y100.0 I67.615 J-18.117
N73 G01 X120.0
...
```

In order to 'decifer' the program and find the arc radius, you have to know what the arc vectors IJ represent. Both represent the distance and direction from the arc start point to the arc center - measured along the X-axis (I-vector) and the Y-axis (J-vector). Start point of the arc in the example is in the block N72.

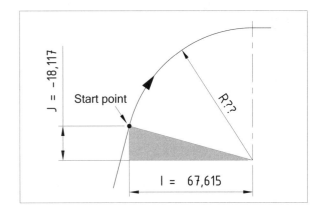

Once you identify the IJ vectors as two sides of a right triangle, the radius R can be calculated using the classic *Pythagorean Theorem*:

$$R = \sqrt{I^2 + J^2}$$

Applying the known vectors I and J, the radius can be calculated:

$$R = \sqrt{67.615^2 + 18.117^2} = 70.00009939$$

Checking the radius in the drawing, it should be **R70**.

Surface Finish

Surface finish is influenced by several factors, such as speeds and feedrates, depth and width of cut, the type of material, the diameter and length of tool, rigidity of setup and a number of other factors.

Typically, adjusting the speed and feed at the machine are the most common methods to improve surface finish, but other measures should also be applied in many cases. For example, using the same size end mill as directed by the program but with a different flute geometry will influence the surface finish. If applicable, a larger cutter radius will also have a positive effect on surface finish. Changing the offset amount for stock allowance between roughing and finishing operations may also improve the surface finish.

Other factors include suitable cutting fluids, direction of cut, geometry of the cutting tool, and others.

Stock Allowance

Typically, stock allowance is the amount of material left for finishing. As the amount of stock has a great deal to do with the final part quality, it is necessary for the operator to set such amount correctly. Take the last program example *(Version 2)*. Using the nominal cutter radius, the finishing offset amount for D51 will be 6.000. It is the D61 offset amount that leaves the stock. Adding stock to the part means the offset has to be greater than 6 mm. There are no magic numbers here. If you set D61 to 6.500 mm, there will be 0.5 mm (about 0.02") left - *per side* - for finishing. Even without any magic numbers, this is the amount of stock often used by operators as a starting point.

Program dimensional data normally uses the drawing dimensions (contour edge), so the exact setting of the offsets is in the hands of CNC operators.

MATERIAL INSPECTION

When delivered to the CNC machine, blank material (stock) allocated for machining is not always what is should be. Programmers often base toolpath motions on supplied material data and do not have the opportunity to check the real material to be used for machining. This is one of the most common frustrations a CNC operator has to face. In addition, the lead time is increased and productivity goes down.

Ideally, all stock material used for a particular batch of machining should be the same. Practically, there are several reasons why material will vary is size and condition. One is that two different sources were used. Another is a preparation of the material did not include quality control. There are other reasons as well, but for the operator, the most important reason is to *know the differences*.

Programmers are sometimes aware of a possible problem and incorporate suitable measures in the program. For example, they may provide an option for one or two facing cuts, using the block skip function (*page 220*). It is common for CNC operators to check a few samples from the supplied material before starting the job.

PROGRAM EVALUATION

Most CNC operators are not part programmers. Yet, their skills are often measured by the output of parts only - the more the better within the same time frame. This approach does not take into consideration that the operator - especially one who is also responsible for machine setup - has to know what the part program actually does. A few programmers offer a brief description of individual operations, many do not. Some companies provide a routing sheet or similar document describing individual operations at each stage of manufacturing. Small machine shops do not have such luxury.

In order to shorten the lead time, the part program is often used as is, without further examination. This approach can be justified for programs that had been used earlier - programs that had been verified. For new programs, it is always a good idea to evaluate the program before using it. Incidentally, it does not matter whether the program was developed manually or by using programming software. Either method can produce its own errors. Errors generated by a computer software should be non-existent, or at least far between, but that is not always the case. Errors generated by manual programming are much more frequent. Knowing what method of programming had been used will make a difference at the machine when the program is evaluated.

Programs Generated Manually

Manual programming, in contrast to programming using software such as Mastercam®, is still a common method in many shops. Manual programming is also used for simple parts, even if computer assisted software is available. Manual programming for CNC lathes is quite common, because of the powerful built-in cycles, which are themselves computer (control) based.

A program developed manually may have syntax errors or logical errors. Syntax errors are detectable by the control, logical errors are not - those are the serious ones.

Syntax Errors

When evaluating a part program developed manually, the first thing you may be looking for is *syntax errors*. The most common one is the letter **O** instead of the digit zero (**0**):

GOO X23.5   ... *letter O is a syntax error*

as compared with

G00 X23.5   ... *digit 0 is slimmer than letter O*

Fortunately, the control system can detect syntax errors, and some operators actually count on it by running the program without any motions. In this respect, syntax errors are easy to find and correct. All non-syntax errors are logical errors that are more difficult to find.

Dimensional Errors

Dimensional errors are *logical* errors. Logical errors are much more difficult to detect in the program. Logical error is a correct syntax with incorrect data. For example, **X50.0** is a perfectly correct program entry, as long as the programmer did not mean **X5.0**, for example. Errors of this type can generally be detected during program proving, using various methods, such as single block, dry run, machine lock, Z-axis lock, etc.

Decimal Point

A missing decimal point has always been at the top of many programming errors of the logical type. In standard programming, all dimensional addresses, including those in degrees, as well as arc vectors, use decimal point. Feedrates generally use decimal points, but there are more exceptions in this area. A brief list of addresses that *do not* use decimal point, at least not for Fanuc and similar controls follows:

- **Address O** *Program number*
- **Address N** *Block number*
- **Address H** *Tool length offset number*
- **Address D** *Cutter radius offset number*
- **Address G** *G-codes (special codes do use decimal point)*
- **Address M** *Machine and miscellaneous functions*
- **Address L** *Repetitive count for cycles and subprograms*
- **Address K** *Only if used for the same purpose as address L*
- **Address P** *A dwell time in milliseconds (pause)*
- **Address S** *Spindle speed (r/min or cutting speed)*
- **Address T** *Tool number*

As always, check the control system manual for exact representation of various addresses.

When evaluating a program, usually from a control display screen or from a printed copy, you should know that, for example, ...:

➡ *In metric units:*

X10.0 is the same as **X10.** as well as **X10000**

X0.1 is the same as **X.1** as well as **X100**

➡ *In imperial units:*

X10.0 is the same as **X10.** as well as **X100000**

X0.1 is the same as **X.1** as well as **X1000**

Note the number of zeros for each unit when the decimal point is absent.

It is not only the missing decimal point that can cause problems, it can also be a decimal point in the wrong place. For example, block numbers and spindle speeds do not accept a decimal point. Its presence in such commands will cause an error at the control.

> **Programming rules for decimal point also apply to setup data input at the control, including offsets**

Calculator Input

A short note on a feature called *Calculator Input*. Some machines, such as routers, hobby mills, and special purpose machines do use the calculator input as a standard input of data that would normally use decimal point. The subject of calculator input is related to programming a decimal point (or not). Program *Version 1* on *page 102* uses the standard method of programming decimal point.

Here is the same program using the calculator input feature, with underlined changes:

```
(*** VERSION 1 ***)
N1 G21
N2 G17 G40 G80       (T01 = 12 MM = IN SPINDLE)
N3 G90 G54 G00 X-8 Y-8 S1200 M03
N4 G43 Z2 H01 M08
N5 Z-3
N6 G01 G41 X2 D51 F200
N7 Y61
N8 G02 X98 Y59 I45.082 J-141.063
N9 G01 Y2
N10 X-8
N11 G40 G00 Y-8 M09
N12 G28 Z-3 M05
N13 M30
%
```

In this particular example, only zero was programmed after the decimal point, so all decimal points were eliminated. Now compare the two following blocks:

`N8 G02 X98.0 Y59.0 I45.082 J-141.063`

with

`N8 G02 X98 Y59 I45.082 J-141.063`

It is easy to see that the first version uses the decimal point in standard way, whereby the second version uses calculator input. Values that are not point-zero (.0) must be written in full and will be evaluated as expected.

Note:
Some G-codes use decimal point as well, for example, G84.1 may be used for rigid tapping cycle. This designation is not a decimal point in the mathematical sense, but a separator which must always be used, even in the calculator input mode.

Positive / Negative Value

Another item to look for during program evaluation is the wrong numeric sign, particularly the *negative* sign. Programs do not use any symbol for a positive number - it is simply assumed that `X10.0` indicates a positive amount. Negative numbers *must* contain the negative sign symbol (dash or minus sign on the keyboard), for example, `X-10.0`.

How can you evaluate all positive and negative entries during program check? The main key is to understand how the part is oriented relative to part zero (program zero). Many engineers and designers who are familiar with CNC technology try to dimension the part in such as way that all - or most - dimensions are located in the *Quadrant I*. In this quadrant, all X and Y values are positive. If X0,Y0 and Z0 are programmed, there is never a sign - zero is neither positive nor negative.

XY signs in each quadrant are illustrated below:

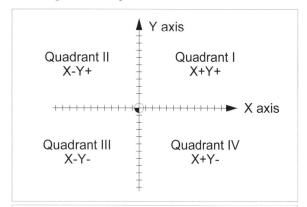

POINT LOCATION	COORDINATE	
	X AXIS	Y AXIS
QUADRANT I	+	+
QUADRANT II	-	+
QUADRANT III	-	-
QUADRANT IV	+	-

If you have a setup sheet showing the part layout, or at least a description of the layout, you can tell whether the machining takes place in a single quadrant or cover more than one. From the illustration, it is apparent that ...:

- **In Quadrant I:**   X = positive   Y = positive
- **In Quadrant II:** X = negative   Y = positive
- **In Quadrant III:** X = negative   Y = negative
- **In Quadrant IV:** X = positive   Y = negative

This is the standard method of *absolute* method of programming (using **G90** command), which always indicates a tool XYZ location from part zero (origin).

In incremental mode of programming (using **G91** command), the motion itself is programmed as a distance and direction from the current tool position.

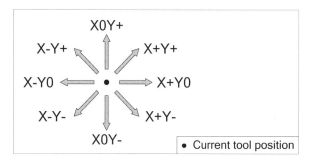

For reference, it is apparent from the last illustration that when viewing a vertical CNC machining center, that motion directions can be easily determined:

- Motion to the left: X-positive
- Motion to the right: X-negative
- Motion away: Y-positive
- Motion towards: Y-negative
- Motion up: Z-positive
- Motion down: Z-negative

Keep in mind, that positive/negative direction relates to the *tool motion*, not the machine table motion.

Programs Generated by Software

You may also encounter a decimal point error in part programs generated by computer software. These errors usually indicate a poorly configured post processor. The purpose of a post processor is to output program in the format compatible with a particular CNC machine. For example, the post processor may output a quarter second well in **G82** fixed cycle as **P0.25** or **P.25**. As most Fanuc controls do *not* accept this input with decimal point, the post processor has to be updated.

It is not uncommon to correct part program generated by computer software at the control. For example, if the programmer forgot to turn on the coolant, there will be no **M08** in the program. It is simple to add it to the program in *Edit* mode, but this should be done as a temporary measure only. Keep in mind that the source of the omission is still erroneous and must be changed.

Part Setup and Tooling

Part setup is a multi-level process. It covers setup of work holding device on the machine table (milling) or preparing the chuck and jaws (turning). It also covers a great amount of various tasks and procedures, such as tooling, offsets, drawing and program evaluation, etc. This handbook was developed for the purpose of providing details about control setup as it relates to a part to be machined. Check various chapters for more details.

PROGRAM INPUT

In the majority of cases, a part program is developed away off the CNC machine (often called *off-line*). In order to machine the part, the control system has to be able to process it. There are two common methods of processing the part program - *internal* and *external*.

Memory Mode

The *internal* method of program processing is the most common. *Memory mode* means that the whole program is entered into the control system memory and processed from there. All changed to the program (if necessary) are done in the memory, using the *Edit* mode.

To load a program to the memory can be done by typing the program manually, which is the least efficient method. Normally, the program is loaded from an external computer (a desktop or a notebook), via a special cable, through the communication ports.

While it makes no difference of how the program is entered and stored in the internal memory of the control system, what matters is that the program uses the internal storage, rather than an external storage.

DNC Mode

DNC method is the *external* method of program processing. The common abbreviation **DNC** may mean either *Direct Numerical Control* or *Distributed Numerical Control*. Both really mean the same thing, but the *direct* type is often used for connection of a computer with a *single* CNC machine, while the *distributed* type is used for connection with *several* CNC machines. In both cases, a suitable software and setup must be done at both ends - at the computer (PC) and the control (CNC). Apart from the external computer, a suitable cable is required between the PC and CNC.

In DNC mode, the program is stored on the PC and sent to the control system in chunks of data at a time. When the data is processed, it is flushed from memory and a new chunk of data is sent over. This is a continuous activity and there is no delay in program execution. Any changes to the program must be made at the PC side, not the CNC side of the operation.

PROGRAM VERIFICATION

CNC programmers make an honest effort to develop their programs error-free. They have several ways to check the program before it reaches the machine shop. If the part program had been developed by a computer software, the verification is different than if the same program was developed manually. Programming software of this kind provides quite accurate visual verification in several forms, as a graphic toolpath displayed on the computer screen. Visually verifying a part program developed manually is much more difficult to do.

Manual Verification

Verifying a program manually means to physically read the program and interpret it. Every operator approaches this task in a unique way, but there are many common methods.

Common methods are too numerous to list, but a short list may provide some ideas what to look for:

- Syntax accuracy
- Program consistency
- Tool usage
- Clearances
- Spindle speeds
- Cutting feedrates
- Depth of cut
- Width of cut
- Optional stops
- Program and setup relationship
- ... and many others

Some operators may ask somebody else to look at the program, with the possible outcome that another person may see what the operator may miss.

If the control system has a graphic option, the toolpath can be shown on the control screen. However, this is not a common option and relating on the manual program verification itself is the only way.

Software Verification

The current software market offers quite a selection of programs that do a visual verification of CNC programs. They vary from almost free to very expensive. Unfortunately, the software cost does not always reflect the quality and power of the software itself. It is quite common to find an expensive verification software that uses solid model type of display, yet it fails to support many programming features. While solid type display is very attractive in many ways, it also requires data related to the stock and tools. Software that is based on a wireframe display is very fast, does not need any special definitions, and in the majority of cases, it does the job very well. In all cases, the software includes an text editor geared to CNC work that can be used to write and/or edit part programs. One powerful and reasonably priced software is **NCPlot**® (*www.ncplot.com*). Most software vendors offer a time limited evaluation version.

> The main purpose of program verification is to detect program errors before they affect the part

PRODUCTION MACHINING

Production machining is the ultimate objective of all CNC work - it is its final phase. After the part had been machined, additional activities may take place, such as inspection, but the CNC work itself is completed.

Production machining starts when all setup, testing and proving had been done. What are the duties and responsibilities of the CNC operator? That largely depends on the type of work assignment. Some operators are not required to do anything else but to load the part, push the *Cycle Start* button, and unload the part when the cycle is completed. This type of work requires the least amount of skill. Operators on this level are common in production of large volume of a single part, where literally thousands and thousands of parts in the same batch have to be machined. In such environment, if the operator requires assistance, there is usually a much more qualified person available, who is generally responsible for overall setup, offsets adjustment, and troubleshooting. Such person is typically responsible for several CNC machines and belongs to the setup and support group.

Responsibilities

In smaller shop or for production with low batches, the CNC operator is often responsible not only for the setup but also for the machine activity during production. Generally, the responsibilities of the operator are:

- **Loading and unloading parts**
- **Loading and unloading programs**
- **Monitoring each part run**
- **Performing periodical part inspection**
- **Checking tools for wear and other flaws**
- **Replacing insert or tool when necessary**
- **Adjusting offsets for optimum part dimensions**
- **Deburring sharp edges if necessary**
- **Interaction with the programmer**
- **... and many others**

A number of additional duties and responsibilities could be added to the list, depending on company requirements. One such responsibility that should definitely be on the list is *'thinking of improvements'*. An experienced CNC operator is at the end of all work done ahead of machining, such fixture design, tool selection, programming, etc. In this position of direct contact with the actual machining, thinking of improvements at all levels will make the same part or similar parts more efficient next time.

PART INSPECTION

Part inspection after the whole batch had been completed is not a constructive approach. As precise as CNC work can be, there are factors that have to be considered - factors that directly influence the part quality. These factors include dimensional accuracy, surface finish quality, overall appearance, ease of handling, etc. Inspecting the part during machining is important, as any changes required can be done immediately. There are methods of preventing a scrap (see *page 109*) that should be applied for any job.

Part inspection at the machine generally requires only a minimum number of inspection tools - typically a vernier (caliper) for general dimensions and micrometer for precise dimensions. Other measuring instruments include a height gage, thread gauges, plug gauges, and a variety of special instruments.

Some CNC machines are equipped with a feature called *'in-process gaging'*, which includes a built-in electronic probe and a special macro program designed to inspect a part automatically while it is still part of the overall setup. Macros programs can also be used to adjust offsets automatically as well.

PROGRAM SOURCES

Part programs can be developed by using different methods - manual method and software based method have been already mentioned. Although the final program should do the required job well regardless of how it was developed, for the CNC operator the actual method of development may mean focus on different features of the program. The two methods already covered can be supplemented by two additional methods - here is a list covering all four methods:

- Manual programming - Standard
- Manual programming - Custom macros
- Computer generated programs
- Conversational programming

CAM Programs and Machining

Although the manual method of CNC programming is still very common in many machine shops, using dedicated CAM programming software, such as Mastercam®, Edgecam®, and many others, has grown at a rapid rate. Software of this kind is usually sold in a modular form, meaning that you only purchase the module you need for the work you do.

What is a difference between machining a part generated by a traditional manual method of programming and a part generated by computer software?

Some CNC programmers and operators falsely assume that a CAM generated program is always perfect or at least always correct. After all, we all know that computers don't make mistakes, right? Well, this assumption is far from the truth. Hardware problems aside, computers are reliable and always give back what the user asked for. Of course, if the initial input is wrong, the final output - *the result* - will be wrong as well. There is an old computer acronym relating to such situations - it is called GIGO - *'Garbage In, Garbage Out'*.

It is true that errors are more frequent in manual programming, but there are plenty of possibilities of making an error when working with a software based programming system. On a small scale, it could be nothing more than a forgotten coolant function (**M08** missing) - you don't ask for it during program development, you don't get in the final program output. Errors can also be of more serious nature - for example, inputting a wrong dimension may be overlooked on a busy computer screen, but the program output will include it nevertheless.

Some errors in a software generated program can be attributed to a flaw in the post-processor. Post processor is an integral part of the CAM software and has one purpose - to format generic toolpath instructions to match the requirements of a particular CNC machine. If you find that you frequently make the same changes in a software generated part program, chances are great that the culprit is the post processor itself (although it could also be an input error). Post processors of today are very sophisticated, and there is no reason to make any subsequent changes to the program after it has been processed by the computer software. Of course, common changes such as speeds and feeds can still be changed, if required, depending on the actual machining conditions.

> Treat each part program equally,
> regardless of how it was developed

Conversational and Macro Programming

Both conversational and macro programming are outside the scope of this handbook. Conversational programming is performed at the machine and any training program reflects that. Macros are basically manual programs with many powerful features. For learning and understanding, check *Fanuc CNC Custom Macros* book for details. It is published by **Industrial Press, Inc.** (*www.industrialpress.com*).

Conversational programming is an attractive method of program generation. The current leader in this field is the Yamazaki Corporation and their *Mazak®* machine tools and controls. Many other control manufacturers (Fanuc included) also offer conversational programming.

What exactly is conversational programming?

Conversational part programming is based on the real-time interaction of a CNC operator with the control system, done right at the CNC machine.

This reality alone should be enough to dispel the myth that manual programming is obsolete. In the area of conversational programming, Mazak has taken the lead early and has been the leader for many years. Fanuc also had an early conversational programming system called FAPT, but it never caught on as a stand alone system. Other controls, Fanuc included, do offer very high quality conversational programming on several models.

PREVENTING A SCRAP

Without a doubt, the major goal of a CNC operator should always be to make any part to drawing specifications in all respects. At the same time, the operator should also make an effort to prevent making a scrap. It is not uncommon to scrap a few small pieces made of a long bar, before making the 'perfect' piece, for example. This leisurely approach often leads to a certain level of complacency, and may prove costly when the number of parts is limited, for example, when a customer provided the exact number of stock material.

Causes of Scrap

There are only a few reasons for scrap to happen in the first place. On the human side, it is either negligence or it is incompetence. On the machine site, it is a hardware or software failure. Needless to say, the human errors are in great majority as causes for scrapped part.

Steps to Eliminate Scrap

There are several methods how CNC programmers and operators can prevent a scrap. Some have to be included in the program, others can be done at the machine directly. Programmers often include extra features in their programs that help to eliminate scrap. Typical methods include:

- Providing a trial cut (test cut), using block skip function or some other method - see *Chapter 21*
- Including *Program Stop* function M00 or *Optional Program Stop* function M01 in the program
- Including messages or comments in the program
- Making a setup sheet with a sketch
- Splitting a tool motion for single block operation
- Programming sufficient clearances

These are just some methods that can be used by the program to make the operator's work easier.

CNC operators at the machine also have a number of ways to prevent a scrap from any cause - which really means to identify a potential problem before it becomes a real problem. The first method is to truly understand the program - to be able to interpret it and know what it does. Mistakes made during tool or fixture setup are also a common cause of scrap. Errors also happen during data entry into the control system. Placing the right settings into a wrong offset register is quite a common cause, as are errors of decimal point entry, in the form of a missing decimal point or a decimal point in the wrong place.

Even after a few parts have been machined successfully, scrap can happen for any number of reasons. Typical errors happen when changing offsets, cutting tools or inserts, modifying program, etc.

Let Offsets Work for You

A verified program is no reason to decrease your attention at the machine. When working with offsets, there are ways to let them work for you. First, look at the following table - it shows how a measured part dimension influences the whole part. The table is best to be used for external or internal contours:

Measurement at machine	EXTERNAL CONTOUR	INTERNAL CONTOUR
IN TOLERANCE	OK	OK
UNDER SIZE	SCRAP	RECUT POSSIBLE
OVER SIZE	RECUT POSSIBLE	SCRAP

Considering what measurement constitutes which result, the operator can take precautions by intentionally manipulating cutter radius offset (**G41** or **G42**). Take the following simplified program as an example - it does nothing more than uses a contour toolpath to machine a small simple rectangle 75 x 50 mm. Concentrate on the concept rather than the part itself:

```
N1  G21
N2  G17 G40 G80 T01 (10 MM END MILL)
N3  M06
N4  G90 G54 G00 X-7.0 Y-7.0 S1000 M03
N5  G43 Z2.0 H01 M08
N6  G01 Z-5.0 F500.0
N7  G41 X0 D51
N8  Y50.0 F150.0
N9  X75.0
N10 Y0
N11 X-7.0
N12 G40 G00 Y-7.0 M09
N13 G28 Z2.0 M05
N14 M30
%
```

The tolerance on the 75 mm width of the part is quite tight, 30 microns plus, nothing minus. Even if you use a brand new cutter, there is a good chance that the width will be right at zero or slightly under. Of course, the width may also end up slightly larger, too. Rather than taking a chance, a simple procedure may be helpful in all situations that require change of offset setting:

1. Identify the normal offset value for D51
2. Estimate the amount of possible error
3. Temporarily increase the offset
4. Cut, measure and adjust offset

Now for the details. For *Item 1*, the answer is 5.000 mm. The end mill has a diameter of 10 mm, so the 'normal' setting of offset 51 will be 5.000. *Item 2* may need a special attention - you don't want to either overestimate or underestimate. Quality of the tool is important. For a new cutter the error should be extremely small - let's assume no more than 10-15 microns. Take this possible deviation into account for *Item 3* - you want to change the offset, so the part will be wider than 15 microns. If stock to be removed allows, you may decide that you want to leave 200 microns on the width.

Offset Adjustment - Example

➡ *What will the temporary offset setting be ?*

Correct answer to this important question is the real key to understanding the concept of offsets.

From the earlier chart it is clear that a measured external width that is greater than the drawing width can be cut again to final size. That means adding a stock to the toolpath. Adding a stock means increasing the offset from 5.000 to - to what? You want to end up with a width that is 200 microns larger - 5.200 mm.

Keep in mind that the 200 microns is distributed over the contour width, so it is only 100 microns per side! That means changing the offset to 5.100, *NOT* to 5.200! It would not be a disaster if you make this mistake, but the main purpose of the process would be lost. The result is that upon completion of *Item 3*, the D51 setting will contain temporary 5.100 mm offset.

Now, to *Item 4*. When the cutting is completed, measure the 75 mm width. One of three possibilities will be the result:

- **ON SIZE**   Width will be exactly 75.2 mm
- **OVER SIZE**   Width will be greater than 75.2 mm
- **UNDER SIZE** Width will be smaller than 75.2 mm

Even if the width is smaller than 75.2 (under size), it still should be greater than the 75.03 maximum size. If it is not, you made a mistake in *Item 2* - estimating the amount of error. Let's look at the three possible outcomes and the offset adjustments, but first you have to decide what the final width should be - for the example, let's aim for the middle of the tolerance, which means 75.015 width.

Measured sizes are examples only:

- **ON SIZE result = 75.200:**

➡ *(75.015 - 75.200) / 2 = -0.0925*

Use 0.093 to match three decimal places. We have to subtract 0.093 from the current setting of the offset 51:

5.100 - 0.093 = 5.007 is the final setting

- **OVER SIZE result = 75.243:**

➡ *(75.015 - 75.243) / 2 = -0.114*

5.100 - 0.114 = 4.986 is the final setting

- **UNDER SIZE result = 75.185:**

➡ *(75.015 - 75.185) / 2 = -0.085*

5.100 - 0.085 = 5.015 is the final setting

One clear lesson can be learned from these examples - the process of establishing the final offset amount is the same, regardless of the actual size of the test measurement. To learn these methods requires practice - paper instructions may give you the necessary start, but actual doing makes a big difference. Try to learn on a scrap piece of material first.

> **The ultimate objective of CNC machining should be to *eliminate* scrap, not to minimize it**

PROGRAM MODIFICATION

When the CNC operator receives a completed program for machining, such program should be totally free of errors and in no need of any further changes. In many cases, such a situation is more optimistic than realistic, with no real blame to place. Most part programs will need some change at the machine (during setup), even if such a change is no more than some minor tweaking of spindle speeds and cutting feedrates.

An experienced CNC operator will be able to harmonize all machining responsibilities with those of a CNC programmer (and to be fair, the reverse is also true). Of course, the most ideal situation would be a 100% cooperation and 100% mutual exchange of 100% of ideas flowing between 100% of CNC programmers and 100% of CNC operators. This is a very worthy objective, but really not realistic to achieve.

What does all this mean to a CNC operator? What changes in a program or related to a program are likely to be required at one time or another? Apart of corrections (if any), here are only some ideas:

- **Changing tool size**
  ... typically from the one used by the program
- **Effect of insert size change on final dimensions**
  ... e.g., R0.8 (R0.0313) to R1.2 (R0.0469) in the same holder
- **Changing offsets**
  ... working with tolerances - preventing scrap
- **Clearances**
  ... above part and upon breakthrough
- **Dwell times**
  ... they are often programmed too long
- **General problem solving**
  ... control alarms - diagnostics
- **Order of operations**
  ... in rare cases, application of tools may need change

When something goes wrong at the CNC machine, the outcome may be either quite positive or very negative. of course, wrong is wrong, regardless of how we look at it. The key question here is what is the exact response from the control system. When a computer software encounters something wrong, it should alert the user. Alarms of many types are inherent part of the computer. Usually, alarms are accompanied by a short message that identifies the alarm number and some explanation. If that is not enough, the control and/or machine manual usually provides additional details.

Any program changes should be discussed with the programmer or supervisor and always documented. Any permanent changes should be done as soon as possible.

MAINTAINING SURFACE FINISH

Apart from maintaining precise drawing dimensions for any machined part, it is equally important to maintain the required quality of machined surfaces. Machined surface of a ceratin quality is required for two purposes:

- Practical functionality
- Visual appearance

Practical functionality may include, for example, a requirement for a certain surface finish to make assembly of two parts easier. More often, high surface finish is required for two matching parts with a tight fit.

Visual appearance, sometimes called *cosmetic* appearance, is strictly for visual purposes, rather than practical purposes. A good looking part is not only more attractive, but also easier to sell.

Surface Finish in a Drawing

An engineering drawing may include special symbols that define surface finish requirements for individual features of a given part. Because of different standards, metric and imperial drawings use different symbols:

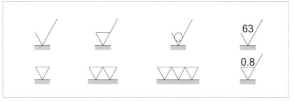

These, and other symbols are used in the drawing to identify surface finish. Some are used for metric drawings exclusively, others for imperial drawings. Interpretation of engineering drawings (blueprint reading) is a subject that covers drafting methods, including symbols. Overall, the CNC process offers control of surface finish quality in three main areas:

- Selection of cutting tool
- Selection of cutting data
- Machining direction

Effect of Cutting Tools

Proper selection of cutting tools is the first step towards achieving desired surface finish. There is a direct relationship between cutter radius and cutting feedrate:

> **Overall quality of surface finish will increase with large tool radius and slow feedrate**

Selection of Cutting Data

This area generally applies to the selection of speeds and feeds, but it also includes width and depth of cut, as well as other changes listed elsewhere. Unless using certain machining cycles, there is not much the operator can do at the machine to change width and depth of cut without some serious re-programming.

Machining Direction

The directions shown below are based on a right hand tool and normal (CW) rotation, using **M03** in the program. In CNC programming for milling, the tool always moves around the stationary part.

G41 LEFT	G42 RIGHT
G42 RIGHT	G41 LEFT
G40 - NONE -	G40 - NONE -

In a small experiment, machine a slot through a short length aluminum part as one cut, where the slot width is equal to the cutter diameter at the depth of 6-10 mm (~0.25 - 0.4 inches). When the part is inspected, each slot side (wall) will have a noticeably different surface finish. One side will be smooth and shiny, the other will be rougher and a bit 'flaky'. The cause of the good and not so good surface finish was the result of the way how cutting edges (flutes) entered and exited the material.

During contour milling, you may have noticed in programs that the tool is always moving on the left side of the contour. There may be special exceptions, but tool moving on the left side of the contour is the most common. The reason programmers choose this method is because on CNC machines it provides much better cutting conditions and surface finish. This method is called *climb milling*, and is one of two methods available:

- **Climb milling** ... also known as **Down** milling
- **Conventional milling** ... also known as **Up** milling

Follow the text and the illustrations for more details.

Climb Milling

In climb milling, the chip thickness is *thin* when the cutting flute leaves the contour, eliminating tool pressure at the contour. In this mode, the tool has the tendency to push down the part toward the table, and for that reason, climb milling is also known as *down* milling. **This method is preferred for CNC contour milling.**

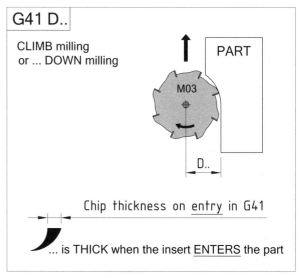

Conventional Milling

In conventional milling, the chip thickness is *thick* when the cutting flute leaves the contour, forcing tool pressure at the contour. In this mode, the tool has the tendency to pull up the part away from the table, and for that reason, conventional milling is also known as *up* milling. **This method should be avoided for milling.**

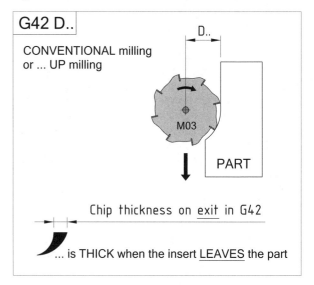

Chatter / Vibration

Every machinist has experienced the sound - sharp, squeaking, a bit eerie, but definitely spelling *trouble*. This is the sound of *chatter* that leaves nothing to imagination - every machinist will know that something is not right with the cutting tool performance. Even if the sound itself were tolerable, its results are not. Chatter leaves a bad surface finish on the part and negatively influences dimensional tolerances. Fortunately, chatter can be controlled to the point of total elimination.

Chatter during machining is a type of an undesired vibration, caused by either the cutting tool or the part, working against each other. The actual cause is not always easy to identify without special equipment, and the actual remedy may take a few trials to succeed.

Causes of Chatter

Chatter is more often present on CNC lathes, particularly during deep hole boring operations. A boring bar can work very well if the ratio of its overhang length to its diameter is within reasonable limits. Normally, the ratio is about 3:1 for steel shank boring bars, and 5:1 (or more) for carbide bars. In practise, it means that a 25 mm steel shank bar (1.0 inch) can be extended about 75 mm (3 inches), or 125 mm (5 inches) for carbide bar. Actual cutting conditions, such as the material type, depth of cut, cutting speeds and feedrates also play a large role. Long bar overhang is one of the most common reasons for chatter to occur. Although the solution is simple - *just shorten the bar* - it is not always feasible, due to the part design. Carbide bar can help, or a special devibrator boring bar can also be an option. Otherwise, a change of setup, including increasing the number of operations may be the last resort.

If boring bar overhang is not the culprit, the problem may be the overhang of the part itself. Part extended from the chuck also has its own limitations. Fortunately, an extended part can be supported by means of tailstock and/or steady rest (or follow rest). Boring operations are not possible with a tailstock support, and the alternative of steady rest or follow rest attachment can provide the solution.

> Always make sure the boring bar is properly set in relationship of the tool tip and the part center line

Another cause for chatter is when machining thin stock, such as tubular stock (pipes) or large thin plates.

Elimination of Chatter

Apart of the ideas already presented, chatter can also be eliminated by the way machining operations are selected prior to boring. For example, if an external tool removes the bulk of material by roughing, it may also leave a thin stock left for boring. Thin stock should always be of concern, as it is another cause of chatter. Rearranging the order of operations (bore first, then turn) may eliminate the chatter.

Cutting depth has also been known to cause chatter, and reducing it may help to eliminate chatter. Keep in mind that cutting depth works together with spindle speeds and feedrates, so they both have to be considered as well.

Although the initial tendency to eliminate chatter is to *reduce* speeds and feeds, often the exact opposite works much better.

There are at least two other factors that should be considered when encountering chatter, particularly on CNC lathes or during face milling. One option is to change the cutting insert. Inserts come with different type of built-in chip breakers. A positive type chipbraker will have a positive influence on eliminating chatter in many situations as it requires less pressure to enter the material.

The other factor that may help in eliminating chatter is the tool nose radius. For most lathe work, nose radius of 0.8 mm (0.0313 inches) is common. Smaller radius will result in less pressure during cutting, but presents problems of its own - possible change of cutting feedrates (in the program) and offset settings (at the machine).

For CNC milling operations, the basic principles apply equally.

COOLANTS

These days, a lot of attention is aimed at dry cutting, basically cutting without coolant. Modern development of cutting tools has brought this possibility into the forefront. In spite of the new trends, most metal removal is not machined dry and some form of coolant is required.

During actual machining, when the cutting tool is in contact with the material - *and that applies equally to conventional machining as well as CNC machining* - most parts will require coolant applied during the actual machining process. Coolant is a necessity for most - but certainly not for all - machining operations. Excluding the 'dry' type of machining operations, coolant is commonly used for the majority of machining applications.

Machining metals is a group that benefits the most from proper application of coolant, but other materials may benefit as well. The only obvious exception to using coolants is when machining parts made of wood or similar materials, in which case the coolant is not only unnecessary but is also not wanted. Materials in this category are rather rare in shops that specialize in metalworking as their main focus of business. In these companies, correct using of coolant - at least for most of machining operations - will provide many numerous benefits.

Benefits of Coolants

The benefits of using coolants during metal machining serves the following purposes:

- ... to cool the tool cutting edge
- ... to lubricate the tool contact area with the work
- ... to flush chips away from the part
- ... to clean machine slides

Cutting tool generates a great amount of heat and friction during a cut. With higher spindle speeds and cutting feedrates, the friction and heat increase. Coolant should always be supplied to the cutting edge of the tool, using various flexible pipes and nozzles.

Each CNC machine tool has a central coolant system, where the coolant mixture is stored. Using the recommended type of coolant, typically a mixture of biodegradable soluble oil and water, the coolant decreases tool wear, extends tool life, flushes out chips, and cools the machined part.

A lack of coolant or a loss of its lubricating capabilities is the common cause of poor surface finish, where the tool 'rubs' against the material rather than cuts it.

Coolant is not recommended with certain cutting tool materials, for example, ceramics. Often, some types of cast iron are machined without the use of a coolant. Coolants may not always be composed of soluble oil and water, they may be cutting oil, lard, mist, etc.

Working with Coolant

Just like tools wear out when working for an extended period of time, so does coolant. Quality of the coolant can be checked by using a tool called *refractometer*. This instrument shows how the light bends when passing through a substance. In terms of coolants, it measures the degree of fluid concentration. A typical ratio of water and soluble oil is 20 : 1, which is not the same as 20% mixture.

One negative effect on coolant is the machine slide lubricant. This oil constantly lubricates machine slides, and while doing it, it also drips into the coolant container, causing a floating film-like surface called *tramp oil*. This floating oil prevents air flow to the coolant. If not attended to, it can cause rancid smell, particularly during extended periods of machine inactivity. The smell is caused by bacteria and can be hazardous when in contact with skin. There are various methods of cleaning the tramp oil in order to extend coolant working life.

When working with coolants, the lesson is clear - take care of the coolant before it smells. Prevention is always preferred solution.

12 MACHINING HOLES

At first look, machining holes may seem rather a simple operation. In one view, it *is* a simple operation. Looking from another view, there are many possibilities that have to be considered, particularly when something does not work the way it should. For the CNC operator, machining holes almost always involves a programming feature called *canned cycles,* also known as *fixed cycles*. Programmers provide these cycles not only because they significantly shorten programming time, but if some changes are necessary, they can be done quickly at the CNC machine.

Understanding the structure of fixed cycles and how they are used in the program is a critical knowledge that every CNC operator should have. After all, if you cannot interpret what the program is telling you, how can you make any changes for possible improvement?

FIXED CYCLES - OVERVIEW

The main focus of this handbook is not aimed at the programming process, it is aimed at the control setup and working with part programs that have been already completed. The programming aspects are presented here only when they are directly related to machining in the shop and when they enhance the operator's overall understanding of this important subject.

DRILLING HOLES

Machining holes on a CNC machining center or a mill involves many different cutting tools and many different machining methods, but no machining method is more common than *drilling*. Many different drills can be applied to machine a single hole, but the programming and machining process always remains the same.

The most common drill for machining small holes is a standard twist drill, with a 118° tool point angle. The length of the drill point has to be always considered, as well as the breakthrough amount for through holes.

In order to drill any hole, it is a common practice to use a spot drill or a center drill first, to start the hole in a precise location, then follow this operation with another, using a drill to machine the hole to desired depth.

The process of drilling a hole typically involves at least four steps:

Step 1   Rapid to XY location
Step 2   Rapid to Z-clearance
Step 3   Drill the hole to depth at feedrate
Step 4   Rapid retract from the hole

The best way to illustrate the process is to use a practical example. Consider the following drawing - it contains only three holes, marked A-B-C for reference. Only drilling is required, so we forget any other tools - a ⌀10 mm drill is all that is needed.

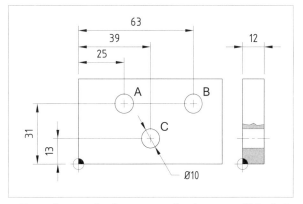

Even the very basic program for the part will include some decisions and some calculations. This is part of the programming process. The drill diameter D is 10mm. The result of the process is the Z-clearance at 2.5 mm above the top of part (practical decision), breakthrough clearance of 1 mm (practical decision), and calculated drill point length P, using the standard formula for 118°:

$P = D \times 0.3$ to be used as $P = \varnothing 10 \times 0.3 = 3.0$ mm

The XY coordinates of each of the three holes are presented in the drawing directly, by ordinate dimensions. The program is in metric units, and since only a single tool is used, there will be no need for a tool change in the program. Although important for actual machining, the speeds and feeds will be presented as *'reasonable'* only. The order of machining will be A > B > C. Brief comments in the program will describe each activity.

115

MACHINING HOLES

Drilling Example

Here is the program listing *NOT* using any shortcuts (known as fixed cycles). This method of programming is called *'long hand'*.

```
N1  G21                             (METRIC)
N2  G17 G40 G80                     (INITIALIZATION)
N3  G90 G54 G00 X25.0 Y31.0 S850 M03 (HOLE A)
N4  G43 Z25.0 H01 M08               (INITIAL LEVEL)
N5  Z2.5                            (FEEDPLANE)
N6  G01 Z-16.0 F175.0               (HOLE A DEPTH)
N7  G00 Z2.5                        (RETRACT)
N8  X63.0                           (MOVE TO HOLE B)
N9  G01 Z-16.0                      (HOLE B DEPTH)
N10 G00 Z2.5                        (RETRACT)
N11 X39.0 Y13.0                     (MOVE TO HOLE C)
N12 G01 Z-16.0                      (HOLE C DEPTH)
N13 G00 Z2.5                        (RETRACT)
N14 Z25.0 M09                       (INITIAL LEVEL)
N15 G28 Z25.0 M05         (RETURN TO MACHINE ZERO)
N16 M30                             (END OF PROGRAM)
%
```

Look at the program carefully and evaluate it as best as you can. There is nothing wrong with the program in terms of machining, all three holes will come out right. The problem is what can be done with such a program at the machine, should some changes be necessary. The answer is - not much. The first observation is that for only three holes, the program is quite long. Now imagine a similar program for three hundred holes. Second observation is that there are too many repetitions - the retract is the same, the depth is the same. Are they necessary?

Consider the motions for first hole:

N3	Rapid to XY location	G00 X25.0 Y31.0
N5	Rapid to Z-clearance	Z2.5
N6	Drill to depth at feedrate	G01 Z-16.0 F175.0
N7	Rapid retract from the hole	G00 Z2.5

For all three holes, the motions in blocks N5, N6 and N7 will be the same. It is this repetition that led towards development of a drilling cycle and other cycles for machining holes. The idea was to specify all data only once and then only those that change. This approach produced a much shorter program, where the information was squeezed - or *canned* - into a small space.

The format of a standard drilling cycle is:

```
G98 G81 X.. Y.. R.. Z.. F..
```

or

```
G99 G81 X.. Y.. R.. Z.. F..
```

Although written at the block beginning, **G98** or **G99** comes into effect during the retract only:

G98	Tool retracts to initial level
G99	Tool retracts to R-level
G81	Drilling cycle
X.. Y..	Hole position
R..	Clearance in Z-axis where feedrate begins (feed plane)
Z..	Z-axis position for depth
F..	Cutting feedrate

Other cycles may have additional address, but in principle, they all works on the same basic structure. Here is the program for the same three holes using **G81** cycle.

```
N1  G21                             (METRIC)
N2  G17 G40 G80                     (INITIALIZATION)
N3  G90 G54 G00 X25.0 Y31.0 S850 M03 (HOLE A)
N4  G43 Z25.0 H01 M08               (INITIAL LEVEL)
N5  G99 G81 R2.5 Z-16.0 F175.0      (HOLE A)
N6  X63.0                           (HOLE B)
N7  X39.0 Y13.0                     (HOLE C)
N8  G80 Z25.0 M09                   (RETRACT)
N9  G28 Z25.0 M05         (RETURN TO MACHINE ZERO)
N10 M30                             (END OF PROGRAM)
%
```

The illustration shows the first hole relationship between the long-hand method and the cycle:

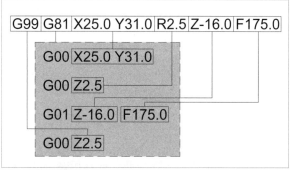

Terminology

Three words have been used that relate to fixed cycles and all relate to the Z-axis. They should be understood well:

- **Initial level**   *The last Z-axis position before the cycle*
- **R-level**   *Z-axis position where feedrate begins*
- **Depth**   *Z-axis position where feedrate ends*

The address R was used for the lack of another letter. Each address can occur only once in a block.

Initial Level and R-Level

As a CNC operator, watch for both the programmed initial level and the R-level, especially the initial level. In the example, the last Z-position before the cycle is in block N4, as Z25.0. The initial level is normally higher than the R-level, but both could be the same. With obstacles between individual holes, make sure the program uses the initial level higher than the highest obstacle.

> **Initial level must always be higher than the highest obstacle between holes**

If there are no obstacles between holes (such as in the example), the initial level makes no difference. In the example, it could have been any position above Z2.5, including Z2.5.

Depth

Also watch for the depth. Drills and similar tools are usually set to their lowest point (tip). This is the point that will reach the programmed depth. If you find that the depth is not correct, check the program first. If the programmed depth is correct, check the setup for tool length offset. Also check if the tool is firmly mounted. It is not unusual for a tool with straight shank to be pushed into the holder under cutting pressure.

Standard twist drill has an angular endpoint. The angle is usually 118°-120° for most materials, but cutting harder materials may require a wider angle, such 135°, and cutting softer materials may work better with angle as small as 90°. If you have to change the drill, you may also have change the programmed depth. For that, the following formula may come handy:

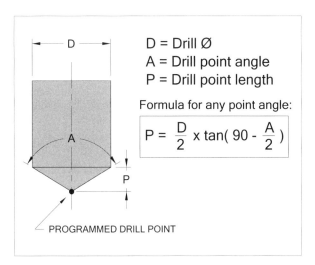

D = Drill Ø
A = Drill point angle
P = Drill point length

Formula for any point angle:

$$P = \frac{D}{2} \times \tan\left(90 - \frac{A}{2}\right)$$

PROGRAMMED DRILL POINT

While the formula can be used for any drill point angle, using a constant is a much faster way. To find the point length P, multiply the drill diameter by the constant.

60°	$P = D \times 0.866$
75°	$P = D \times 0.652$
80°	$P = D \times 0.596$
82°	$P = D \times 0.575$
90°	$P = D \times 0.500$
100°	$P = D \times 0.420$
110°	$P = D \times 0.350$
118°	$P = D \times 0.300$
120°	$P = D \times 0.289$
135°	$P = D \times 0.207$
150°	$P = D \times 0.134$
180°	$P = 0.000$

In the example,

∅10 mm drill x 0.3 = 3 mm

Spot Drilling + Counterbore + Countersink

Spot drilling is an operation where a small short tool make an indentation in the material to guide the drill that follows into a precise location. Traditional center drill can do the same job, but is more difficult to calculate and it cannot cut a standard chamfer. Spot drill, having the drill point angle of 90° can cut a chamfer on holes up to about 25 mm or 1 inch.

The cycle in the program uses **G82** command and add a dwell (pause) into the cycle:

G98 G82 X.. Y.. R.. Z.. P.. F..

or

G99 G82 X.. Y.. R.. Z.. P.. F..

While all other data have the same meaning as for the **G81** cycle, the dwell address P requires some attention. Dwell in fixed cycles is always programmed in *milliseconds* (1 ms = 1/1000th of a second). This is one area where the operator may significantly improve the machining efficiency. Unfortunately, many dwell times are programmed too long. One second does not sound that long, but is it really necessary? In order to be able to answer that question, you should understand what the dwell is for in fixed cycles.

The major reason for dwell is to cleanup the bottom of the hole. Dwell is typically programmed for spot drilling, counterboring and countersinking. In order to cleanup the bottom of the hole, a single revolution of the spindle should do the job. Considering that spindle override may be as low as 50% of the programmed speed, two revolutions become the minimum. The simple formula to calculate minimum dwell in *milliseconds* is:

$$\text{Minimum dwell} = \frac{60 \times 1000}{\text{rpm}} = \frac{60000}{\text{rpm}}$$

Typically, the result is doubled and rounded to a more 'attractive' number. For example, if the machine spindle speed is programmed as

```
S1250 M03
```

Minimum dwell is **60000 / 1250 = 48 ms**. Doubling and rounding the result will result in programmed dwell of 100 ms, or **P100** in the fixed cycle. That is 10% of one second. Always keep a basic rule in mind:

> **Dwell time INCREASES as spindle speed DECREASES**

Both cycles **G81** and **G82** can be shown graphically:

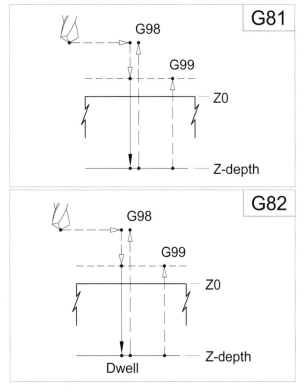

Peck Drilling

Peck drilling is another name for drilling with interruptions. Another description is *deep hole drilling* or *chip breaking*. There are two fixed cycles for peck drilling available. The first cycle is **G83**:

```
G98 G83 X.. Y.. R.. Q.. Z.. F..
```

or

```
G99 G83 X.. Y.. R.. Q.. Z.. F..
```

The addition of the Q-address and the change of cycle number make it a peck drilling cycle. The other cycle has the same format, but function differently. It is the **G73** cycle:

```
G98 G73 X.. Y.. R.. Q.. Z.. F..
```

or

```
G99 G73 X.. Y.. R.. Q.. Z.. F..
```

The graphic representation shows the differences between the two cycles:

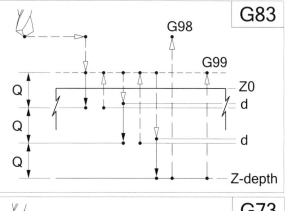

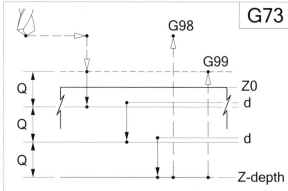

As you see, the difference is only in the retract. While **G83** retracts to the R-level after each peck, the retract in **G73** is only minimal, shown as the *'d'* dimension. In both cycles, the **G98** or **G99** is effective only on the last retract. It is easy to change the cycle number at the machine - no other changes are necessary, unless you also want to change the *Q* amount of peck depth.

*For the non-programmable **d** dimension see page 251.*

G73 is mainly used to break chips, while **G83** is used for standard peck drilling. Coolant reaching the drill tip is important, and **G83** is better, unless coolant through the spindle is available.

Changing Q-depth

The most common change of a peck drilling cycle at the control is changing the Q-amount - the depth of each peck for the same Z-depth. Make sure the Q-amount is reasonable for cutting and consider the number of pecks:

> Smaller Q-amount = Number of pecks INCREASES
> Larger Q-amount = Number of pecks DECREASES

Take the rule loosely - a very small amount of change may not result in the number of pecks, the change has to be significant. However, you can easily calculate the number of pecks. The first item to know is the total distance the drill will travel. This distance is always calculated between the R-level to the Z-depth. Once the distance is known, divide it by the number of pecks required and that will be the new Q-amount of peck depth. Be careful in rounding - always round up. Here is a practical example:

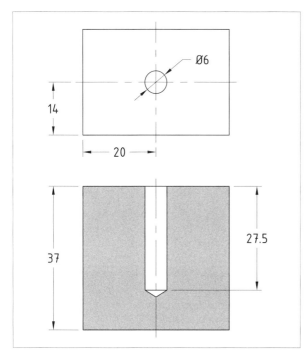

The cycle block in the program contains all data:

```
G99 G83 X20.0 Y14.0 R2.5 Z-29.3 Q8.0 F200.0
```

Before editing the program to increase or decrease the peck depth, you should know how many pecks will the program produce as is. During machining, you can count the pecks, but applying math is much more useful.

First, figure out the total travel between the R-level and the Z-depth. That calculation has to include the drill point length, which is

P = 6 x 0.3 = 1.8

Full diameter depth is 27.5, R-level is 2.5 mm above part zero, so the total travel will be the sum of all three:

Total travel = 2.5 + 27.5 + 1.8 = 31.8

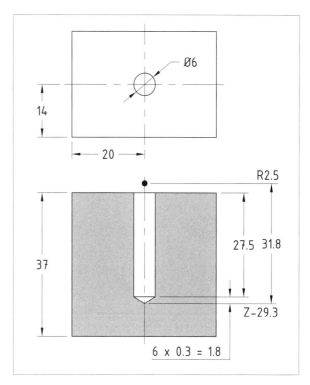

Programmed Q-depth is 8 mm and the number of pecks can be calculated:

Number of pecks = 31.8 / 8.0 = 3.975 = 4 pecks

You find that five pecks would work better, so you calculate the new Q-depth:

Q-depth for 5 pecks = 31.8 / 5 = 6.36

As the result is less than three digits, the number can be used as calculated. Now, suppose you want *nine* pecks for some hard material:

Q-depth for 9 pecks = 31.8 / 9 = 3.53333

Although mathematically correct rounding would be **3.533**, the minimum Q-depth must be **3.534**. Practicably, **Q3.6** would be more likely. In imperial units, the rounding will be to four decimal places, again, always rounding upwards. Also keep in mind that the Z-depth will never be exceeded, regardless of the Q-depth amount.

TAPPING OPERATIONS

A hole that was drilled earlier can accept another operation, such as boring, reaming or tapping. Let's look at tapping first. Tapping is a combination of forming and cutting. The tap has to be selected strictly for its size and the hole has to be drilled in such a way that there is enough material left for the tap. The drill selected for tapping is commonly known as the *tap drill*.

Tap Drill

Programmers will use a tap drill most suitable for the job. There are numerous tap drill charts available, supplied through vendors, manufacturers and other sources. You can also find a typical chart in the *Reference* section of this handbook, on page 271.

A modified example from the drilling section can be used as a sample of what to look in the program for a tap.

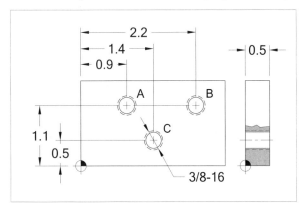

The program shows a tool **T03 - 3/8-16 TAP**. The tap drill program is not listed, as it does not bring any new information. The typical (most common) size of the tap drill is ⌀5/16 (0.3125), based on standard charts.

```
(G20)                           (IMPERIAL)
...
(T03 - 3/8-16 TAP)
N22 T03                         (TOOL CALL)
N23 M06                         (TOOL CHANGE)
N23 G90 G54 G00 X0.9 Y1.1 S600 M03    (HOLE A)
N24 G43 Z1.0 H03 M08            (INITIAL LEVEL)
N25 G99 G84 R0.3 Z-0.7 F37.5    (HOLE A)
N26 X2.2                        (HOLE B)
N27 X1.4 Y0.5                   (HOLE C)
N28 G80 Z1.0 M09                (RETRACT)
N29 G28 Z1.0 M05   (RETURN TO MACHINE ZERO)
N30 M30                         (END OF PROGRAM)
%
```

The format of a tapping cycle is very similar to **G81**:

G98 G84 X.. Y.. R.. Z.. F..

or

G99 G84 X.. Y.. R.. Z.. F..

G74 has the same format.

There are several features in a typical tapping program that the CNC operator should be aware of:

- **G84 is used for right-hand taps with M03 spindle rotation - right-hand tap must be used**
- **G74 is used for left-hand taps with M04 spindle rotation - left-hand tap must be used**
- **Spindle speed is *selected* for the machining conditions**
- **Feedrate is always *calculated* (see below)**
- **R-level has been increased to allow extra space for acceleration with high feedrates**
- **Z-depth should always guarantee full diameter thread**
- **Taps vary in geometry - select the most suitable one**

Feedrate Calculation

Since tapping is a cutting operation combined with forming, the shape of the tap has to be formed for the thread to be accurate. That means maintaining the thread pitch. The great majority of taps have only a single start, so the pitch and lead of the thread are the same. Remember, that feedrate for tapping (or threading) always considers the lead of the thread, not the pitch. This is not an issue in tapping, but it is very important for threads with more than one start (multi start threads). To review:

- **PITCH** is the distance between corresponding point of two adjacent threads
- **LEAD** is the axial distance measured over one revolution

Tapping feedrate is a relationship between the programmed spindle speed and thread lead (pitch for normal tapping). Metric threads are always defined by pitch, for example, **M10x1.5** is a 10 mm thread with a pitch of 1.5 mm. Standard V-shape imperial threads have the same form but are defined with the number of threads per inch (TPI).

Here are two related formulas for calculating tapping feedrate *F* with known spindle *S* and either pitch *P* or number of threads per inch *TPI*.

$$F = S \times P$$

This formula is always used for metric threads, but it can also be used for imperial threads, which requires calculation of the pitch first:

$$P = 1 / TPI$$

This is an extra step, which is not necessary. A better way is to use the TPI directly:

$$F = S / TPI$$

In the example, the feedrate was calculated using the last formula:

F = 600 / 16 = 37.5

NOTE - *It is not unusual to change spindle speed at the machine to create optimal conditions. Changing spindle speed for tapping is no exception, as long as the feedrate is changed accordingly*

For example, you may find that the programmed spindle speed *S600* can be changed to *S700*. It is not enough to change the address *S* only, but the feedrate *F* has to change as well:

S600   F = 600 / 16 =   F37.5
S700   F = 700 / 16 =   F43.75   (F43.7)

In the less likely event, the reverse of the above statement is also thru - if you change the feedrate, you must also adjust the spindle speed. For the number of threads per inch *TPI* use this formula:

$$S = F \times TPI$$

For pitch *P*, use this formula:

$$S = F / P$$

For example if you want to change the feedrate of F37.5 to F35.0 the spindle speed will change to:

S = 35.0 x 16 = S560

which is the same as

S = 35.0 / 0.0625 = S560

Troubleshooting

Threads made by tapping can create problems just like threads machined on a lathe. The problems can be quite numerous, so it is always important to categorize them and narrow down the cause. Generally, tapping problems can be filed in four categories:

- **Tap breakage**
- **Stripped thread**
- **Dimensional problems**
- **Surface finish problems**

Tap Breakage

Tap will most likely break, if the tap itself is too large or the pre-drilled hole is too small. Dull taps are more likely to break than new taps. In the program, check if the XY coordinate positions of the tap drill and the tap match. Check if the spindle speed and feedrate relationship is correct. Sometimes a different type of lubricant may solve the problem.

For blind holes check the depth of the drilled hole. Most drawings allow certain flexibility for the hole depth, as long as it does not break through the material. Tapped depth in blind holes should be checked periodically, so the full diameter tap depth is maintained. Removing chips prior to tapping may be another way to eliminate tap breakage.

Stripped Thread

Finding causes for a stripped thread may not always be easy, but trying the remedies suggested here should always be the first step. Dull threads, wrong alignment, hard material, poor setup - they all may contribute to the problem.

One solution is to *underfeed*. Take 2-5 percent off the calculated feedrate and try again. An even more innovative solution, particularly for very fine threads, may be to underfeed on the way *into* the material and overfeed on the way *out*. The problem with this solution is that tapping cycle uses only one feedrate for both motions. Writing a simple macro is the best solution. Since macros belong to manual programming on a much higher level, the following macro will be provided with minimum explanation.

It is written for a feedrate of 80% in and 120% out. This percentage will most likely require some degree of experimentation, but has been proven to work well.

Macro Example

Macro call example for imperial units, where the argument T is the number of *threads per inch* (TPI):

```
G65 P8020 R0.4 Z0.25 S750 T36

O8020 (SPECIAL TAPPING MACRO - IMPERIAL)
(R = #18 = FEEDPLANE)
(Z = #26 = TAP DEPTH)
(S = #19 = SPINDLE SPEED - RPM)
(T = #20 = NUMBER OF THREADS PER INCH =====)
#3003 = 1
G00 Z[ABS[#18]] S#19 M03
#3004 = 7
G01 Z-[ABS[#26]] F[#19/#20*0.8] M05
Z#18 F[#19/#20*1.2] M04
#3004 = 0
M05
M03
#3003 = 0
M99
%
```

Macro call example for metric units, where the argument T is the thread *pitch*:

```
G65 P8020 R3.0 Z6.5 S750 T0.75

O8020 (SPECIAL TAPPING MACRO - METRIC)
(R = #18 = FEEDPLANE)
(Z = #26 = TAP DEPTH)
(S = #19 = SPINDLE SPEED - RPM)
(T = #20 = THREAD PITCH ===================)
#3003 = 1
G00 Z[ABS[#18]] S#19 M03
#3004 = 7
G01 Z-[ABS[#26]] F[#19*#20*0.8] M05
Z#18 F[#19*#20*1.2] M04
#3004 = 0
M05
M03
#3003 = 0
M99
%
```

The 3000-series variables disable or enable single block and feedrate overrides. Take these examples for what they are - only examples that can be adapted.

Dimensional Problems

Adjustment in tapping speed (don't forget to change feedrate, too) is one possible approach. Using coated taps will prevent metal sticking to the thread, and so will a suitable lubricant. Changing the tap drill size slightly will also change the percentage of the thread depth. Using high helix tap geometry minimizes pressure. These are all possible solutions to consider.

Surface Finish Problems

Rough surface finish is generally the result of the tap pushing itself into the material rather than cutting it smoothly. Changing tap geometry often eliminates the problem of rough threads. Also, check the tapping head - some designs allow torque adjustments. Small tap drill may also contribute to poor surface finish, If the CNC machine has rigid tapping capability, use rigid tapping rather than the traditional tension/compression type tapping heads. Rigid tapping always surpasses regular tapping, but the machine tool has to support this feature. Underfeeding a couple of percent may also help.

Graphical Representation

Both G84 and G74 tapping cycles can be represented graphically, similar to other cycles described in this section. Watch for spindle rotation and correct tap selection.

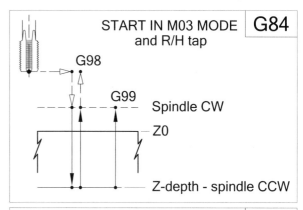

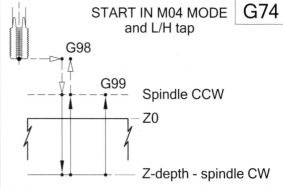

Rigid Tapping

Rigid tapping does not much change in the program, but the tap is installed is a solid holder (rigid holder). Spindle synchronizes the tapping. There are many benefits, but the CNC machine has to support this feature.

BORING OPERATIONS

Boring - or more accurately - *single point boring* is basically a sizing operation. It adds *roundness* and *cylindricity* to the hole, improves its surface finish, and with a micro attachment on the boring bar, hole size can be adjusted very accurately.

All remaining cycles are defined as *boring* cycles, starting with **G85**. Their differences are often subtle. A short list below illustrates the order of basic boring motions, programming format and brief motion details.

G85 Cycle

G85   Feed to depth > Feed out

    G98 G85 X.. Y.. R.. Z.. F..
    G99 G85 X.. Y.. R.. Z.. F..

Feeding out with a regular boring bar may actually damage the surface due to the release of tool pressure.
This is the preferred fixed cycle for REAMING

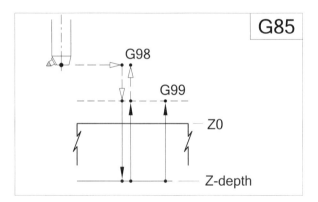

G86 Cycle

G86   Feed to depth > Stop > Rapid out

    G98 G86 X.. Y.. R.. Z.. F..
    G99 G86 X.. Y.. R.. Z.. F..

A very useful fixed cycle for rough or utility type boring. When the tool retracts at a rapid rate, the spindle does not rotate, and the relief of tool pressure will leave a straight vertical line on the hole surface.

In fact, **G81** *cycle that is closely associated with drilling can also be used for rough or utility type boring. The difference is that it will leave a helix on the hole surface, since the spindle is always rotating.* **G86** *or even* **G81** *cycles are often used when the hole has to be round, but surface finish is not critical.*

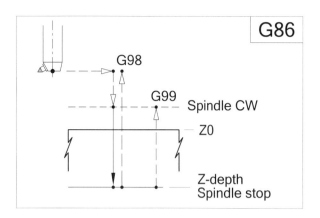

G87 Cycle

G87   Backboring (** G98 MODE ONLY **)

    G98 G87 X.. Y.. R.. Q.. Z.. F..

Apart from the requirement of a special tool, this cycle starts from the bottom of the hole and feeds up. It is used mainly when the backbore diameter is only marginally larger that the existing hole. **G98** *mode only.*

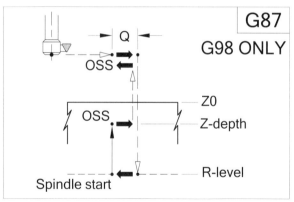

Note the OSS abbreviation - OSS stands for 'oriented spindle stop'. Also note the **Q-address** *- this is amount of the bar shift. Oriented spindle stop and the shift and their effect on tool setup is described in detail for* **G76** *fixed cycle on page 124.*

G88 Cycle

G88   Feed to depth > Stop > Manual operation ...

For all practical purposes, ignore the fixed cycle **G88**. *It is used for very special purposes only and appears in part programs extremely rarely (if ever).*

There is no need for illustration.

G89 Cycle

G89   Feed to depth > Dwell > Feed out

```
G98 G89 X.. Y.. R.. P.. Z.. F..
G99 G89 X.. Y.. R.. P.. Z.. F..
```

G89 *cycle does everything the* **G85** *cycle does, but requires a dwell P at the hole bottom.*

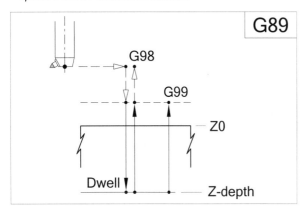

G76 Cycle

G76   Fine boring with shift

```
G98 G76 X.. Y.. R.. Z.. P.. Q.. F..
G99 G76 X.. Y.. R.. Z.. P.. Q.. F..
```

This is the king of the boring cycles and should be understood well, as its setup is not like setup of other tools.

G76 is the most commonly used boring cycle, particularly for work where the hole quality matters. It is often called a *fine boring cycle*, since it does not leave any residue on the finished hole diameter. Two key elements in this cycle is the *shift* and the *oriented spindle stop* (OSS).

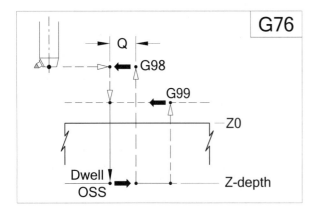

OSS and Shift

Oriented spindle stop (OSS) is a position of the spindle that is consistent all the time. It is a fixed position. Normally, it is used for alignment of tool with the ATC arm during tool change. It is also used in **G76** (and **G87**) cycle. When you are setting up a boring for a cycle with a shift, you have to know *where* the oriented position is, and *which direction* the shift takes place. Machine specifications provide that information.

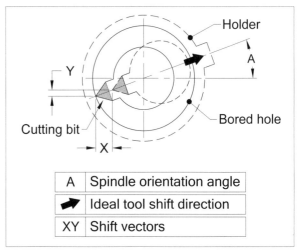

In the illustration, note that the tip of the boring bar always has to move into a clear area, still within the hole. For that reason, check the program for the Q-value - it is the amount of shift. For **G76** cycle, there is no need to program that amount more than 0.25 mm or about 0.01 inches. One of the methods used to find the direction is to specify the Q-value quite high, then watch the direction on the screen, as *Distance-To-Go*. Of course, the test will be done in the air, and several repetitions are often necessary. Don't forget to change the original, rather small amount.

That concludes the chapter on fixed cycles. There are many references in the handbook that relate to machining holes using fixed cycles. Also keep in mind that if you see the address *K* (or *L* for older controls) while any cycle is active, it means the number of repetitions. This selection is often used in incremental mode. Any fixed cycle is cancelled by the **G80** command or *any* motion command, such as **G00**, **G01**, etc.

For a shift using *I* and *J* in **G17** plane, see *page 250*.

DRILLING FORMULAS

In drilling operations, including spot drilling and center drilling, the tools have a known shape, only their sizes change. When machining holes, a number of useful formulas can make the job much easier. Even small changes to the program may require calculations or charts. In this section you will find a number of formulas that will make your job much easier.

Center Drills

The *L* dimension in the following illustrations is a suggested depth of cut to achieve effective diameter *E*. Feel free to modify the suggested values. *D1* and *D2* are standard sizes.

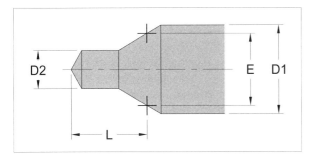

D1	Pilot Ø - D2	E	L
3.15	1.25	2.50	2.98
4.00	1.60	3.00	3.61
5.00	2.00	4.00	4.63
6.30	2.50	5.00	6.49
8.00	3.15	6.00	8.21
10.00	4.00	8.00	10.22
12.50	5.00	10.00	12.48
16.00	6.30	12.00	13.54
20.00	8.00	16.00	17.73
METRIC - ISO 866 / DIN333A			

Number	D1	D2	E	L
# 1	0.125	0.047	0.100	0.106
# 2	0.188	0.078	0.150	0.163
# 3	0.250	0.110	0.200	0.219
# 4	0.312	0.125	0.250	0.269
# 5	0.438	0.188	0.350	0.382
# 6	0.500	0.218	0.400	0.438
# 7	0.625	0.250	0.500	0.538
# 8	0.750	0.312	0.600	0.651
IMPERIAL - ANSI B94.11M type A				

Number	D1	D2	E	L
# 1	3.175	1.194	2.600	2.700
# 2	4.775	1.981	3.800	4.100
# 3	6.350	2.794	5.100	5.600
# 4	7.925	3.175	6.400	6.800
# 5	11.125	4.775	8.900	9.700
# 6	12.700	5.537	10.200	11.000
# 7	15.875	6.350	12.700	13.700
# 8	19.050	7.925	15.200	16.500
METRIC equivalent of ANSI B94.11M type A				

Spot Drills

Spot drill have a 90° tip angle, a thin web and no land on edges. They are used for spot drilling holes up to ⌀25 mm or ⌀1.0 inch. The full diameter of the spot drill is not used for machining, only the angular tip.

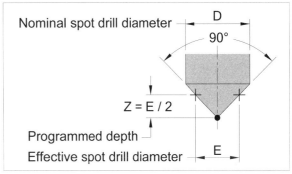

Many holes benefit from the 90 tip angle, and programmers use the spot drill not only to accurately mark a hole position but to chamfer the hole as well.

The formula below calculates the Z-depth:

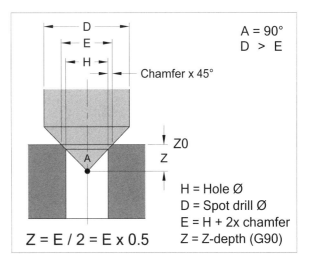

Spot drilling of large holes will not allow a chamfer, but the Z-depth can still controlled:

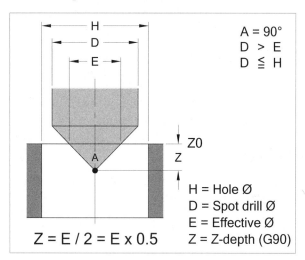

A = 90°
D > E
D ≦ H

H = Hole Ø
D = Spot drill Ø
E = Effective Ø
Z = Z-depth (G90)

$$Z = E / 2 = E \times 0.5$$

Twist Drills

The most common drilling tool is a twist drill, either HSS or HSS with cobalt, vanadium, and other elements that increase their strength and tool life. Calculations for a drill Z-depth include:

- Through holes
- Blind holes

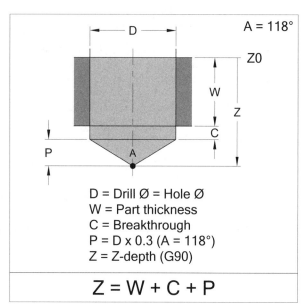

A = 118°

D = Drill Ø = Hole Ø
W = Part thickness
C = Breakthrough
P = D x 0.3 (A = 118°)
Z = Z-depth (G90)

$$Z = W + C + P$$

Use these formulas if a programmed hole needs an adjustment on the depth. Changing the program should always be the last resort, unless the program is obviously wrong. Always check your setup, particularly tool length offset, check if the tool is firmly mounted, and check if the wear offset is set properly (if available).

Formulas are great time savers, but being able to understand formulas allows you to work with them more intelligently. You will also be able to make your own formulas.

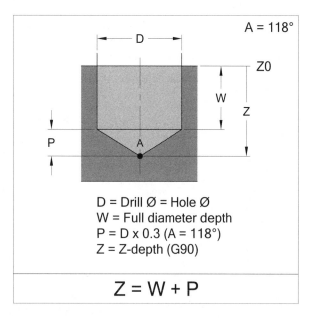

A = 118°

D = Drill Ø = Hole Ø
W = Full diameter depth
P = D x 0.3 (A = 118°)
Z = Z-depth (G90)

$$Z = W + P$$

13 MULTIPART SETUP

The basic concepts of multi-part setup were described in the earlier chapter *'Work Offset Settings'*. Study this chapter first - it covers the basic mathematical concepts and practical procedures of establishing part zero at the machine, during actual setup. This chapter will further expand the subject to apply to two or more parts set on the machine table to be machined as a group.

MULTIPLE PARTS

What is a multipart setup? As the question suggests, when more than one part is set within the machine work area, the setup qualifies as a multi-part setup. Parts can be set in the same fixture or they can be independently set at various positions of the machine table.

Understanding standard setup for a single part and being able to use work offsets is a prerequisite for working with multiple parts.

In most machining applications, there is only one part set on the machine table at any time. Any one CNC machine tool equals a single part setup. That is the usual method of setting many CNC machining centers. If the part occupies a large area of the machine table, this is quite normal. In cases there is a large table area unoccupied, a setup for more than one part might be worth considering. certain conditions apply.

Setup Conditions

Setting more than a single part on the machine table is a very attractive possibility, providing certain basic conditions are satisfied:

- All parts must be located within the machine travel limits
- All parts have to be set in a fixed position
- Part program must support multiple parts
- Each fixed position must be common to all parts machined at that location
- Distance between part zero locations measured along an axis may or may not be known

The first three items follow common sense - they refer to an actual part position within a specific machine work area and the necessity of part program support (work offsets and machining).

Working within limits should be quite easy to visualize - if the tool cannot reach the part, there is no machining that can take place. As for the second item, having all parts located in the same fixed position is a fundamental requirement of any CNC setup. All this would not be possible, if the program did not contain the change of work offsets, tool changes, machining, etc.

The second item requires a fixed position for each part located on the table, each with its own part zero (part origin). In this case, two possibilities emerge:

- Distance between part origins *is not* known
- Distance between part origins *is* known

Depending on which of the two options is the case, programmers will handle it in different ways. There are several programming methods available, such as:

- **Work offset change G54 - G59 and extended set G54.1**
- **Local coordinate system G52**
- **Offset setting by program G10**
- **Using macros G65**

The following sections will illustrate various options.

ORIGIN DISTANCES UNKNOWN

As an example, take three identical machinist's vises and see how they can be located on the machine table. One common option is to set each vise in a position that is not related to any other position.

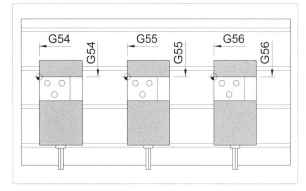

The schematic illustration shows three vises set on the machine table with distances between them unknown.

127

Even if the vises appear to be in a row, there is always a difference between their positions in both axes. Never assume that the table slots guarantee accuracy of the setup. The Y-axis settings may be close, but not very likely to be exactly the same (although possible).

The most common programming method for this kind of setup is to use multiple works offsets. There is really no other alternative. Probing macro is a possibility, but it is not available to most machine shops.

A sample offset entry may look something like this:

00	EXT				
X	0.000				
Y	0.000				
Z	0.000				
01	G54	02	G55	03	G56
X	-839.528	X	-624.883	X	-412.067
Y	-383.712	Y	-384.567	Y	-382.631
Z	0.000	Z	0.000	Z	0.000
04	G57	05	G58	06	G59
X	0.000	X	0.000	X	0.000
Y	0.000	Y	0.000	Y	0.000
Z	0.000	Z	0.000	Z	0.000

The three offset entries are visually emphasized.

Program Data

In the following drawing are three holes that have to be machined on three vises located in a row on the machine table.

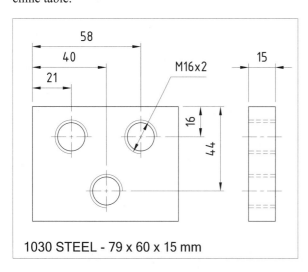

1030 STEEL - 79 x 60 x 15 mm

Setup such as this can be used for a large volume of parts, to minimize frequent tool changes. Although the setup of three fixtures takes longer than a setup of a single fixture, the extra time can still be justified in saving the extra tool changes on a large volume of parts.

In the program, the programmer will most likely store the hole locations in a subprogram and apply three standard operations for this type of work:

- **Spot drill** ⌀ 25 mm or ⌀ 0.75 inch spot drill
- **Drill** ⌀ 14 mm or ⌀ 35/64 inch tap drill
- **Tap** M16x2 plug tap

Actual machining will make three holes on each vise, before a tool change:

```
N1 G21              (=== USING WORK OFFSETS ===)
N2 G17 G40 G80 T01
N3 M06              (T01 - SPOT DRILL TO SPINDLE)
N4 G90 G54 G00 X21.0 Y-16.0 S800 M03 T02
N5 G43 Z2.5 H01 M08
N6 G99 G82 R2.5 Z-8.5 P150 F175.0 L0
N7 M98 P1001                    (LEFT VISE)
N8 G55 M98 P1001                (MIDDLE VISE)
N9 G56 M98 P1001                (RIGHT VISE)
N10 G80 G54 Z25.0 M09
N11 G28 Z25.0 M05
N12 M01

N13 T02
N14 M06             (T02 - TAP DRILL TO SPINDLE)
N15 G90 G54 G00 X21.0 Y-16.0 S750 M03 T03
N16 G43 Z2.5 H02 M08
N17 G99 G81 R2.5 Z-20.2 F220.0 L0
N18 M98 P1001                   (LEFT VISE)
N19 G55 M98 P1001               (MIDDLE VISE)
N20 G56 M98 P1001               (RIGHT VISE)
N21 G80 G54 Z25.0 M09
N22 G28 Z25.0 M05
N23 M01

N24 T03
N25 M06             (T03 - TAP TO SPINDLE)
N26 G90 G54 G00 X21.0 Y-16.0 S600 M03 T01
N27 G43 Z5.0 H03 M08
N28 G99 G84 R5.0 Z-18.0 F1200.0 L0
N29 M98 P1001                   (LEFT VISE)
N30 G55 M98 P1001               (MIDDLE VISE)
N31 G56 M98 P1001               (RIGHT VISE)
N32 G80 G54 Z25.0 M09
N33 G28 Z25.0 M05
N34 M30

O1001               (SUBPROGRAM FOR THREE HOLES)
N101 G90 X21.0 Y-16.0
N102 X58.0
N103 X40.0 Y-44.0
N104 M99
%
```

There is **G54** command in blocks N10, N21, and N32. It is included as part of better structure, but it can be omitted, as it is repeated at the beginning of each tool.

Note that there absolute mode **G90** is used, and there is no cancellations of data between vises. The CNC system will calculate the distance-to-go automatically.

In order to understand what happens at the control, consider the distance the tool will travel between the *last hole of the left vise* and the *first hole of the middle vise*:

| Left vise | G54 | X-839.528 | Y-383.712 |
| Middle vise | G55 | X-624.883 | Y-384.567 |

➥ *The last hole of the left vise in X-axis is at:*

X-axis: -839.528 + 40.0 = -799.528

The first hole of the middle vise in X-axis is at:

X-axis: -624.883 + 21.0 = -603.883

The X-difference is the X-travel:

X-axis travel: 799.528 - 603.883 = **195.645 mm**

➥ *The last hole of the left vise in Y-axis is at:*

Y-axis: -383.712 - 44.0 = -427.712

The first hole of the middle vise in Y-axis is at:

Y-axis: -384.567 - 16.0 = -400.567

The X-difference is the Y-travel:

Y-axis travel: 427.712 - 400.567 = **27.145 mm**

These calculation are not necessary - they are included here for better understanding of the subject presented.

Setup Process

There is no different setup for multiple fixtures than there is for a single fixture. In the example of three vises, setup each vise exactly the same way as you would for a single vise. Check the program, and look up work offset in the range of **G54** to **G59** or even in the extended range of **G54.1 P1** to **G54.1 P48**, if the programmer has that option available.

Typically, the offsets for three vises would be **G54** for the first vise, **G55** for the second vise, and **G56** for the third vise, as shown here. The standard range of six work offsets can accommodate six fixtures. For more than six, you will either need the extended set of work offsets or the **G10** command (if distances are known).

ORIGIN DISTANCES KNOWN

The same three vises used in the previous example, can also be mounted on a single fixture. In this case, the distance between individual origins is known. There are other setups of multiple parts where distances between the origins are known. In all cases, the options the programmer has are much greater.

The three-vise setup on a fixture will form a single unit. Only one work offset is normally required, for example, the **G54** offset to the left-most vise location. Do not expect the Y-axis distances to be the same - even on a single fixture, the distances will vary. The drawing below shows an example of a single fixture setup with three vises mounted in a row - **G54** is set at the machine:

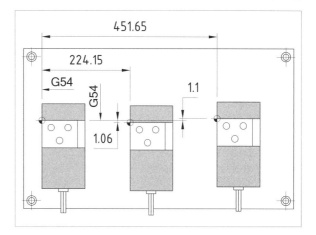

Setup with known dimensions between fixtures is best handled in the program by using the local coordinate system **G52**.

The **G52** command is used to shift the current coordinate setting by a programmed amount (known difference). Any one work offset has to be in effect (**G54** in the example), for the local coordinate system to work. Also, the programmer has to know the distances between individual fixtures. That information may be available in the fixture drawing or by measuring it at the machine. Once the distances are known, they will remain the same for each subsequent setup. Only the work offset setting (**G54**) will be different with each setup, when the fixture is removed and later placed on the machine table again.

In the illustration, and the following program example for the same three holes, note a very important feature:

> **G52 shift amount is always based on the current work offset**

Program Data

There is only one work offset in the example program - **G54**. All known distances are included in the program.

```
N1  G21                     (=== USING G52 SHIFT ===)
N2  G17 G40 G80 T01
N3  M06              (T01 - SPOT DRILL TO SPINDLE)
N4  G90 G54 G00 X21.0 Y-16.0 S800 M03 T02
N5  G43 Z2.5 H01 M08
N6  G99 G82 R2.5 Z-8.5 P150 F175.0 L0
N7  M98 P1001                       (LEFT VISE)
N8  G52 X224.15 Y-1.06   (SHIFT TO MIDDLE VISE)
N9  M98 P1001                     (MIDDLE VISE)
N10 G52 X451.65 Y1.1      (SHIFT TO RIGHT VISE)
N11 M98 P1001                      (RIGHT VISE)
N12 G52 X0 Y0          (NORMAL G54 IN EFFECT)
N13 G80 Z25.0 M09
N14 G28 Z25.0 M05
N15 M01

N16 T02
N17 M06               (T02 - TAP DRILL TO SPINDLE)
N18 G90 G54 G00 X21.0 Y-16.0 S750 M03 T03
N19 G43 Z2.5 H02 M08
N20 G99 G81 R2.5 Z-20.2 F220.0 L0
N21 M98 P1001                       (LEFT VISE)
N22 G52 X224.15 Y-1.06   (SHIFT TO MIDDLE VISE)
N23 M98 P1001                     (MIDDLE VISE)
N24 G52 X451.65 Y1.1      (SHIFT TO RIGHT VISE)
N25 M98 P1001                      (RIGHT VISE)
N26 G52 X0 Y0          (NORMAL G54 IN EFFECT)
N27 G80 Z25.0 M09
N28 G28 Z25.0 M05
N29 M01

N30 T03
N31 M06                    (T03 - TAP TO SPINDLE)
N32 G90 G54 G00 X21.0 Y-16.0 S600 M03 T01
N33 G43 Z5.0 H03 M08
N34 G99 G84 R5.0 Z-18.0 F1200.0 L0
N35 M98 P1001                       (LEFT VISE)
N36 G52 X224.15 Y-1.06   (SHIFT TO MIDDLE VISE)
N37 M98 P1001                     (MIDDLE VISE)
N38 G52 X451.65 Y1.1      (SHIFT TO RIGHT VISE)
N39 M98 P1001                      (RIGHT VISE)
N40 G52 X0 Y0          (NORMAL G54 IN EFFECT)
N41 G80 Z25.0 M09
N42 G28 Z25.0 M05
N43 M30

O1001             (SUBPROGRAM FOR THREE HOLES)
N101 G90 X21.0 Y-16.0
N102 X58.0
N103 X40.0 Y-44.0
N104 M99
%
```

Note the **G52** cancellation at the end of each tool. This is necessary in order to start each tool the same way.

Using Work Offsets

When the distances between origins are not known, using work offsets makes sense. When the distances *are* known, using work offsets does not make sense. Yes, work offsets can be used in such cases, but that means more work during actual part setup. This is unnecessary work and non-productive work.

Using Local Coordinate System

The program example used the method of work offset shift. Most CNC machines have a control feature called *Local Coordinate System*, associated with the **G52** command. Using this feature in the program is much preferred to the previous method, using work offsets.

The main difference is that all known dimensions between fixtures are entered in the program and not at the machine, during setup. Once verified, the program can be used over and over, anytime. If you are editing an existing program, do not forget to cancel the **G52** setting at the end of each tool, using the following block:

```
G52 X0 Y0
```

There is no need to cancel the **G52** command before any new **G52** is used. The control system will calculate the differences automatically.

Using G10 Setting

Fanuc and similar controls offer the **G10** command as means to set various data through the program. This subject will be covered in more detail, starting on *page 226*.

Here is an example of **G10** for the first tool, using the same fixture as for the previous example:

```
N1  G21                   (=== USING G10 SETTING ===)
N2  G17 G40 G80 T01
N3  M06              (T01 - SPOT DRILL TO SPINDLE)
N4  G90 G54 G00 X21.0 Y-16.0 S800 M03 T02
N5  G43 Z2.5 H01 M08
N6  G99 G82 R2.5 Z-8.5 P150 F175.0 L0
N7  M98 P1001                       (LEFT VISE)
N8  G10 L2 P2 X-627.275 Y-393.68 (MIDDLE VISE)
N9  G55 M98 P1001                 (MIDDLE VISE)
N10 G10 L2 P2 X-399.775 Y-391.52 (RIGHT VISE)
N11 G55 M98 P1001                  (RIGHT VISE)
N12 G80 G54 Z25.0 M09
N13 G28 Z25.0 M05
N14 M01
```

Other tools use the same method of programming. In this program example, **P2 = G55** work offset.

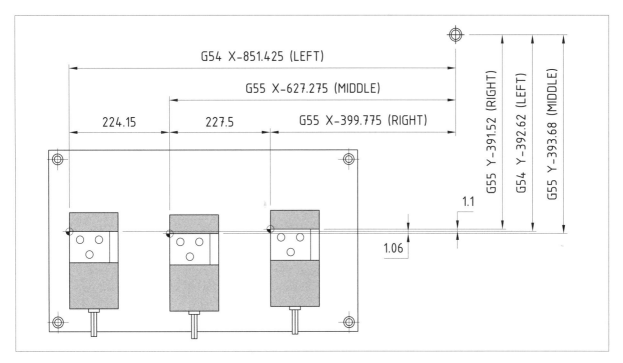

What does the program actually tell us? First, the **G54** work offset for the left vise is entered at the machine, during setup - in this case, it is *not* a known dimension. This dimension may or may not be known during programming. After all, the fixture will not always be in exactly the same position after each setup, but in some setups it will be known. In the example, the **G54** work offset is *not* known. If the dimension *is* known, you would see in the program a block containing `G10 L2 P1 X.. Y..`, followed by **G54**. The *L2* indicates a selection of work offset to distinguish it from other offsets.

As the **G54** setting is *not* known in the example, it has to be set at the machine, using standard edge finder methods. In the program, the P-address always identifies reference to the actual work offset number, and *P2* identifies the work offset used as the *second* available, therefore **G55**. What may be confusing is that **G55** is used for *both* the middle *and* the right vise setting. That is true in both the illustration at the top and the program sample presented. The reason for that is how the **G10** command is defined:

> XYZ setting of the G10 command is always measured from the machine reference point (machine zero)

The most important aspect of the **G10** command is that it actually *replaces* the work offset setting in the control. There can only be one **G55** at a time.

In a summary, the **G52** and **G10** commands can be compared. If you compare the **G52** setting of the last example, with the **G10** method used in this example, there is a significant difference:

- **G52** ... *is a shift distance from the current work offset*
- **G10** ... *replaces the current work offset setting*

Whether the program uses the command **G52** or **G10**, there is never a need to make more than one work offset setting at the machine. If the reference position of the fixture is also known, there is no need to make any work offset settings at all.

Using Custom Macro

Custom macro feature, such as *Fanuc Custom Macro B* is a special control option. Unless the CNC machine you are using has this option installed, you cannot take any benefits from it.

A special custom macro can also be developed if there is an additional need for more than the six standard work offsets or the extended set of 48 additional work offsets. In either case, only a single work offset is required, and the macro program takes care of the rest. This is strictly a programming job and no operator's interference is required. The subject of macros is beyond the scope of this handbook - you may want to check *Fanuc CNC Custom Macros* at **www.industrialpress.com** for more details.

EXTENDED WORK OFFSETS

In the text presented so far, there were many references to the common six work offsets in the range of **G54** to **G59**. These are standard work offsets, available on just about every CNC machining center.

While the six work offsets are more than enough for the majority of work done in smaller machine shops, there are cases when more work offsets are needed, particularly in larger companies. One such case is when a dedicated set of a dozen or more CNC machines is used to complete a single part that requires many operations. Take, for example, an automobile engine. There are many operations required for full completion of this complex part, including many faces. Large manufacturers will be more efficient if they design a full line of individual cells (CNC and other machines), with each cell dedicated to a certain machining operation.

Some of these operations will require no more that the six standard work offsets, while others will require more than six. This is where the optional extended set of work offsets comes in. It is optional, because you have to pay to have it installed.

Available Range

Fanuc and similar control systems offer additional forty eight (48) work offsets. Combined with the six standard work offsets, the number of work offsets is staggering *fifty four!*

As expected, there must be some difference in the program between the standard and the extended work offset range. Indeed, there is.

Standard vs. Extended Work Offset Range

In the CNC program, the standard range of six work offset is recognizable by the G-code:

- **G54 to G59 is the standard work offset range**

The extended set uses **G54.1** command, followed by the P-address:

`G54.1 P1` to `G54.1 P48`

▸ *Example:*

G54 = the first standard work offset

G54.1 P1 = the first extended work offset (above **G59**)

Incidentally, **G54.1 P..** is the same as **G54 P..**, etc.

Z-AXIS AND MULTIPART SETUP

So far, the examples shown in this chapter have focused on the top view of X-axis and Y-axis of the setup. Even if the same fixtures (such as vises) are of the same make, it is to be expected to have differences, however small, relating to the X and Y axes. This situation has been the focus of all previous sections in this chapter.

In actual multi-part setup, the Z-setting between individual fixtures is also an important issue to be addressed.

Variable Height

Just as you cannot expect absolute precision along the X and Y axes for multiple part setup, you cannot expected it in the Z-axis either.

The Z-axis offset settings control the height of the part and the machined depths. Assuming the part program is correct and that it takes into consideration the setup variations, it still comes down to the actual and physical setup at the CNC machine, using offsets.

Even with all parts having the same thickness, some deviation of the Z-axis settings is to be expected. With a single part, it is only a matter of the *tool length offset* setting. With multiple parts used in a single setup, each part has to be accurate, which means each part has to have a unique setting in the program. Tool length offset has been described in a separate chapter, starting on *page 87*.

The first application will be the same part in three vises, with minor deviation in height. For clarity, the fixture (vise) is not shown, only the important dimensions.

Tool Length Offsets

Tool length is normally measured for a single part; you measure one tool length for each tool on the same part. This does not change for multiple part setup. In the illustration, the part in the *left vise* was measured.

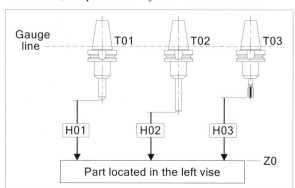

The measurement method used here is the so called *touch-off* method. This is not the most efficient method, but one of the most common, particularly in small machine shops. The principles described here apply to any other method of tool length offset measurement.

For the example, three tool length offsets have been entered into the offset registry. Although the illustration shows a column for wear offset, some controls have only one column only for the geometry offset (as measured).

OFFSET		
No.	Geometry	Wear
001	-368.345	0.000
002	-312.794	0.000
003	-357.820	0.000
...	...	...
...	...	...

When three parts are setup in three different vises, the tool offset may or may not be the same for all three parts. A good bet is that they will *not* be the same. What are the options to make the adjustments for the part height differences?

The first answer might be more tool length offsets. In theory, that would solve the problem, but only in theory. Consider not only three parts but sixteen - for example, four parts on each face of a four-sided tombstone, common in horizontal machining. Not only that the setup would take a lot of time, you may actually run out of the available offsets.

Better solution is to use the Z-setting of each work offset. Look at the three parts (illustration is simplified for clarity) with some minor height variations.

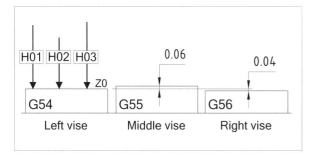

Although the illustration is not to scale, it clearly shows that the middle part is 60 microns higher and the right part is 40 microns lower.

These adjustments will be placed in the work offset of Z-axis setting, which is normally zero:

00	EXT				
X	0.000				
Y	0.000				
Z	0.000				

01	G54	02	G55	03	G56
X	-839.528	X	-624.883	X	-412.067
Y	-383.712	Y	-384.567	Y	-382.631
Z	0.000	Z	0.060	Z	-0.040
04	G57	05	G58	06	G59
X	0.000	X	0.000	X	0.000
Y	0.000	Y	0.000	Y	0.000
Z	0.000	Z	0.000	Z	0.000

Note that the **G54** setting in the Z-axis is 0.000 for the left vise - this is where the part was measured and offsets entered into the tool length offset registry. For the middle and right parts, the control system will make all necessary adjustments.

Note - Even if the part stock is exactly the same height, the adjustments may be necessary for deviations between individual vises

Using EXT offset

The two work offset adjustments shown above control each fixture separately. That is not always the case, as some adjustments apply to all fixtures equally. For example - for some reason, you change parallels in all three vises and replaced them with new ones that are 3 mm narrower. All three parts are now lower by 3 mm.

What are the options?

One option is to change the tool length offset for each and every tool by 3 mm:

- **Tool T01 / H01** ... *from* -368.345 *to* -371.345
- **Tool T02 / H02** ... *from* -312.794 *to* -315.794
- **Tool T03 / H03** ... *from* -357.820 *to* -360.820

In all three cases, the change was the same - 3 mm. Since the parts are now lower than before, the original distance measured will be 3 mm larger. There is nothing wrong with this method in mathematical terms. What is wrong is its inefficiency. For a large number of tools, to change all offsets is time consuming and prone to errors.

The solution is to use the **EXT** (*External* or *Common*) offset. All you have to do is to put the amount of 3 mm and its directional adjustment (minus) into the **EXT** offset, which is part of the work offset settings:

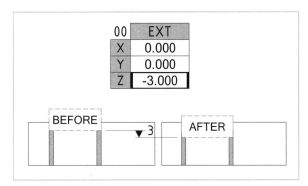

For more details on the **EXT** work offset, see *page 85*.

PROGRESSIVE SETUP

Suppose that a certain part has five operations before it is fully completed. Design permitting, each operation requires a specific (progressive) setup. If the part is small enough to fit on the machine table, a total of five different fixtures will be required. The core of the program will be to make connections between the fixtures by using various offsets. For setup like this, study this chapter carefully - the information is there - just make sure you apply it well.

SUMMARY

This chapter has presented various setting options applied to locating several parts on the machine table at the same time. As expected, there are variations of these settings, depending on the exact work situation. Also, there are applications of these methods that have not been mentioned. Make sure you understand the examples as presented - then you will be able to apply the knowledge to similar settings, those that are not described in this handbook.

14 CUTTER RADIUS OFFSET

All modern CNC controls for machining centers, lathes, and other machine types, provide a feature called *'cutter radius compensation'* or *'cutter radius offset'*. The basic purpose of cutter radius offset is to compensate for the cutting tool radius. The cutter radius offset feature of the control system is beneficial to both CNC programmers and CNC operators.

BENEFITS

CNC programmers benefit from this feature by practically disregarding the actual radius of the cutting tool at the time of program development. That allows dimensional data to be taken directly from engineering drawing dimensions. Programmers still have to consider the tool size when determining clearances and approach towards the part contour, but most calculations are greatly simplified.

At the CNC machine, one major benefit of the cutter radius offset is that the CNC operator can select the cutter with a certain degree of freedom, and is not constrained by the program itself. The other reason is that the cutter radius offset allows to adjust (fine tune) the size of the cutter to achieve dimensions within certain tolerances, typically caused by the tool wear.

PROGRAM DATA

Normal use of the cutter offset is for two-dimensional cutting motions, and the three program commands used are all preparatory G-codes:

G40	Cutter radius offset to the LEFT of cutting direction
G41	Cutter radius offset to the RIGHT of cutting direction
G42	Cutter radius offset mode CANCEL

On CNC milling machines and machining centers, it is necessary to program the number of the offset registry, where the actual offset amount will be entered during part setup. This is achieved with the address *D*, followed by the offset number.

Address D

Similar to address *H* for tool length offset in **G43** mode, cutter radius offset also requires offset number within the same range, using the **G41/G42** command. In this case, the letter used for the cutter radius offset number is the address *D*.

The H-offset for tool length is generally the same as the tool number, for example, for tool number six (T06) the tool length offset will be H06. For the D-offset that is not necessarily the same. The reason is that there are three offset memory types available on modern CNC systems.

Offset Memory Types

Fanuc identifies their offset memory types (offset registers) as *Type A*, *Type B*, and *Type C*. As the following table shows, more options are available with the higher types than with the lower types:

Offset memory Type A				
		Offset No.	Offset	
		01	0.000	
		02	0.000	
		03	0.000	
		...	...	

Offset memory Type B				
	Offset No.	Geometry	Wear	
	01	0.000	0.000	
	02	0.000	0.000	
	03	0.000	0.000	
	...	...	...	

Offset memory Type C				
Offset No.	H-offset		D-offset	
	Geometry	Wear	Geometry	Wear
01	0.000	0.000	0.000	0.000
02	0.000	0.000	0.000	0.000
03	0.000	0.000	0.000	0.000
...	...	...	...	...

Entering particular offset data is always related to the programmed offset number. Practical setup examples are shown later in this chapter.

Equidistant Toolpath

When you set the work offset such as **G54**, the part X0Y0 represents the point that corresponds with the center line of the cutting tool. For example, if the **G54** is measured as X-625.0 Y-413.0, and the tool is moved to the position of **X36.0 Y24.0**, it will be 589 mm in the X-axis and 389 mm in the Y-axis from machine zero. The centerline of spindle will be located at this point of the contour:

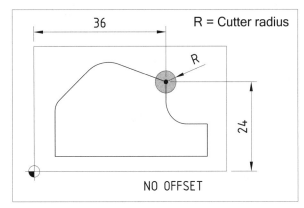

With cutter radius offset in effect, the tool will be shifted - or *offset* - by its radius in such a way that the cutter edge - *not its center* - will be in constant contact with the programmed contour. This offset toolpath is called *equidistant toolpath*.

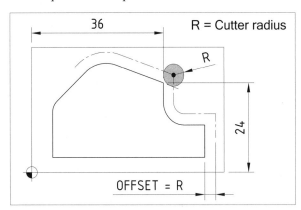

If the cutter radius were included in the program, it would have to match exactly the cutter radius used for machining. Such a situation is very difficult to achieve, which means the programmed radius and actual radius may not match and the part would not be machined to drawing specifications. Being able to adjust cutter radius *at the control* is one of the most important features of modern CNC machines. Such adjustment can be done not only during setup but also during production.

Cutting Direction

The cutting direction *LEFT* or *RIGHT* is always related to the cutter direction moving along a contour. This is a very important observation. When you are viewing the XY motions of a vertical CNC machining center, you will see that the machine table moves left and right along the X-axis, and towards you or away from you along the Y-axis. In fact, many machines even show positive and negative arrows on the machine slides, to indicate the motion direction. In programming - *and program interpretation* - there is only one very important rule:

> **Direction of tool motion LEFT and RIGHT always refers to the right-hand tool moving around a stationary part**

This rule is the basis of **G41** command being to the *left* and **G42** being to the *right* - left or right of the contour, viewed into the cutting direction. To be even more precise, the commonly used rule of *G41 = Left* and *G42 = Right* is true for **right hand** cutters only - the rules for left-hand cutters are reversed.

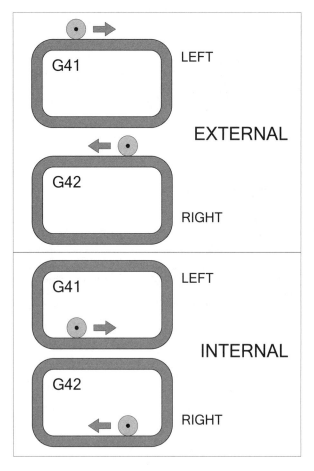

WORKING WITH RADIUS OFFSET

Study the following program segment carefully - it shows the first few blocks of a tool that uses cutter radius offset for contouring:

```
...
(T06 = 18 MM END MILL)
N43 T06
N44 M06
N45 G90 G54 G00 X-12.0 Y-12.0 S950 M03 T07
N46 G43 Z2.0 H06 M08
N47 G01 Z-5.0 F125.0
N48 G41 X3.0 D56 F200.0
...
```

Interpreting the part program correctly is one of the most important skills of a CNC operator. You don't have to be able to write programs, but you have to be able to interpret them. Let's look at the six numbered blocks above - what features relate to the part setup?

- *Feature 1*   Tool selection + Tool number   18 mm
- *Feature 2*   Work offset   G54
- *Feature 3*   Tool length offset + H-address   G43 H06
- *Feature 4*   Cutter radius offset + D-address   G41 D56

All these features are critical. Starting with *Feature 1*:

The tool selection is often described as a comment, either at the top of the program along with other tool descriptions, or - as in this case - at the beginning of each tool. Program comments are enclosed in parentheses and the example tool is specified as an end mill with 18 mm diameter. The same comment also identifies the tool as tool number six, which is confirmed in the program block N43. Block N43 will force the control system to search for tool registered as tool number six, unless the tool is in the stand-by position already. The following block N44 will bring this tool into the spindle and it becomes the *current* tool or *active* tool.

Feature 2 identifies which work offset the program - or this particular tool - uses. As is common for many jobs, the **G54** work offset is used in this program as well.

> Always check for the work offset
> Do not assume it is always G54

Feature 3 is in block N46 - **G43 H06**. This is the tool length offset, having the same number as the tool. Using the same offset number as the tool is practical, as there is usually only one tool length offset for each tool.

Feature 4 - and the main focus of this chapter - is the cutter radius offset. The offset direction is specified to the left of the contour - as **G41**. When working with CNC milling machines and machining centers, you will encounter mostly **G41** - **G42** will not be as common. The reason for this is the construction of CNC machines, combined with tooling used. For milling, the preferred way of moving the tool around contour in *climb* milling mode (also known as the *down milling*). This mode provides the best machining conditions and surface finish quality. In climb milling mode, the cutting produces a chip that is thick at the cutter entry into the material, but thin when the cutter leaves material.

To achieve climb milling mode with a standard right-hand cutter, the programmer develops the toolpath in the direction that is to the *left* of the contour machined. Preparatory command **G41** is used for this purpose.

D-offset Number

Just as the tool length offset requires offset number, cutter radius offset requires a number as well. Using the T06 from the example, it would make sense to program:

```
T06 - H06 - D06
```

However, that is not the numbering used in the program example. While the T06 and H06 do correspond to each other, the D-address uses *D56* as the offset number rather than the expected *D06*. The reason is that many control systems offer *Memory Offset Type A* as standard, the other two being optional. Looking at the table on page 135, you will see that *Type A* provides only one column for both tool length and cutter radius offset. As only one entry is possible for each offset number, tool length and cutter radius numbers must be different.

In order to make some order to this requirement, programmers add an arbitrary number to the tool number, such as 30, 40 or 50. This addition is strictly at the discretion of the programmer. Generally, the number added should be greater than the number of tools. As many controls have at least 99 available offsets, splitting the range in the middle offers a solution that many programmers use. That is why the numbering is the example is:

```
T06 - H06 - D56
```

Memory offset *Type B* is also shared and will require adding an arbitrary number same as *Type A*. Only the highest level control systems use memory offset *Type C*, where T# = H# = D#. The other items, namely *Geometry* and *Wear* offsets will be explained as well.

Look-Ahead Type Offset

Before explaining the *'Look-Ahead'* type of cutter radius offset in the section title, let's look at the example program once more, this time from a totally different perspective - *this is manually generated CNC program*:

```
...
(T06 = 18 MM END MILL)
N43 T06
N44 M06
N45 G90 G54 G00 X-12.0 Y-12.0 S950 M03 T07
N46 G43 Z2.0 H06 M08
N47 G01 Z-5.0 F125.0
N48 G41 X3.0 D56 F200.0
...
```

Suppose that you have entered only the six numbered blocks above into the control (no others), and want to test the program at the control. You have also entered the cutter radius as 9.000 in the offset registry number 56. Now the question - think before reading further:

• **What are the XY coordinates of the cutter when all six blocks shown have been processed (executed)?**

Let's evaluate - block *N45* brings the tool to a position of *X-12.0 Y-12.0* - no cutter radius offset is in effect at this time. After the Z-depth is reached in block *N47*, the cutter radius offset is called. **G41** indicates position to the left, and *D56* is linked to the offset setting of 9.000 in the control. Seems easy - so where is the tool center? What are the XY coordinates when block *N48* is completed? Here is a simple illustration:

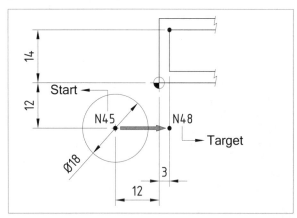

The programmed motion is along X-axis only, from X-12.0 to X3.0 - from the *start* point to the *target* point. The distance between the two points is 12 + 3 = 15 mm.

• **Where will the cutter center be, when the target is reached in block N48?**

The quick answer is that it will be 9 mm short of the target. After all, the travel distance *(Distance-To-Go)* is 15 mm, cutter radius is 9 mm, so the travel will be shortened by 9mm to actual *Distance-To-Go* of 6 mm.

There is a catch, however. Yes, this is exactly what *should* happen - the cutter *should* travel only 6 mm, but that will **not** be the case. The cutter will actually travel full 15 mm and its center position after block N48 is executed will be X3.0 Y-12.0. So, what is the problem?

The problem is that the control does not 'know' what is the left position. **G41** specifies cutter radius offset to the left, but that statement is always tied up to the direction of motion. This direction is the missing piece - there is *no information* in the six blocks as to what direction the cutter should be left of! Consider this illustration:

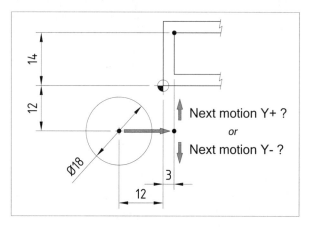

In order to establish the cutting direction, the next tool motion must be known. Here are the two possible options illustrated:

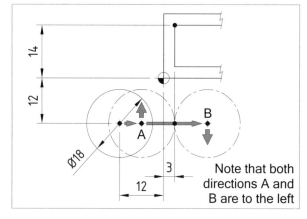

Both directions *A* and *B* are to the left of the target - the actual tool position will depend on the *next* motion programmed. Will it be *A* or *B*?

If the next position is *not* available, the control system cannot provide correct solution and the cutter center will be at the target location - *uncompensated*. This is exactly the same situation, as if the cutter radius offset were not programmed at all:

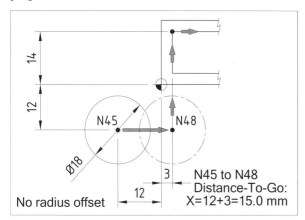

No radius offset
N45 to N48 Distance-To-Go: X=12+3=15.0 mm

Control systems are designed with a feature called *'look-ahead'* type of cutter radius offset. Because of computerized processing, the control can 'look ahead' in the program, and 'see' what is the next motion and apply the cutter radius offset as required. In fact, while the cutter radius is in effect, the control will constantly look ahead. Looking ahead means that the system will store the next motion in a buffer (temporary memory), *before* actually processing that block.

How does all that affect the program example? If you really want to test only a partial program, you must also include block *N49*, that includes the next motion - not six block shown, but seven blocks.

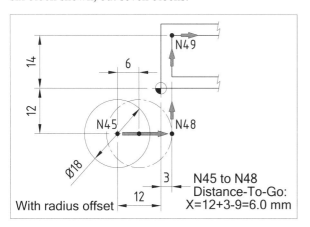

With radius offset
N45 to N48 Distance-To-Go: X=12+3-9=6.0 mm

Although this example was intended mainly to illustrate the subject of cutter radius offset, when testing, it is always a good idea to follow a simple rule:

> **Test only complete program and make sure all tools and offsets are set before testing**

To conclude - from the last illustration, the first motion is to the **X3.0** target location, followed by motion in the *positive* Y-direction, to the target of **Y14.0**. The minimum program to test has to include this next motion:

```
...
(T06 = 18 MM END MILL)
N43 T06
N44 M06
N45 G90 G54 G00 X-12.0 Y-12.0 S950 M03 T07
N46 G43 Z2.0 H06 M08
N47 G01 Z-5.0 F125.0
N48 G41 X3.0 D56 F200.0
N49 Y14.0
...
```

Now the test will work just fine, but keep in mind the last rule - test the program after the *complete* program is loaded and *all* setting are done.

INITIAL CONSIDERATIONS

The title of this section may seem unnecessary, particularly if you consider that the cutter radius setting at the machine control requires to input the radius amount, and that's it. In principle, that is true - just input the *proper* - that is *suitable* - radius amount and all is done. Well, not quite.

The key words in the previous statement are *proper* and *suitable*. What is a *proper* - or *suitable* - radius amount to be input and registered as the offset setting?

Cutter Diameter

If the cutting tool selected for the previous program example is an 18 mm end mill, it only means that the *nominal* diameter of this tool is 18 mm. The elusive word *nominal* in this sense means *normal, assumed*; it means *as per catalogue*, it means *ideal*, and so on with similar meanings. What it does *not* mean is that it is an actual, real, perfect diameter.

Cutting tools (such as end mills) are not all manufactured to the exact standards. For many different reasons, the expected diameter of 18 mm may be slightly more or slightly less. This has to be taken into consideration.

Effects of Machining

Even if the cutting tool - such as our 18 mm end mill - is 'perfect', at least in the sense of its actual diameter, there are other factors that have to be taken into account. These factors include two possible and common problems - *tool deflection* and *tool wear*.

Tool *deflection* happens when the cutting tool deviates from its axis of rotation during machining. Cutting tools that have a large ratio between the tool diameter and the tool length (*i.e.,* long tools with small diameter) are more likely to have deflection related problems than short tools with relatively large diameter. Deflection can also happen when the tool takes a heavy cut. In this case, the tool is pushed away from its centerline by the cutting forces. There are some other causes of tool deflection, but regardless of the actual cause, as long as the tool can remove material, it is usable.

Tool *wear* is a situation where the cutting tool becomes progressively smaller as it removes material, particularly after it is used for machining of large number of parts. As long as the tool wear amount is very small, there is no need to replace the tool, possibly quite an expensive one. At the same time, the decreasing tool size, even very small, may have an significant effect on the final dimensions of the machined part.

In both cases - and providing the problem is not excessive - the CNC operator can adjust the cutter radius offset setting in order to maintain dimensional precision. This adjustment means *increasing* or *decreasing* the currently set offset amount.

Cutter radius offset can be even adjusted *before* the first part machining takes place, for the single reason of *preventing* a scrap.

Scrap Prevention

Whatever the actual cause may be, a part that is removed from the production batch because it does not meet specifications is called a *scrap*. Scrap is always an undesirable situation, and one that can be very expensive. Causes of scrap can be numerous, such as using the wrong tool, setup error, program error, tool breakage, and many others. For the purposes of this chapter, we will evaluate causes of scrap due to erroneous use of cutter radius offset and look for ways to prevent a scrap from happening. In this sense, scrap can be defined as:

> Part is considered a scrap if its measured dimensions are outside of specified tolerances

Study the following table:

	External	Internal
ON size	No corrective action	No corrective action
OVER size	Recut is possible	SCRAP
UNDER size	SCRAP	Recut is possible

This concept is easy to illustrate:

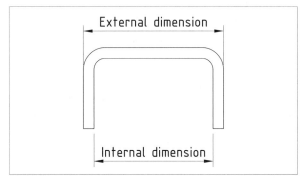

Two closed contours above represent external and internal cutting. From the top table, it is apparent that if either dimension is *within* tolerance *(ON size)*, there is no need for corrective action - the part is as per drawing.

ACTUAL OFFSET SETTING

Without a doubt, the CNC operator is responsible for maintaining accurate drawing dimensions on the machined parts. Maintaining dimensional tolerances starts with the part program. It is absolutely necessary for the machine operator to know whether the program uses nominal drawing dimensions or dimensions somewhere within the tolerance high and low range. For the examples, the above two contours will have some dimensions:

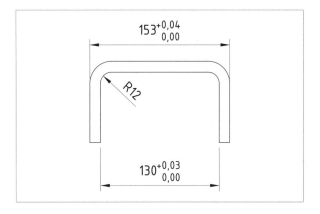

Most manual programs are developed using only the nominal dimensions - XY coordinates do not include any tolerances at all. That is the common case and will be used for further evaluation purposes. We will consider the program to be correct and functioning properly.

The drawing used for testing contains two linear dimensions that have to be maintained during machining. One is external, the other internal. The internal radius is included for another reason - tool selection. Other dimensions are not required for the offset settings.

Programmed Tool Size

For machining the external contour, an end mill of almost any diameter could be selected by the programmer. That is not the case for the internal contour. The smallest internal radius of the contour is 12 mm, which means that a tool *smaller* than 24 mm in diameter should be selected. In theory, a 24 mm cutter can also be used, but this tool would produce a very poor finish and most CNC programmers would not even consider it.

For the examples, these tools will be used:

- T01 - external contour ...   ∅25 mm end mill - D51 offset
- T02 - internal contour ...   ∅20 mm end mill - D52 offset

Each offset setting will be discussed independently.

External Offset Settings

As expected, the programmer used the overall part dimension of 153 mm for calculating the contour points. Offset D51 is used in the program, and its amount must be set at the machine control. What should that amount be for a new, never used, ∅25 mm end mill?

As most CNC machines require this offset to be the amount of the tool *radius*, the setting be:

D51 = 12.500   ... *applies to manually generated programs*

With a new tool, rigid setup and well designed program, the contour, when measured, should be exactly 153.000 mm. As this dimension is within the specified tolerance, all seems to be well. A good argument for this offset setting may be that as the cutter wears out slightly, it becomes smaller and the dimension will still fall within tolerance, as it will be slightly increased. There could be a problem, however - what happens, for example, if the tool is very slightly rotating off its center line?

In this case, the tool effective diameter increases, removes more material, and the size will be too small.

When setting up the part and entering the initial offset, *you don't know what the measured dimension will be*. That is reason enough to make sure that the dimension will be a little *larger* - remember, larger external size can be recut, if necessary. If larger size is the objective, what should the initial radius offset setting be?

The dimensional tolerance is up to 40 microns over the nominal size, but not smaller than the nominal size. Now the all important question many CNC operators may ask:

Should I **increase** or **decrease** the 12.500 setting?

Various resources offer many different ways, often relating to the larger or smaller imaginary tool size, or to the length of actual travel (distance-to-go). While there is nothing basically wrong with these descriptions, they get a bit complicated when it comes to external vs. internal contour and a number of different tolerance ranges.

Yet, there is a very simple way to determine the offset adjustment. Instead of focusing on the tool, it focuses on the stock - do you want to *add stock to the contour* or do you want to *remove stock from the contour*?

> To ADD stock to the contour,
> INCREASE the current offset amount

> To REMOVE stock from the contour,
> DECREASE the current offset amount

The simplicity of these two definitions is that they apply equally to both *external and internal* contours.

For the initial setting, it may be a good choice to 'add stock' to the contour and make it slightly larger. Keep in mind that the tolerance of 40 microns is over the part width, but the offset setting has to selected per side.

> Cutter radius offset must always be applied per side

The initial offset setting can be anywhere between 12.500 and 12.520. If you start close to 12.520, you will have a better chance to recut if necessary. If you start closer to 12.500, you will not need to adjust the offset due to tool wear as often. A good choice for this example may be:

D51 = 12.505   ... *applies to manually generated programs*

This will add 10 microns to the overall width (5 microns per side), for the theoretical dimension of 153.010.

In summary, for this example, it is better to start with an offset setting being rather too large than too small.

Internal Offset Settings

Same as for the external contour, the programmer used the overall part dimension of 130 mm for calculating the contour points. This contour also has a positive tolerance, this time of 30 microns over the part width. In spite of the tolerance direction being positive, the approach to setting the initial offset is much different.

The objective of machining this internal contour is to make it slightly larger than its nominal size. While we were 'adding' stock to the external contour to make it larger, we must 'remove' stock for the internal contour to achieve the same result.

The default 'normal' setting for offset 52 would be:

`D52 = 10.000` ... *applies to manually generated programs*

10 mm is the cutter radius. Based on the two previous statements, we have to *decrease* the offset amount for the final machining. Initially, the setting can be larger, for example 10.200, so a measurement can be taken, difference found, and offset adjusted. Various scenarios will be presented later in this chapter.

In this case, the range of offset setting should be between 10.000 and 9.985. If you start close to 10.000, you will have a better chance to recut if necessary. If you start closer to 9.985, you will not need to adjust the offset due to tool wear as often. A good choice for this example may be:

`D52 = 9.990` ... *applies to manually generated programs*

This will add 20 microns to the overall width (10 microns per side), for the theoretical dimension of 130.020.

Conclusion

The main objective in preventing a scrap is to make sure the part can be measured, offset adjusted, and the cut repeated with the new offset setting. Although this method can be used in most cases, also keep in mind that the cutting condition for the recut are not the same as for the original cut. Width of cut will most likely be quite different and cutting forces will be different. With these variables, your final dimension may be slightly off and the surface finish may also suffer to a small degree.

Always use common sense when setting offsets. There is no magic involved, just plain understanding of the way cutter radius offset actually works.

USING OFFSET MEMORY TYPES

At the beginning of this chapter, three unique offset memory types have been described. The fastest way to find out which of the three types your control supports is to open the offset screen and count the number of columns used for offset data entry:

- **ONE column**   ... offset memory Type A
- **TWO columns**   ... offset memory Type B
- **FOUR columns**   ... offset memory Type C

Regardless of which type your control has, it is always the program that determines the offset number. In the next three illustrations, one for each type, a program segment with the offset number will be shown, along with the method of offset entry.

Geometry and Wear Offsets

Before discussing individual types of memory, it is important to understand the terms *Geometry* and *Wear* offsets. These terms are part of the offset vocabulary and apply to both tool length offset and cutter radius offset. They also apply to lathe applications.

Geometry offset is the offset initial setting. This setting is usually established during part setup, but it can also be set off-machine and transferred to the offset via electronic device, such as a chip reader that will load setting done away from the machine.

Wear offset does not always have to refer to the physical wear of the cutting tool. Wear offset is nothing more than an adjustment applied to the geometry offset. Having both geometry and wear offset means that the original settings are always available and any adjustment can be viewed independently.

Modern CNC lathes have both types as standard, but that is not the case for offsets used in milling applications. Lathe offsets are described in a separate chapter.

Offset Memory - Type A

The first thing to know about *Type A* offset is that both geometry and wear are combined. That means there is no distinction between them and only one offset entry is available. Because of the lower cost of controls using this feature, this is a very common offset type.

To work with *Type A* offset means that the initial offset amount is registered in the offset number specified by the program, but any wear adjustments are applied to this original setting. Let's look at a few examples.

The following illustration shows *Type A* offset with two settings - one for the tool length, the other for the cutter radius. This offset type is called *shared* offset:

Offset No.	Offset
...	...
06	-213.687
...	...
56	8.000

The above example represents offset settings for tool *T06*. Tool length offset 06 (*H06* in the program) is used for reference only, to show the difference in offset entry. As the program also calls for cutter radius offset *D56*, the suitable cutter radius is entered in offset 56. In the example, 8 mm radius represents a 16 mm end mill.

The program will include blocks for both types:

```
...
N61 T06
N62 M06
N63 G90 G54 G00 X-10.0 Y-10.0 S950 M03 T07
M64 G43 Z2.0 H06 M08
N65 G01 Z-6.75 F200.0
N66 G41 X6.0 D56
N67 Y120.0
...
```

Type A - Offset Adjustments

As both geometry and wear offsets are rolled into a single offset entry, there is only a single place where radius adjustments can be made, whether for the geometry or the wear offset - that is offset number 56. If the cutter radius offset has to be adjusted, it will be either increased or decreased.

→ Example 1 - Radius has to be increased by 0.035 mm:

The initial setting of 8.000 mm has to be changed by 0.035 mm up, to the new amount of *8.035* mm.

Offset No.	Offset
...	...
56	8.035
...	...

→ Example 2 - Radius has to be decreased by 0.02 mm:

The initial setting of 8.000 mm has to be changed by 0.02 mm down, to the new amount of *7.980* mm.

Offset No.	Offset
...	...
56	7.980
...	...

→ Example 3 - Current radius needs an increase of 0.01 mm:

The current offset setting is 7.980 mm and has to be adjusted further by increasing it by 0.01 mm. Using the **+Input** key (incremental offset) just enter 0.01 and press **+Input** - the new offset setting will be *7.990* mm.

Offset No.	Offset
...	...
56	7.990
...	...

In all cases, make sure you understand the adjustment method described earlier in this chapter.

Also note that in all three cases, the new offset represent correct radius setting, but the initial setting is not available.

Offset Memory - Type B

Offset memory *Type A* is a shared offset. Offset memory *Type B* is also a shared offset but with a difference:

On the offset registry display, there are *two* columns. Although similar to *Type A*, this offset allows the adjustment to be placed in the *Wear* offset column, while keeping the original *Geometry* offset intact. The same examples as for *Type A* will be used for *Type B*.

Type B - Offset Adjustments

→ Example 1 - Radius has to be increased by 0.035 mm:

The initial offset setting of 8.000 mm is now the *Geometry* offset that will remain the same, while the adjustment amount of 0.035 mm will be ***added*** to the current setting of the *Wear* offset.

Offset No.	Geometry	Wear
...	...	...
56	8.000	0.035
...	...	...

→ Example 2- Radius has to be decreased by 0.02 mm:

The initial offset setting of 8.000 mm is now the *Geometry* offset that will remain the same, while the adjustment amount of 0.02 mm will be ***subtracted*** from the current setting of the *Wear* offset.

Offset No.	Geometry	Wear
...	...	...
56	8.000	-0.020
...	...	...

→ Example 3 - Current radius needs an increase of 0.01 mm:

The current wear offset setting is -0.02 mm and has to be adjusted further by increasing it by +0.01 mm to be -0.010 as new setting.

Offset No.	Geometry	Wear
...	...	...
56	8.000	-0.010
...	...	...

For the last example, or for any update, use the **+Input** key (incremental offset) and just enter the offset change amount, such as 0.01 and press **+Input** - the new offset setting will be *-0.010* mm.

> **+INPUT key will always ADD the specified amount to the current setting**

Changing Geometry Offset Only

Some CNC operators ignore the *Wear* offset for the *Type B* or *Type C* and change the *Geometry* offset the same as for the *Type A*, where there is no choice. While this method is correct mathematically, it is not recommended for the offsets where *Geometry* and *Wear* amounts are separate.

Offset Memory - Type C

The three examples used for the previous two types will be used for the *Type C* as well, with very little actual change. The main difference between *Type B* and *Type C* is that the T-offset and the D-offset can have the same number, for example T06 and D06. Otherwise, the radius amount entry and adjustments for *Type C* are exactly the same as those for *Type B*:

Example 1 - Radius has to be increased by 0.035 mm:

Offset No.	H-offset		D-offset	
	Geometry	Wear	Geometry	Wear
...	...	...	...	...
06	-213.687	0.000	8.000	0.035
...	...	...	...	...

Example 2 - Radius has to be decreased by 0.02 mm:

Offset No.	H-offset		D-offset	
	Geometry	Wear	Geometry	Wear
...	...	...	...	...
06	-213.687	0.000	8.000	-0.020
...	...	...	...	...

Example 3 - Current radius needs an increase of 0.01 mm:

Offset No.	H-offset		D-offset	
	Geometry	Wear	Geometry	Wear
...	...	...	...	...
06	-213.687	0.000	8.000	-0.010
...	...	...	...	...

PROGRAMS FROM COMPUTER

So far, the examples have been all generated by manual programming methods. What are the differences - if any - when working with a program that had been generated by a computer software, one that commonly ends with letters CAM, such as Mastercam® or Edgecam®?

Mastercam Example

In a CAM software, the cutter radius offset is automatic. Here is a brief description of how Mastercam handles cutter radius. From the main selection, there are three selections related to cutter radius offset:

- **Compensation type**
- **Compensation direction**
- **Tip comp**

Compensation type has five selections, described shortly. *Compensation direction* can be set to *Left* for **G41** or *Right for* **G42**, and *Tip comp* applies only to spherical cutters and is not important here.

Depending on which of the five options the CNC programmer selects, the output will be different.

Computer

This is the default setting. In this case, the software will compensate for the cutter radius automatically by the radius of the selected cutter. There will not be any **G41** or **G42** in the program, and quick fine-tuning at the control is not possible. Programmers use this setting for roughing operations and for contouring where tolerances are not critical.

Control

This setting simulates - note the word *simulates* - a toolpath that corresponds to the manual methods of programming. The program output will generate **G41** or **G42** and the D-offset number. In this case, the typical D-offset amount will be set at the machine to the *actual cutter radius*. All changes are done by adding to or subtracting from the original radius or current offset setting.

➥ *Note that this is NOT the preferred programming method as it only simulates the toolpath*

Wear

Programmers working with computer assisted software prefer this method. Toolpath corresponding to the actual toolpath can be displayed and error detection is improved. Backplot or verification of the toolpath is accurate. This method will in effect duplicate the *Computer* method, with one major difference. There *will* be **G41** or **G42** output in the program, along with the selected D-offset number. In this case, the typical D-offset amount will be set at the machine to *zero*. All changes are done by adding to or subtracting from zero or current offset setting.

➥ *Note that this is the most common output from computer generated programs*

Reverse wear

As the name suggests, this selection will change all **G41** commands to **G42** and all **G42** commands to **G41** in the program output. It will *not* change the programmed toolpath direction. D-offset setting is the same as for the *Wear* option.

Off

Off selection disables cutter radius offset altogether. There will be no **G41** or **G42** in the program and the tool will follow the geometry as defined. Programmers use this option when the cutter centerline must follow the defined geometry. *Left* or *Right* compensation direction is irrelevant in this case.

> Always check what method of cutter radius offset has been used in the program

SOLVING PROBLEMS

A smoothly running program is always the objective of CNC programmers and operators. When something goes wrong, the most important step is to find the cause of the problem. Problems can be related to the program or to the setup, even as a combination of both.

Cutter radius offset is a complex subject that is not always understood in its entirety, and this lack of knowledge may hold back prompt solution. Fanuc controls have several alarm codes (error or fault codes) that relate specifically to errors related to cutter radius offset. These will be explained shortly, along with their explanation in Fanuc manuals.

Alarm Number 041

For the longest time, Fanuc alarm 041 (#41) has been the most common error related to cutter radius offset. This alarm happens during program run and displays the following message on the system screen:

> **INTERFERENCE IN CRC**

First, **CRC** is an abbreviation for *'Cutter Radius Compensation'*. The word 'interference' simply means that there is something in the way of making a successful calculation of the compensated toolpath. There is an explanation in the control manual:

> *Overcutting will occur in cutter compensation C. Two or more blocks are consecutively specified in which functions such as the auxiliary function and dwell functions are performed without movement in the cutter compensation mode. Modify the program. Although the cause of the alarm is specified, it by no means the only possible cause. Let's look at an expanded program used earlier, this time containing this error:*

```
...
N61 T06
N62 M06
N63 G90 G54 G00 X-10.0 Y-10.0 S950 M03 T07
N64 G43 Z2.0 H06 M08
N65 G01 Z-6.75 F200.0
N66 G41 X6.0 D56    (CUTTER RADIUS OFFSET IS ON)
N67 Y120.0
N68 X200.0
N69 S1500           (*** NO AXIS DATA ***)
N70 G04 P500        (*** NO AXIS DATA ***)
N71 Y6.0
N72 X-10.0
N73 G40 G00 Y-10.0 M09   (OFFSET CANCELED)
N74 G28 Z2.0 M05
N75 M01
```

Cutter radius offset mode starts in block N66 and is canceled in block N73. Keep in mind that all modern controls are of the 'look-ahead' type, where the control 'looks ahead' for the next axis direction. If the next block does *not* contain axis data (X and/or Y), the control will make one more attempt to look to the block after the one that did not contain axis data. If that block does not contain axis data either, the alarm is issued. If only one block does not contain axis data, the radius offset is calculated properly. This is definitely a program error.

Some control models may allow more than two consecutive blocks with no data, but the best is not to include more than one, if necessary.

Insufficient Clearance

An overcutting alarm will also occur when the cutter radius cannot fit into the space provided by the program. For example, you have a program in your hands where tool T03 that starts like this:

```
...
(T03 - 20 MM DIA END MILL)
N21 T03
N22 M06
N23 G90 G54 G00 X-4.0 Y-4.0 S900 M03 T04
M24 G43 Z2.0 H03 M08
N25 G01 Z-8.0 F250.0
N26 G41 X3.0 D53
N27 Y14.0 F125.0
...
```

The program looks fine, and no obvious error is immediately noticeable. Yet, there will be an alarm issued by the control system. Look at the following illustration:

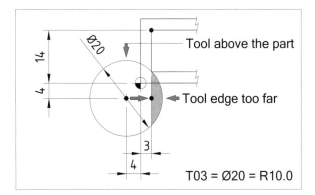

When the program and the illustration are connected, there are two issues that need addressing:

- **The tool is located *above* the lower corner of the part**
- **The tool reaches *past* the target point**

The first issue is a poor programming decision, but not an error that control can detect, therefore no alarm will result. Considering this issue separate from the second, when the tool plunges to the `Z-8.0` depth, it will actually remove material. In some cases it may be acceptable, in others not. Plunging in the Z-axis in this example is not a good choice - just look at how close the tool is to the corner of the part. By moving the Y-position lower (into Y-negative direction), the problem can be eliminated.

Much more serious problem is the second issue. From the `X-4.0Y-4.0` start point, the tool is moved in X-axis only, to `X3.0` in block N26. While the numbers may look alright at a glance, careful evaluation uncovers the error.

As the approach to the contour is along the X-axis only, the clearance in the Y-axis is less important, as long as it does not interfere with the part material (see earlier comment). Now, consider the actual tool motion in X:

- Start point   X-4.0
- Target point   X3.0
- Distance between start and target   7.0 mm
- Cutter radius   10 mm (D53)

These four numbers represent the only data the control can and will work with. What happens next involves some internal calculations to find *distance-to-go*, the length of motion to the target point:

Distance-to-go = 7.0 total distance - 10.0 radius = -3.0 mm

The actual *distance-to-go* would be negative 3.0 mm in block N26 - *if the control allowed it*. It is this last comment that is the reason for the alarm - the control will *not* allow it.

Some programmers and operators have difficulty understanding the reason for the alarm - after all, why the motion could not be in the opposite direction? The answer is in the alarm itself:

'Overcutting will occur in cutter radius compensation'

The actual message may be somewhat different, but its meaning is the same. *By moving the tool in the opposite direction, a collision is possible - therefore, overcutting can occur* - this is the core of the message.

The fact that there is no collision (overcutting) possible, is in the human knowledge, not in the CNC environment. Keep in mind that even if collision cannot happen in this case, it can happen in other cases (see the next section for details).

The solution is quite simple:

> **In G41/G42 mode, the *Distance-To-Go* must be greater than the cutter radius**

Corrected program will need *larger* clearance in block N23, to reflect the cutter 10.0 mm cutter radius:

```
...
(T03 - 20 MM DIA END MILL)
N21 T03
N22 M06
N23 G90 G54 G00 X-9.0 Y-9.0 S900 M03 T04
M24 G43 Z2.0 H03 M08
N25 G01 Z-8.0 F250.0
N26 G41 X3.0 D53
N27 Y14.0 F125.0
...
```

Again, consider the actual tool motion in X:

- Start point   X-9.0
- Target point   X3.0
- Distance between start and target   12.0 mm
- Cutter radius   10 mm (D53)

The actual *distance-to-go* will be 2 mm in block N26:

Distance-to-go = 12.0 total distance - 10.0 radius = 2.0 mm

The illustration shows the program correction that addresses both issues:

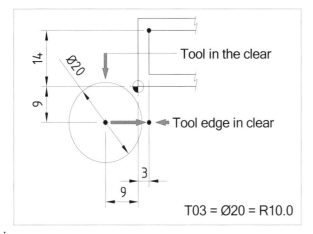

The root of the problem that was illustrated as insufficient clearance can also occur during actual cutting.

Cutter Radius Too Large

During a cut with cutter radius offset **G41/G42** in effect, the same *'overcutting will occur ..'* error will happen if any inside radius of the contour is smaller than cutter radius. The drawing is followed by the program:

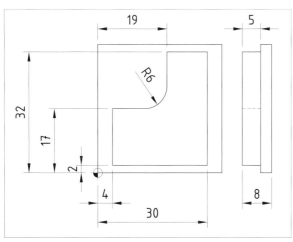

Only the relevant data are shown in the drawing. The program example will make the finishing contour only, using ⌀ 16 mm end mill:

```
...
N33 T07                (T07 = 16 MM END MILL)
N34 M06
N35 G90 G54 G00 X-10.0 Y-10.0 S900 M03 T08
N36 G43 Z2.0 H07 M08
N37 G01 Z-5.0 F300.0
N38 G41 X4.0 D57
N39 Y17.0 F170.0
N40 X13.0
N41 G03 X19.0 Y23.0 I0 J6.0
N42 G01 Y32.0
N43 X30.0
N44 Y2.0
N45 X-10.0
N46 G40 G00 Y-10.0 M09
N47 G28 Z-10.0 M05
N48 M01
```

The most important block of the program is the corner radius in block N41. This is an *inside* arc, which means the cutter radius will be fully located inside of the arc during machining. A careful study of the program itself will reveals no errors - the program is correct. Setting a ⌀ 16 mm end mill, registering the **D57 = 8.000** will precede the program run - now, the setup is correct.

During machining, the control will stop at block N39 and enters alarm condition. The message will again be along the lines of *'Overcutting will occur in cutter radius compensation'*. This is the same message as in the previous example, with a difference. The difference this time is that overcutting will really occur - *if the control allows it*. Internally, the control system still has no way of 'knowing' whether physical material exists or not - it only evaluates individual tool positions and how they relate to the cutter radius. This is consistent with the previous example.

What exactly happens here? Cutter radius offset on modern CNC systems is the *'look-ahead'* type. Its purpose is to detect errors such as the one in this example. While the program is correct, the tool selection is not.

Overcutting Error Message

Whenever the overcutting message occurs, the reason for it is always the same - the cutter radius cannot fit into the space provided in the program.

In the example, a tool that has a radius of 8 mm tries to cut an inside arc that has the radius of only 6mm. Since the larger tool cannot fit into the space provided, an alarm (error) condition is the result.

Without the alarm, overcutting would actually happen and the result would be scrapped part, broken tool, or both. Modern controls prevent this possibility from happening - the alarm may not be welcome, but it is better than the alternatives.

Understanding this problem is the key to being able to fix it. In this case, the drawing radius is given and the only solution is to select a cutter whose radius is smaller than the smallest inside radius of the contour. In the example, any cutter with a radius *smaller* than 6 mm can be selected - do not select a tool radius that is the same as the inside corner radius.

Hopefully, the verbal explanation was sufficient, but a few illustrations should also help. During machining, the vector (imaginary line) from the center of the cutter must always be perpendicular to the end point (target point) specified in the program - here are some examples:

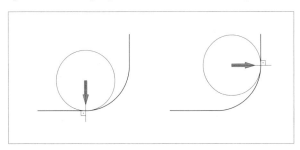

The above illustration shows how it should be.

The next illustration shows what would happen when the cutter radius is larger than the inside contour radius:

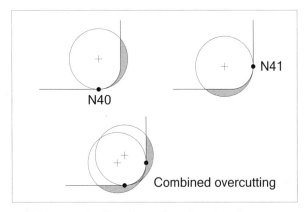

This example also shows that changing the program itself would not solve the problem. After all, the program is correct, only the tool selection is wrong. Always try to find the real cause of the problem - that is the only way to solve it.

Contour Start / End Error

Apart from the error of large tool radius being used for a small radius of the part, common errors relating to cutter radius offset can be found at the beginning or the end of the contour. These errors are generally related to starting the radius offset tool late or cancelling it too early.

Most programmers apply the Z-axis motion to the required depth, then apply cutter radius offset. Consider the following simplified program listing:

```
N1 G21
N2 G17 G40 G80 T01
N3 M06
N4 G90 G54 G00 X-15.0 Y-15.0 S1350 M03 T02
N5 G43 Z2.0 H01 M08
N6 G01 Z-5.0 F250.0
N7 X0
N8 G41 Y55.0 D51 F150.0
N9 X77.0
N10 Y0
N11 X-15.0
N12 G40 G00 Y-15.0 M09
N13 G28 Z-5.0 M05
N14 M01
```

The error is in blocks N7 and N8. A casual look may not reveal the problem, but when you evaluate the program carefully, you will see the error is in applying the cutter radius *too late*:

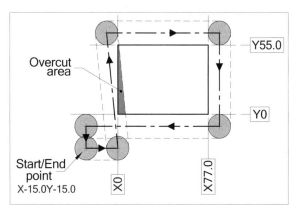

To correct the program is to use the **G41** command in block N7:

```
N6 G01 Z-5.0 F250.0
N7 G41 X0 D51
N8 Y55.0 F150.0
N9 X77.0
```

As the illustration shows, the result of such error - *if not detected* - is a severe overcutting that will certainly result in a scrapped part.

An error very similar to this one can also occur at the end of the contour, when the cutter radius offset is programmed too early. Using the same simple example illustrates the error:

```
N1 G21
N2 G17 G40 G80 T01
N3 M06
N4 G90 G54 G00 X-15.0 Y-15.0 S1350 M03 T02
N5 G43 Z2.0 H01 M08
N6 G01 Z-5.0 F250.0
N7 G41 X0 D51
N8 Y55.0 F150.0
N9 X77.0
N10 Y0
N11 G40 X-15.0
N12 G00 Y-15.0 M09
N13 G28 Z-5.0 M05
N14 M01
```

Block N11 contains cutter radius offset cancellation while the bottom edge is still being completed. This is an example of programming the cancellation *too early*:

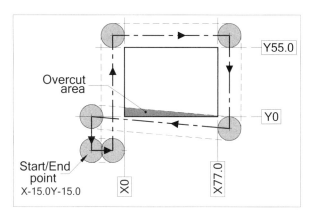

The program after correction will move the **G40** command to the next block N12:

```
N10 Y0
N11 X-15.0
N12 G40 G00 Y-15.0 M09
N13 G28 Z-5.0 M05
```

In this example, overcutting will also cause the part to be scrapped.

➧ *In both cases shown, there will be no alarm at the control, so it is very important to watch for these types of motions during part setup and program proving*

CHANGE OF TOOL DIAMETER

One of the main benefits of a program using cutter radius offset is that the CNC operator can use a *different* tool diameter than the one selected by the programmer. In principle, it means that if the offset amount represents the actual cutter radius, the part will be machined correctly. This is true, *in principle*.

There are at least two main factors to be considered before you change the cutter diameter:

- Available clearance
- Smallest inside radius of the part

Both have been covered in this chapter. To make the offset adjustment easier, you can use simple formulas that apply to both changes of cutter diameter - from larger to smaller and from smaller to larger. Changing tool diameter means selecting a new tool that is smaller or larger than the tool originally programmed.

New Diameter is Smaller

If the new cutter diameter is *smaller* than the one programmed, the change may affect some speeds and feeds, but it could also negatively affect width and depth of cut that is not easily changed in the program. On the other hand, change to a smaller cutter does not require any clearance change. At worst, the clearance may be a too large for efficient machining.

Evaluate this program example for T03:

```
...
(T03 - 20 MM DIA END MILL)
N21 T03
N22 M06
N23 G90 G54 G00 X-9.0 Y-9.0 S900 M03 T04
M24 G43 Z2.0 H03 M08
N25 G01 Z-8.0 F250.0
N26 G41 X3.0 D53
N27 Y50.0 F125.0
...
```

This is the final version of the corrected program listed at the bottom of *page 146*.

➧ *Example*

The part program calls for a 20 mm end mill diameter, but you only have a 16 mm diameter available. The adjustment process is a simple one - find the difference between the two diameters and adjust the current setting of the D53 offset.

If the previous offset amount was the cutter radius (D53 = 10.000), then the new setting will also be cutter radius - *the new one* - D53 = 8.000:

New radius = 8.000

Old radius = 10.000

Adjustment = 8.0 - 10.0 = -2.0

... therefore

Modified offset D53 = 10.0 + -2.0 = 8.000

If the previous offset amount was zero (D53 = 0.000), then the new setting will be the difference between the old radius and the new radius of each tool and will be *added* to the old offset setting:

New radius = 8.000

Old radius = 10.000

Adjustment = 8.0 - 10.0 = -2.0

... therefore

Modified offset D53 = 0.0 + -2.0 = -2.000

New Diameter is Larger

If the new cutter diameter is *larger* than the one programmed, the changes to be considered are much more important. Just as before, the change may affect some speeds and feeds, as well as width and depth of cut. The main impact of changing a smaller diameter to a larger one is - again - in the two factors mentioned several times - clearances and smallest inside radius.

Consider the earlier program example for T03:

```
...
(T03 - 20 MM DIA END MILL)
N21 T03
N22 M06
N23 G90 G54 G00 X-9.0 Y-9.0 S900 M03 T04
M24 G43 Z2.0 H03 M08
N25 G01 Z-8.0 F250.0
N26 G41 X3.0 D53
N27 Y50.0 F125.0
...
```

Now consider that the ⌀ 20 mm end mill has been replaced by a ⌀ 25 mm end mill. What changes do have to be made? The first change relates to the cutter radius offset setting in the control.

If the previous offset amount was the cutter radius (D53 = 10.000), then the new setting will also be cutter radius - *the new one* - D53 = 12.500:

New radius = 12.500

Old radius = 10.000

Difference = 12.5 - 10.0 = +2.5

... therefore

Modified offset D53 = 10.0 + 2.0 = 12.500

One other change is required to prevent control alarm. The distance between start point X-9.0 and target point X3.0 is 12 mm. That is not enough to accommodate 12.5 mm offset, and the X-9.0 must be changed by more than 0.5 mm, for example to X-11.0:

```
N23 G90 G54 G00 X-11.0 Y-9.0 S900 M03 T04
```

If the previous offset amount was zero (D53 = 0.000), then the new setting will be the difference between the old radius and the new radius of each tool and will be *added* to the old offset setting:

New radius = 12.500

Old radius = 10.000

Difference = 12.5 - 10.0 = +2.5

... therefore

Modified offset D53 = 2.500

Units Conversion

The program calls for a ⌀ 20 mm end mill, but all you have is a ⌀ 0.75 **inch** end mill. It is always more complicated to switch imperial tools for metric or vice versa. In both cases, convert the new tool to the units of the program - in this case, inches to millimeters:

0.7500 inches = 19.050 mm

For all practical purposes, you are now working with a 19.05 mm end mill diameter. If the offset setting for the 20 mm end mill was 10.000, the offset for the 19.05 mm end mill (0.75 inch) will be 9.525.

15 TOOLS FOR CNC LATHES

One of the main differences between CNC milling and turning machines is the application of tools. While tools for milling applications are of the *rotary* type, standard turning tools are *stationary*. Rotary tools can be part of so called mill-turn machines, but they are for milling, not turning or boring.

In this chapter, tools for the basic two-axis configuration of CNC lathes (rear type) will be used.

PROGRAMMMING LATHE TOOLS

When you receive a program for CNC lathes, the tool number will appear slightly different than the similar tool number for milling - it will have *four digits*, rather than two:

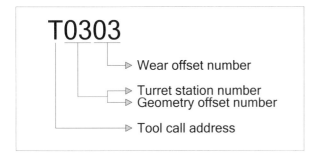

The T-address is also the command for automatic tool change - *there is no* **M06** *in the lathe program*.

The first pair identifies the turret station number, in which the tool is located. It *also* identifies the **GEOMETRY** offset number for the tool. This is the offset where the measured dimensions between the tool tip and part zero are entered.

The second pair identifies the **WEAR** offset number for the current tool. In most programs, the wear offset will have the same number as the geometry number, but that does not have to be the case in every lathe program. Machining that requires two or more wear offsets for the same tool will have different numbers, for example:

`T0313`, `T0323`, *etc.*

You may also see in some lathe programs that the first tool may have the wear offset zero, for example:

`T0100`

This programming method requires a short explanation - tool 01 is used with no wear offset.

Zero Wear Offset

Zero wear offset is sometimes programmed for the first tool only, but - *if used* - it is often programmed for all tools in the program.

The geometry offset is set from the machine zero position, which also means the first tool starts from machine zero position. Take this program beginning:

```
N1 G21 T0101
N2 ..
```

The moment this block is processed, the CNC system will activate two settings simultaneously:

- **Sets the current GEOMETRY offset 01**
- **Applies the current WEAR offset 01**

If the wear offset is set to zero, there will be no problem. If the offset is negative or set to a very small positive amount, there will also be no problem. The problem may occur if the positive wear offset is too large - in this case, *overtravel* will occur. This is one reason why many lathe programs start with:

```
N1 G21 T0100
N2 ..
```

There is another reason to start with zero wear offset.

Feedrate Error

An error code informing the operator that *feedrate is missing* can come up as an unpleasant surprise the moment the *Cycle Start* button is pressed for the first part. If the block `N1 G21 T0101` is programmed and the wear offset is stored with an amount other than zero, the *'missing feedrate'* alarm is quite possible. The reason for this unexpected problem is the system setting of the default motion - is it **G00** or **G01**?

151

In the absence of a motion command, usually at the program beginning, the control system will use the default selection stored in the system parameters. For safety reason, many controls are set to **G01** mode, linear interpolation that requires feedrate. In order to perform the adjustment of the wear motion, the tool has to move. If the motion is in **G01** mode, it also requires feedrate. If no feedrate is stored in the system, alarm will be the outcome, regardless of whether the wear offset stores a positive or negative amount.

Programming feedrate with the tool call would not only look a bit absurd, it may not work anyway. Feedrate on lathes is in *mm/rev* or *in/rev*, and a rotating spindle is required. That is not the case at the top of the program.

Prevention

To prevent either of the two problems from happening, it has to start at the programming level. The key is to start with zero wear offset and apply the offset during the first motion, after the spindle is rotating:

```
N1 G21 T0100
N2 G96 S350 M03
N3 G42 G00 X.. Z.. T0101 M08
N4 ..
```

This method will eliminate both problems.

The next chapter *Lathe Offsets* describes practical settings and applications in details, starting on *page 161*.

TOOL NUMBERS

Even the last example shows that more than a single T-address can be entered for one tool. In fact, two or three (even more) are not unusual. Take this simple program that does no more than cuts a front face:

```
N1 G21 T0100
N2 G96 S350 M03
N3 G41 G00 X185.0 Z0 T0101 M08
N4 G01 X17.0 F0.015
N5 G00 W2.5
N6 G40 X300.0 Z50.0 T0100
N7 M01
```

Note that there are *three* entries of tool 01. What would most lathe operators do, if the tool referred to in the program were in turret station 05, for example? Some would remove the tool from station 05 and place it in station 01. Others would change the three numbers in the program. There is nothing wrong with this method, providing *all* *T01..* are changed to *T05..*.

Forgetting even a single one will cause a tool change, which may happen in a dangerous place.

> When editing tool numbers,
> make sure all are edited correctly

TOOL TURRET

Unlike machining centers, CNC lathe do not have a magazine that can hold dozens or even hundreds of tools. Lathe tools are placed into a turret, which can usually hold eight, ten, twelve, or fourteen tools, depending on the size of the machine.

Turret is designed in such a way that it can hold all three types of lathe tools - external tools, internal tools, and tools that are used only at the machine centerline.

The arrangement of tools is very flexible in order to provide the best combination for any machining job. A common 12-station lathe turret with typical tool holders is shown below:

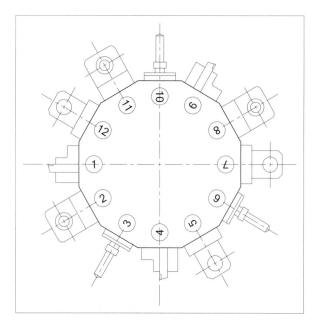

Both right-hand and left-hand tools can be setup in the turret. CNC turning centers that support milling also include motorized stations on the turret, but the basic design is the same. Turning centers with two spindles will also have two turrets, one oriented right to left, the other left to right. Overall design of each turret will be the same.

TOOL HOLDERS

Selecting a suitable tool holder is an important part of setup, regardless of the CNC machine. For lathes, most tool holders are design to accept indexable carbide inserts. Only a few centerline tools are made of HSS.

Standards

Companies that manufacture tool holders and inserts - and there is a large number of them - compete with each other worldwide. Some of the big names (in no particular order) are Kennametal®, Sandvik®, Iscar®, Seco®, Valenite®, Widia®, Sumitomo®, Stellram®, and a number of others. Although there are variances in several minor areas, all companies adhere to certain standards.

Without standards, it would be impossible to match the tools with CNC machines. There are two standards, one for the metric (ISO) system and another one for the imperial (ANSI) system. Only United States and Canada (to some degree) use the ANSI system, all other countries using the ISO system. Even in the two North American countries, emergence of ISO system is noticeable. **ANSI** is an abbreviation for *American National Standards Institute*. **ISO** is an abbreviation for *International System of Standards*. For commercial reasons, there is very little difference between the tools, with the main exception of the units (*mm* versus *inches*).

Identification

Every tooling company uses catalogues to offer their products to the customer. These catalogues should be of interest to both CNC programmers and operators. Apart from the listed tools, an important part of such catalogues is the technical or reference section. There, a table that typically spans over two large pages lists the naming and numbering conventions of cutting tools and inserts (described in the next section).

Tool identification is quite consistent between manufacturers, but the standard allows minor differences to identify features unique to a particular manufacturer.

Note that the tool identification is closely related to the identification of an insert that can be used with the tool. Tables in the tooling catalogues are commonly presented in two versions, one for the ISO system and the other for the ANSI system. It would be pointless to include such a table here, so a typical example of tool holder identification will be used for reference.

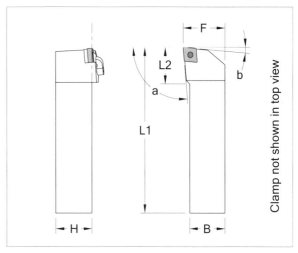

The tool illustrated above is probably the most common turning tool in CNC lathe work. The linear dimensions are self-explanatory - the angle *a* is the angle at the front, called the *lead angle*, the *b* angle is the *back angle*. Depending on the actual tool design, there may be some additional definitions.

The actual catalogue may identify this tool as:

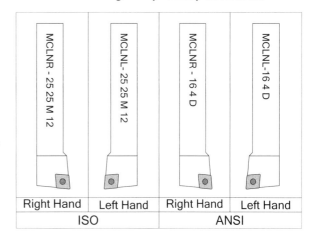

Each letter and each digit has a specific meaning, often with many alternatives. The example presented here represents an accurate description of a particular tool from a particular tooling company; the best approach to tool and insert identification is the tooling catalog from the manufacturer of the tool you will be using.

Tool features such as overall length, length of each side, important angles, hand of the tool, even the shape of the insert are included in the tool designation. Once you understand the basic concept, what appears as a cryptic code now, will make a lot more sense.

ISO and ANSI

Have a look at the initial tool holder identification, illustrated on the previous page. When this tool type is defined by ISO and ANSI standards specifications for a right hand tool, many similarities and some differences (other than the units) can be compared:

ISO

Both shank dimensions B and H are given directly in millimeters and so is the length of the insert edge. Letters *M, C, L, N R*, and the second *M* represent a selection from an identification chart.

MCLNR - 25 25 M 12

	Description	Actual data
M	Insert holding method	M = Top clamp
C	Insert shape	C = 80°
L	Lead angle	L = 95°
N	Insert clearance angle	N = 0°
R	Hand of tool	R = Right hand
25	Shank height	B = 25 mm
25	Shank width	H = 25 mm
M	Tool length	M = 150 mm
12	Insert edge length	12 = 12 mm

ANSI

ANSI identification is a bit shorter. Although there are two numbers, they do not provide any direct dimension: 16 means **16 x 1/16 = 1.0** inch. The shank is assumed to be square. ANSI uses inscribed circle to dimension the insert size: 4 means **4 x 1/8 = 0.5** inches.

MCLNR - 16 4 D

	Description	Actual data
M	Insert holding method	M = Top clamp
C	Insert shape	C = 80°
L	Lead angle	L = 95°
N	Insert clearance angle	N = 0°
R	Hand of tool	R = Right hand
16	Shank size (16 x 1/16)	B = 1.0" H = 1.0"
4	Insert inscribed circle Ø	4 = 4/8 = 0.5"
D	Shank tool length	D = 6 inches

Some options can only be interpreted from a chart.

Metric and Imperial Specifications

As expected, the metric specifications are easier to interpret than the imperial specifications. In both systems, there are several entries that cannot be interpreted without some sort of a *key*. The key in this case is located in a chart that can be found in any tooling catalogue. Note that the first letter (*M* in the example) may have different meaning for each manufacturer.

Insert shapes will be described in the next section.

Gage Insert

A tooling catalogue that lists various cutting tools for CNC turning, contains one important column - this column is identified by the heading *gage insert*. Here is a typical entry from a metric catalogue:

MCLNR2525M12   CN..120408

The catalogue listing uses no spaces in the tool designation, but the actual tool usually does. The **CN..120408** is an insert identification. As you will learn in the next section, this insert has a radius of 0.8 mm. It has been selected as the gage insert, which means tool length of 150 mm is measured over the insert with 0.8 mm radius. Changing the insert to one with a smaller or larger radius will not make the length 150 mm.

Catalogue listing for imperial units will be similar:

MCLNR164D   CN..432

The measurement of the 6 inch tool length is accurate only if the insert has a corner radius of 1/32 = 0.0313.

In both examples, the **CN..** represents the insert designation with **CN** being fixed first two letters, followed by other two letters that vary, depending on the exact insert selection. The letter *'C'* identifies the insert as 80° diamond shape.

Other Tool Holders

There are many tool holders available for CNC lathes. For general reference, the following illustrations will show the shapes of various lathe tools based on selections available in a typical tool catalogue. Note that many designs are unique to a particular manufacture, even patented. Tool identifications have been omitted, as they also vary to some extent from one tooling company to another.

Each illustration will show tool holders that belong to the same general category.

The first group of tools represents a number of tool holders for external machining (turning). Out of the twelve holders below (right hand, left hand, and neutral), only five different inserts shapes will be used - 2 of C, 5 of D, 3 of V, 1 of S, and 1 of T. This ratio roughly represents the usage of each tool shape.

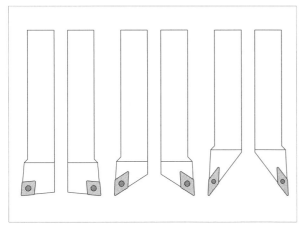

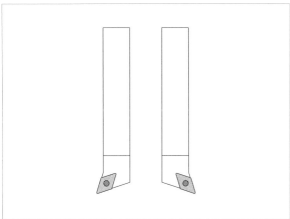

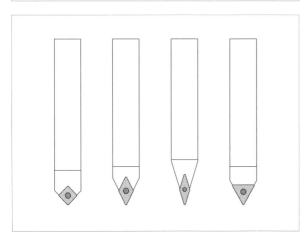

Boring Bars

Boring bars use the same inserts as turning and other tools, and their identification is very similar to the one described so far. One additional feature of boring bars is the bar shank diameter identification.

➤ ISO Example

A metric boring bar identified as **A25RMCLNR12** has a ⌀ 25 mm shank - *A25R* identifies the bar as a steel bar with through coolant (designation *A*), *25* is the bar shank diameter in millimeters, and the letter *R* is the bar length of 200 mm (taken from the identification chart). The remainder of the identification number is the same as for external turning tools.

➤ ANSI Example

An inch boring bar identified as **A16TMCLNR4** has an ⌀ 1.0 inch shank - *A16T* identifies the bar as a steel bar with through coolant (designation *A*), *16* is the bar shank diameter in 1/16ths of an inch: **16 x 1/16 = 1.0** inch, and the letter *T* is the bar length of 12 inches (taken from the identification chart). The remainder of the identification number is the same as for external turning tools.

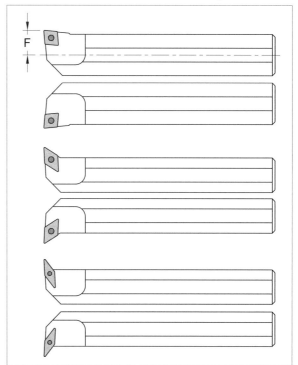

Note the dimension *F* - it is measured between the bar center line and the insert tip. This dimension is important in determining minimum diameter depth.

TOOLS FOR CNC LATHES

Threading and Grooving Tool Holders

Samples of grooving and threading tools are shown below. Boring bars accept all short grooving and threading inserts. The last illustration shows two types of deep grooving tools which can also be used for part-off. The actual tool design varies from one manufacturer to another, but the concepts remain the same.

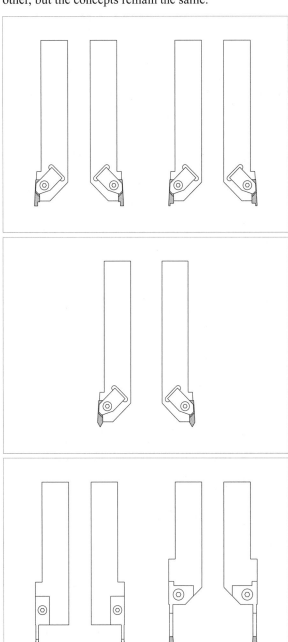

CUTTING INSERTS

A number of references have been made to the inserts that are used in various cutting tool holders. This section will address important details that every CNC programmer and machine operator should know.

Many tools designed for milling use carbide inserts. Many are unique, and some are even the same for milling cutters and turning tools. Usually, there is a lower lower focus on inserts in milling than similar inserts in turning. The main reason is that lathe inserts influence the final size of the part much more than in milling operations.

Insert Shape Identification

Standard insert identification is the same for ISO and ANSI inserts. Just like **MCLNR2525M12** or **MCLNR164D** are the most common external tool holders, **CNMG120408** or **CNMG432** are the most common inserts. The first letter indicates the angle of the insert used for cutting:

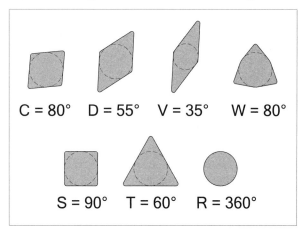

These inserts are only a few of the most common insert shapes. The shapes in the upper row are used more commonly than those in the lower row, with the exception of the triangular insert.

Note that both types *C* and *W* have an 80° cutting angle. The *C-type* is stronger, but has only two cutting edges on one face, whereby the *W-type* is not as strong but has three cutting edges on one face. Depending on the insert geometry, one or both faces can be used for machining.

Insert identification follows established standards, and ISO and ANSI are both very similar. The *'C'* insert shape will be used for the examples, as it fits the tool holders described in the earlier sections.

You will notice that the two specifications described here are practically the same - apart of the coding itself. First, look at the insert description, then the actual meaning of each specification for the given insert. Both insert samples are illustrated:

CNMG - 12 04 08

	Description	Actual data
C	Insert shape	C = 80°
N	Relief angle	N = 0°
M	Tolerance	M = Based on a table
G	Insert style (type)	G = Double sided+Hole
12	Insert size	12 = 12 mm
04	Insert thickness	04 = 4 mm
08	Corner radius	08 = 0,8 mm

CNMG - 432

	Description	Actual data
C	Insert shape	C = 80°
N	Relief angle	N = 0°
M	Tolerance	M = Based on a table
G	Insert style (type)	G = Double sided+Hole
4	Insert size	4 = 4/8ths = 0.5"
3	Insert thickness	3 = 3/16ths = 0.1875"
2	Corner radius	2 = 2/64ths = 0.0313"

Compare the two methods of insert identification - look at the common features of the two types and find the features that are different.

Common Features

Even a very quick look at the ISO and ANSI specifications will show that the meaning of the initial four letters is identical for both standards:

- **C Insert shape** - *see previous page for other selections*
- **N Relief angle**

Depending on the insert manufacturer, a large number of relief angles can be incorporated into their inserts. There is one neutral (N=0°), the others are positive. The type P=11° is one of the most common.

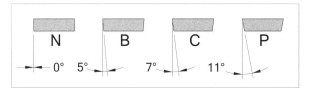

- **M Tolerance**

The insert tolerance is listed in the table provided by the insert manufacturer. This table is quite comprehensive, but for most lathe operations, it does not present a major area of focus.

- **G Style of insert**

The fourth letter tells a lot about the insert itself. Identification indicate presence or absent of a hole, single or double sided design, straight hole or a hole with countersink, and other details. The illustration shows only some of the styles as a general reference. The letter 'G' in the example identifies the insert as double sided, with a hole and built-in chip breaker on both sides of the insert.

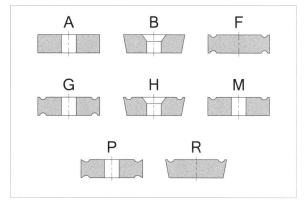

In both ISO and ANSI specifications, the numbers define the size of the insert.

ISO Unique Identification

In the ISO identifications, there are no ambiguities. In case of the **CNMG 12 04 08**, the first pair of numbers **12** is the edge length rounded to an integer. The second pair **04** is the insert thickness and the last pair **08** is the corner radius. All dimensions are in millimeters, and numbers lower than 10 must be prefixed with a zero (4 = 04).

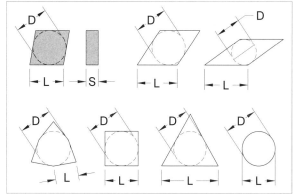

For the *CNMG* type of insert, the edge length *L* and the inscribed circle *D* are very close because of the narrow 5° front angle. The distance is much large for the types *D* or *V*. For example, 55° insert **DNMG 15 04 08** has the actual edge length 15.5 mm (no radius applied).

The corner radius *R* 08 is 0.8 mm and applies to the radius between two cutting edges:

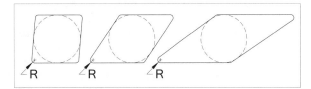

ANSI Unique Identification

The insert identification letters, such *CNMG, DNMG,* and others are the same for ANSI system as for the ISO system. The difference is in the numbers that follow. The size of insert specified as 432, for example, has three parts:

Digit 1	Inscribed circle (IC) diameter in 1/8ths
	ex.: **4**/8ths = 0.5 inch IC
Digit 2	Insert thickness in 1/16ths
	ex.: **3**/16ths = 0.1875 inch thickness
Digit 3	Corner radius in 1/64ths
	ex.: **2**/64ths = 1/32 = 0.0313 inch radius

Grades, Coating and Chip Breakers

Virtually all inserts are listed in catalogues as having been of certain grade. Grade is generally associated with the insert coating. Their purpose is to allow cutting or harder materials and extend overall tool life. These features are non-standard and tooling companies use their own special identification to both.

Cutting inserts are also supplied with or without a built-in chip breaker. Chip breakers deform the chip buildup to the point of breaking it. Those inserts without a built-in chip breaker use a special external shim that is located between the clamp and the top surface of the insert. For some materials, a chip breaker is not needed.

Chip breakers that are built into the insert have been pressed at the time of manufacturing, usually along with various ridges and dents, to further improve the chip forming. The best resource of the most suitable insert and its characteristics is the tooling company catalogue, in most cases available either on-line only or on-line and downloadable as PDF file (requires free *Adobe Reader®* utility - available from *www.adobe.com*).

CHANGING AN INSERT

During production, a worn out cutting insert will have to be changed for a new insert to continue machining. There are two issues to be considered:

- **Insert that has to be replaced with the same radius**
- **Insert that has to be replaced with different radius**

During CNC machining operations it is quite normal to replace worn-out cutting inserts for a specific tool with at least one new insert or a whole set of new inserts. When it comes to changing inserts, it is important to understand that no two inserts are exactly the same when it comes to precision machining. Even if the replacement insert has the same parameters as the original insert, wear offset adjustment will most likely be necessary.

Offset adjustments in this case are rather minor and an experienced CNC operator will handle them with ease.

Corner Radius Change

It is also not unusual to replace one insert with another of the same size, but with a *different corner radius*. A smaller radius may be necessary for a corner at a shoulder, a large radius may be necessary for heavier cuts or better surface finish.

When replacing the same basic insert with a different corner radius (tool nose radius), the originally set geometry offset is no longer valid and has to be adjusted either directly or via wear offset setting. The following illustrations show the *theoretical* amount of change required for *C, D* and *V* shapes of insert in both ISO and ANSI units.

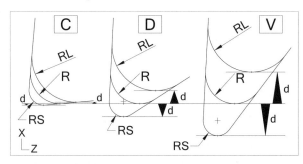

In the following tables, '*d*' is the positive of negative difference from the gage insert with the radius '*R*'. '*RL*' and '*RS*' are larger or smaller radius respectively, in relation the to radius '*R*'. Only the X-dimensions are shown, and the '*d*' is amount per side, *not* on diameter.

In practice, the wear offsets for both X and Z axes will need adjustment. The amounts shown are theoretical.

ISO Example

Three standard shapes with three standard radiuses:

	R	RS	RL
All styles C, D, V	0.08	0.04	1.2
'd' for C-style	0.0	-0.041	+0.041
'd' for D-style	0.0	-0.345	+0.343
'd' for V-style	0.0	-0.843	+0.838

ANSI Example

Three standard shapes with three standard radiuses:

	R	RS	RL
All styles C, D, V	0.0313	0.0156	0.0469
'd' for C-style	0.0	-0.0016	+0.0016
'd' for D-style	0.0	-0.0136	+0.0135
'd' for V-style	0.0	-0.0332	+0.0330

In both cases, notice that the error increases significantly when the tool angle decreases. Illustration and table show change of inserts in the same tool holder.

TOOL CHANGING

One important program feature is the tool clearance in Z-axis during tool changes. For safety reasons, always check if the clearance is sufficient for the tools actually used. The programmer may assume different tool length than the one in the actual setup. Although CNC lathes are designed with sufficient clearances for standard tools, double checking is always worth the time.

During tool change, all tools mounted in the turret are in the working area of the machine. If all tools have approximately the same length, common clearance from the front face (Z0) of 25 mm (1.0 inch) or more is sufficient. The problem may arise when a long tool follows a short tool. This situation can be compounded when the long tool is located in the next turret station to the short tool. Alternating short-long-short tool is always advisable. The following illustration shows an example of a long boring bar (T02) that follows a short tool (T01). If the program contains sufficient clearance for the longest tool, all tools will have safe clearance.

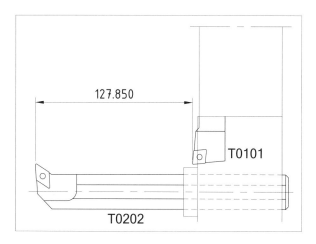

The following program segment shows the last few blocks of the turning tool and a tool change to the boring bar.

```
(TURNING TOOL)
...
N34 G40 G00 X250.0 Z25.0 T0100
N35 M01

(BORING BAR)
N36 T0200
N37 G96 S300 M03
N38 G00 G42 X.. Z.. T0202 M08
...
```

The clearance of 25 mm (1.0 inch) in block *N34* may not be safe during tool change. The block should consider the boring bar overhang of about 128 mm (about 5 inches). Changing the block N34 by rounding the difference of **127.85 - 25 = 102.85 = 110.0 mm** should provide safe tool change. There is no need to be exact, as long as the clearance accommodates the bar overhang:

```
N34 G40 G00 X250.0 Z110.0 T0100
```

Tool clearance in the X-axis is also important, but to a lesser degree. It usually takes place above the part diameter. Just watch for possible contact of the tool with the tailstock during tool changes.

Alternating long and short tools in the turret will increase the clearance behind the part. In some cases, another option may be useful - controlling the direction of the turret rotation. If the machine supports this feature, the manufacturer will provide two M-functions, one for clockwise and one for counterclockwise rotation. Most turrets will rotate automatically, using the shortest distance between tools. In the absence of the special functions, arrangement of tools in the turret can also be controlled.

The remainder of this chapter is related to cutting tools and the chuck, as well as lathe jaws.

LATHE CHUCK

When working with very long and very short tools from the same program, there is possible danger that the long tool may hit the machine wall when the short tool is very close to the chuck. In this case, a change in setup will most likely be required.

Another tool motion to watch is when the tool cuts very close to the jaws. In this case, the best way is to work in single block mode and make frequent use of the feedhold button, while checking the *distance-to-go* amount on the control monitor.

Other two items relate directly to the chuck.

Chuck Clamp and Unclamp

All lathe chucks can be set to hold the part on the outside or on the inside, suing suitable jaws. If the part is clamped outside, the direction of clamping is *towards the centerline*. If the part is clamped inside, the direction of clamping is *away from the centerline*. For this purpose, the operation panels contains a two-way switch that can be set to *chucking outside* or *chucking inside*.

Chuck Pressure

Chuck pressure indicates the amount of force placed upon the chuck jaws to hold a part. Initial pressure of a chuck is set to 'normal', which is a setting the manufacturer considers safe and applicable to most jobs.

Most CNC lathes come with a chuck pressure gage, indicating safe limits. Although it may be tempting to decrease the chucking pressure in order to prevent part deformation, use care - there are usually other options available. Using mandrels and wrap-around jaws are just two of such options.

> **Improper setting of chuck pressure can be hazardous**

When facing to centerline or close to it, the spindle rotates at a very high speed and can easily reach the highest speed available. To clamp the speed down, the program should contain the **G50** command, for example, **G50 S2500**. The spindle will increase gradually to 2500 r/min and remain at that speed to the end of cut. This feature applies to **G96** constant cutting speed mode only.

SELECTION OF JAWS

Interchangeable jaws on a lathe (manual or CNC) belong to two general groups:

- Hard jaws
- Soft jaws

As the descriptions imply, hard jaws are used in situations where the grip strength and rigidity are of prime concern. Soft jaws are typically used for machining operations in their final stages, such as semi-finishing or finishing. Regardless of the types, jaws should always be set so they run concentric. They should also be attached to the chuck safely always using all bolts. Always watch for jaws that protrude too far out.

Hard Jaws

For most lathe roughing work, hard jaws are the best choice, as they provide strong grip. These jaws have serrations that assist in holding the part firmly, but also make an indent in the material. Their depth is fixed.

Soft Jaws

Boring soft jaws is one of the responsibilities of the CNC lathe operator. There are many methods that can be used to bore jaws (manually, MDI, program,...). The most important part of boring soft jaws is to make correct diameter, - see illustration of the effects with large, exact and small bore sizes.

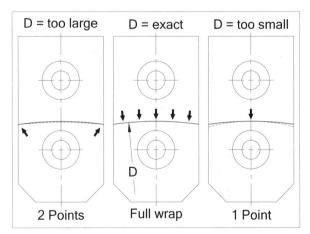

Other important criteria of boring soft jaws is their depth and internal corner radius. Either one set incorrectly will affect setup safety and part quality.

16

LATHE OFFSETS

Offsets applied to CNC lathe applications serve the same purpose as offsets applied to CNC milling applications, applications for wire EDM, router, etc. Due to the individual machine designs, there will be some expected differences. The focus of this chapter is on offsets and their applications in CNC turning, applied to the two main axes, X and Z for a rear lathe orientation.

In order to understand various offsets and their applications on CNC lathes, you have to understand the basic machine layout and geometry of common cutting tools. Although there are many different lathe designs in industry, the principles shown on a standard two-axis lathe apply to all other designs as well, including multi-axis and multi-purpose turning centers, including those with live tooling, sub-spindle, and other features of modern technology of turning centers.

LATHE GEOMETRY

The word *'geometry'*, as it relates to any CNC machine tools, is generally used to describe the relationship between two reference points. All relationships between the *cutting tool* (tool tip reference point) and the *part zero* (program zero reference point) have to be set at the control before any machining can be done and the CNC program processed. Although each CNC machine has its own unique features in terms of axis definition and orientation, the machine geometry is always defined in consistent way.

The basic geometry are two dimensions between the face of the headstock and the turret face:

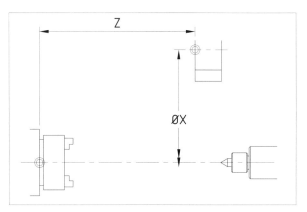

These are factory dimensions and cannot be changed. Note that both the chuck and the tailstock are ignored. A chuck is attached to the headstock face, and different chuck models can be attached to the same headstock. The same applies to different tailstock sizes that can be mounted on the lathe bed.

Both dimensions play a significant role in some very special situations, such as when a complete machine model is used for simulation. In general practice, and in smaller shops in particular, these dimensions are used mainly for reference purposes.

Machine Zero

Machine reference point or *machine zero* or *home position*, they all refer to the same location at the machine - its *origin*. This origin is defined by the machine manufacturer and for normal purposes, this location is *fixed*.

In theory, all tool motions programmed should originate from this fixed machine zero location. For practical reasons, this is not the case, but the concept behind the subject is important.

On CNC lathes (including both *front* and *rear* lathe types with two+ axis models), the machine zero is located at the furthest distance from the lathe centerline in each axis (X-origin of the lathe) and also at the furthest distance from the headstock face (Z-origin of the lathe). These dimensions are factory fixed, are known, and can be verified.

From the illustration at left, the turret position indicated machine zero for a rear type CNC lathe. It is important to understand, that any attempt at a tool motion further away from machine zero position will result in an overtravel condition.

Even the machine zero is not very practical for general tool setup, and programmers do not use it. The only reference in the program to machine zero is the **G28** command - machine zero return:

`G28 U0 W0`   ... *incremental specification*

or

`G28 X.. Z..`   ... *absolute specification*

Part Zero

Setting the zero for a lathe part has been covered earlier in a separate chapter *(see page 71)*. Most CNC lathe setups benefit from the *Z0* at the front face of the finished part. The *X0* location is the same for all lathe work - it is always the spindle centerline.

All initial lathe offsets are based on the part zero. The offsets of this kind are called *Geometry* offsets, and establish the distance and direction between the part zero and the tool reference point. This reference point is nothing more than a fixed point on the cutting tool itself. You may think of it as a tool zero point.

Tool Reference Point

Strictly speaking, the tool reference point can be anywhere, just as a part zero can be (almost) anywhere. On the other hand, the most practical point location is such that can be easily used during setup. This is no different from milling tools, and using a point of reference at the tool tip makes most sense. There is a difference in a different area, and that is the *shape* of the cutting tool. While all milling tools can be represented by a circle, this is not the case for lathe tools. More on lathe tools, see *page 151*.

Types of Cutting Tools

CNC lathe tools can be divided into three categories:

- **External tools**
- **Internal tools**
- **Centerline tools**

Apart of their physical orientation, external and internal tools cover common machining operations such as turning, boring, grooving, threading, and others. Centerline tools are those that do all their work at *X0* - they include drills, taps, reamers, and similar tools. The following three illustrations show the typical tool reference point for each cutting tool type (schematic illustrations only):

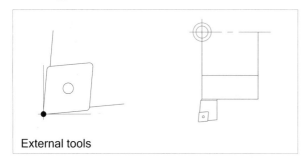
External tools

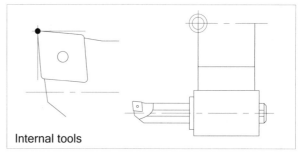
Internal tools

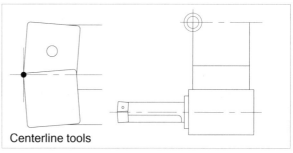
Centerline tools

Now, that the two reference points have been identified, it is important to see how they work together. Each on its own is important, but their mutual relationship is the real objective.

In the part program, there is no direct reference to either of these points. As a CNC lathe operator, you have to *know* where the part zero is and where the tool reference point is. A setup sheet or a tooling sheet may be used for this purpose, a standard company policy may be another resource. Interpreting a programmed dimension also helps. For example, if a dimension from the front face to the nearest shoulder is specified as 20 mm in the drawing, and you will see `Z-20.0` in the program, you will have a key setup related information.

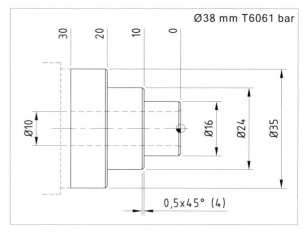

This drawing will also be used with tolerances later.

WORK SETUP

Setting up the relationship between part zero and the tool tip (reference point) is done with the **GEOMETRY** offset of the control system. The program tool definition is a four-digit number that identifies:

- Turret station number
- Geometry offset number
- Wear offset number

There is no other reference to the tool setup in the program. Interpretation of the four digits is as follows - tool T03 is used as an example:

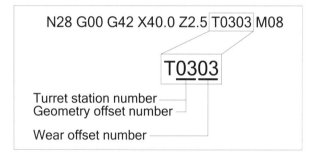

The first pair of digits applies to both the turret station number and the *geometry* offset number. The second pair is the number of the *wear* offset.

Definitions

Geometry offset is the distance measured from the tool tip reference point to part zero along the X-axis as a diameter and along the Z-axis as a distance. In both cases, the direction will be negative for rear type CNC lathes and turning centers.

Wear offset is used for fine adjustments from the original geometry settings. It is used to maintain sizes, including tolerances.

Note - Making all adjustments in the geometry offset is not recommended. Mathematically, the results are the same. Practically, this is not the best method. With any input mistake, the original setup amounts will be lost and have to be recreated. Keep the original settings intact.

At the CNC machine, offsets are entered via control panel keyboard, using the **INPUT** or **+INPUT** keys:

- **INPUT** *key enters* absolute *value and replaces the current setting*
- **+INPUT** *key enters* incremental *value and updates the current setting*

Geometry Offset

Geometry offset is usually set from machine zero, but this is not mandatory. There are different ways of how to set the X and Y offsets, and CNC operators find many ingenious ways. Many modern lathes have a feature that sets the offset automatically. Manually, each axis is measured individually.

The process of measuring the X-geometry offset can be described in a few steps:

- Start from machine zero
- Make sure machine position shows X0.000 Z0.000
- Place a piece of stock to the chuck and clamp it
- Move both axes close to the part, in the front of it
- Turn the spindle on to a reasonable spindle speed
- Move the tool a little below the stock diameter
- Using handle or jog mode, make a cut in Z- of about 6 mm (0.25 inch), enough to place a micrometer there
- Do not move the tool but write down the X-position shown on the screen
- Move the tool to a safe position and stop the spindle
- Measure the diameter just machined

The *Geometry* offset is the sum of the previously shown X-axis position and the diameter measured. For example, starting from machine zero with T03, you have cut the stock at a position shown as X-297.560 and measured the stock diameter as 136.724. The sum of

297.560 + 136.724 = 434.284

will be input in *Geometry* offset 3 with the amount of **-434.284**. The amount must be negative, as the direction of measurement is from tool point to part zero.

Geometry offset for the Z-axis, where the finished front face is the part zero, is often done by touching the front face and adding a face-off amount. For example, the Z-position from machine zero was shown on the screen as Z-685.392. If an earlier tool faced off the part to size, the measured amount is the *Geometry* offset amount of **685.392**, that will be input in *Geometry* offset 3. The display will look like this:

GEOMETRY				
No.	X	Z	R	T
G 001	0.000	0.000	0.000	0
G 002	0.000	0.000	0.000	0
G 003	-434.284	-685.392	0.000	0
G 004	0.000	0.000	0.000	0

If a face cut is expected, just add the extra amount.

Other tools are set in a similar way. Some controls do not have the prefix *G* in front of the offset number. There are two other columns:

- **R-column**
- **T-column**

The R and T Columns

In the column under the heading of **R** is a space that is reserved for the *radius* of the tool nose. Radius of the tool applies to turning and boring tools and is effective only if **G41** or **G42** cutter radius offset is programmed.

The T-column is a column that is reserved for the *tip number* of the tool nose. This setting is also effective only if **G41** or **G42** cutter radius offset is programmed.

Both *R* and *T* settings work together, and only in cutter radius offset mode. The radius provides the *amount* of compensation, the tool tip number provides the control system with information about the *position* of the nose radius center in relation to the tool reference point.

The T-setting is often misunderstood. For those operators who also work on CNC machining centers, they know there is no such column on milling controls. The reason is simple. In milling, the work offset XY is always set to the centerline of the spindle. As all tools used in milling can be represented by a circle, the center of the circle (*i.e.*, center of the tool) is *always* equally distant from the contour machined, in cutter radius offset mode.

The same principle works for the lathes as well. After all, the tool nose radius is part of a circle whose center has to be equally distant from the contour turned or bored. Now for the problem - unlike in milling, *the tool tip reference point of a turning/boring tool is not at the center of this circle.*

What the T-column entry provides is a space for an arbitrary tip number that identifies the tool tip orientation. These numbers are determined by Fanuc and every CNC operator has to understand them well.

Here is an example of a turning tool and a boring bar:

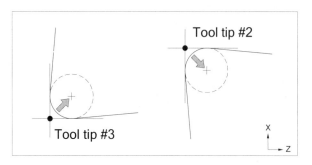

The one and only purpose of the tip number is to provide the control information about the center point location, so the correct toolpath adjustment can be made.

> **Incorrect tip number setting will result in inaccurate dimensions**

Other tool tip numbers for a rear lathe are shown in the illustration:

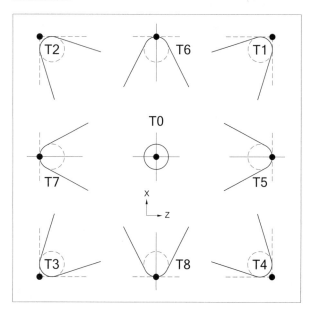

For those who have a front lathe, the numbers are inverted along the X-axis:

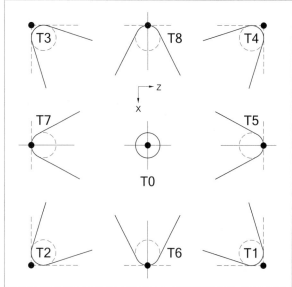

For the external tool T03 used in this example, the final geometry offset will be:

GEOMETRY

No.	X	Z	R	T
G 001	0.000	0.000	0.000	0
G 002	0.000	0.000	0.000	0
G 003	-434.284	-685.392	0.800	3
G 004	0.000	0.000	0.000	0

Wear Offset

Once the *Geometry* offset is set, the *Work* offset screen can be viewed:

WEAR

No.	X	Z	R	T
W 001	0.000	0.000	0.000	0
W 002	0.000	0.000	0.000	0
W 003	0.000	0.000	0.800	3
W 004	0.000	0.000	0.000	0

Although the layout is exactly the same for the wear offset as it is for the geometry offset, there is one noticeable change in offset W003. Wear offset 003 shows the same *R* and *T* values that were set earlier in the *Geometry* offset. Keep in mind this simple rule:

> **R and T values will always be the same in *Geometry* offset and *Wear* offset**

While you can adjust the X and Z settings in either offset, changing the R or T in one, will automatically make the same change in the other.

DRAWING AND PROGRAM

The drawing presented for the example is intentionally very simple. It represents a part that requires all three types of cutting tools in the same program. In the program itself, five tools are used:

- T01   External tool   ... *turning tool - rough*
- T02   Centerline tool   ... *drill*
- T03   External tool   ... *turning tool - finish*
- T04   Internal tool   ... *boring bar*
- T05   Part-off   ... *part-off tool*

Although the tool numbers themselves are not important, what is important is the distribution of the tools in the turret. A short tool (turning) is followed by a long tool (drill) which is followed by a short tool (turning), followed by a long tool (boring), followed by one more tool that is short (part-off). This short-long-short-long setup minimizes the interference problem that may be caused by tools of various lengths.

The part program for all five tools is the program as received in the machine shop, and is listed next. For all practical purposes, the program is correct:

```
(BASIC PROGRAM - NO TOLERANCES)
(38 MM ALUMINUM BAR)
(EXTEND BAR 40 MM FROM FRONT FACE OF JAWS)

(T01 - FACE AND ROUGH TURN OD)
N1  G21 T0100
N2  G96 S200 M03
N3  G00 G41 X40.0 Z0 T0101 M08
N4  G01 X-1.8 F0.2
N5  Z2.5
N6  G00 G42 X40.0
N7  G71 U2.5 R0.3
N8  G71 P9 Q16 U1.5 W0.125 F0.25
N9  G00 X10.0
N10 G01 X16.0 Z-0.5
N11 Z-10.0
N12 X24.0 C-0.5
N13 Z-20.0
N14 X35.0 C-0.5
N15 Z-34.0
N16 U5.0
N17 G00 G40 X100.0 Z75.0 T0100
N18 M01

(T02 - 9 MM DRILL)
N19 T0200
N20 G97 S1600 M03
N21 G00 X0 Z3.0 T0202 M08
N22 G01 Z-35.0 F0.35
N23 G00 Z3.0
N24 X100.0 Z50.0 T0200
N25 M01

(T03 - FINISH OD TURN)
N26 T0300
N27 G96 S220 M03
N28 G00 G42 X40.0 Z2.5 T0303 M08
N29 X10.0
N30 G01 X16.0 Z-0.5 F0.12
N31 Z-10.0
N32 X24.0 C-0.5
N33 Z-20.0
N34 X35.0 C-0.5
N35 Z-34.0
N36 U5.0
N37 G00 G40 X100.0 Z75.0 T0300
N38 M01
```

```
(T04 - BORING BAR - R0.4)
N39 T0400
N40 G96 S200 M03
N41 G00 G41 X16.0 Z2.5 T0404 M08
N42 G01 X10.0 Z-0.5 F0.12
N43 Z-31.5 F0.15
N44 U-3.0
N45 G00 Z2.5
N46 G40 X100.0 Z50.0 T0400
N47 M01

(T05 - 4 MM WIDE PART-OFF TOOL)
N48 T0500
N49 G97 S1800 M03
N50 G00 X40.0 Z-35.0 T0505 M08
N51 G01 X-0.8 F0.1
N52 G00 X40.0
N53 X100.0 Z75.0 T0500
N54 M30
%
```

Each tool has to be set individually, in the turret station, as programmed. Experienced operators find that sometimes it is easier to change the program than move a tool from one turret station to another. Changing the tool number is always faster than resetting the tool.

Changing Tool Numbers

Say you have the 4 mm part-off tool in station 7 - but the program calls for station 5. Changing the program is fast, but be careful. There are four entries related to the tool number - they *all* have to be changed. Required changes are underlined:

```
(T05 - 4 MM WIDE PART-OFF TOOL)
N48 T0500
N49 G97 S1800 M03
N50 G00 X40.0 Z-35.0 T0505 M08
N51 G01 X-0.8 F0.1
N52 G00 X40.0
N53 X100.0 Z75.0 T0500
N54 M30
%
```

The tool number in the comment is not critical, but leaving it intact can be confusing and should be changed as well. Change of the offset number should also match the tool number, to follow common approach.

After modification, the program will use:

- **T07**   ... instead of **T05**
- **T0700**   ... instead of **T0500**
- **T0707**   ... instead of **T0505**

Keep in mind one very important rule:

> **T.... in the program will cause a tool change**

What the statement means is that if you forget to make a change on only one tool number - or make an error in the entry - the control will force an actual tool change. If the tool happens to be at the end of a long bore, for example, the consequences can be severe.

Bar Extension

Extension of a tool is also called tool *overhang* or tool *projection*. The extension generally applies to long tools, particularly boring bars. Boring bars must always be set with the extension long enough to reach the programmed depth *and* include some clearance as well. Longer bars are acceptable, but as in all cases for boring bars, watch the ratio of the extended length to the bar diameter.

The general rule for steel shank boring bars is 3:1 (three times diameter), for carbide bars 5:1 (five times diameter). Actual extension that does not cause vibration is a matter of choosing the right insert, speeds and feeds, depth of cut and, as always, experience.

MANAGING TOLERANCES

As presented, there are no tolerances specified for any dimensions in the given drawing.

In this section, we will look at various tolerances that may be applied to the dimensions for specific engineering requirements. Tool T03 - OD finishing - will be used for the evaluations. That is the reason the finishing did not use **G70** finishing cycle. Had the cycle been used, the process of working with tolerances would be the same.

Three types of tolerances will be evaluated:

- **Diameters only**
- **Shoulders only**
- **Both diameters and shoulders**

Some tolerances presented here may appear a bit extreme (or at least unusual), but they have been carefully selected for educational purposes.

It is likely that you would find some tolerances used more often than others. The objective will be the same for all variations - to set the offset in such a way that the dimension when measured will be somewhere in the middle of the tolerance range specified.

Tolerances and Program Data

Adjusting sizes to match the drawing requirements means adjusting offsets. Before you adjust an offset, you should always know how the programmer handled tolerances in the program.

Some CNC programmers take a dimension and program some 'middle' amount of the tolerance.

For example:

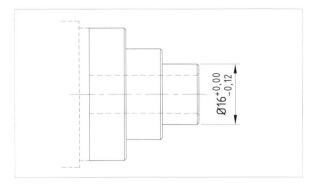

The nominal diameter of 16 mm has a tolerance of zero above the diameter and 120 microns below. That means the in-tolerance range is 15.880 to 16.000 mm. If the programmer did not use the nominal dimension, the program entry for the diameter may look like this:

```
N26 T0300
N27 G96 S220 M03
N28 G00 G42 X40.0 Z2.5 T0303 M08
N29 X10.0
N30 G01 X15.93 Z-0.5 F0.12
N31 Z-10.0
...
```

Many experienced programmers do not use this approach and always program nominal dimensions from the drawing, such as

```
N30 G01 X16.0 Z-0.5 F0.12
```

in this case.

This approach does not require any program changes, if the tolerances are revised for whatever reason.

The division of responsibilities is quite well defined:

> CNC programmer is responsible for correct data,
> CNC operator is responsible to maintain tolerances

This definition can be demonstrated in an example.

Tolerances on Diameters - Part A

When a drawing includes dimensional tolerances, evaluate the tolerances carefully - for example:

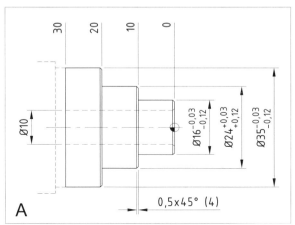

A

You may disagree with the way the drawing tolerances are dimensioned, but it is a legitimate method. Can the program be used as presented so far? It includes a wear offset 03, which can be adjusted to obtain the tolerances required. While that is true, consider the dimensions:

∅ 16 mm has a range of 15.970 to 15.880 mm

∅ 24 mm has a range of 24.030 to 24.120 mm

∅ 35 mm has a range of 34.970 to 34.880 mm

For example, you choose a wear offset 03 amount of fifty microns (0.050 mm). For the diameters ∅ 16 and ∅ 35, the offset amount would be negative fifty micros (-0.050), but for the ∅ 24 it would be positive fifty micros (0.050) resulting in these 'ideal' dimensions:

∅ 16 mm will be machined to ∅ 15.950 mm

∅ 24 mm will be machined to ∅ 24.050 mm

∅ 35 mm will be machined to ∅ 34.950 mm

The obvious problem is that you cannot put two different amounts under one offset entry - you need *two separate wear offset numbers!*

Entering offsets in the program are programmer's responsibility and the program should have another offset, let's say wear offset 13, programmed as T0313. Note that the first two digits identify the turret station and geometry offset - they must remain the same.

Where exactly in the program is the additional offset to be placed? This is a very important question, and a wrong answer may result in a scrapped part. The tool T03 is repeated next, for reference:

```
(T03 - FINISH OD TURN)
N26 T0300
N27 G96 S220 M03
N28 G00 G42 X40.0 Z2.5 T0303 M08
N29 X10.0
N30 G01 X16.0 Z-0.5 F0.12
N31 Z-10.0                       (FIRST DIAMETER)
N32 X24.0 C-0.5                  (FIRST SHOULDER)
N33 Z-20.0                       (MIDDLE DIAMETER)
N34 X35.0 C-0.5                  (SECOND SHOULDER)
N35 Z-34.0                       (LAST DIAMETER)
N36 U5.0
N37 G00 G40 X100.0 Z75.0 T0300
N38 M01
```

The underlined T-words show the common approach to applying a wear offset - comments have been added:

- Call tool - no wear offset ... block N26
- Apply wear offset with a motion ... block N28
- Cancel wear offset when finished ... block N37

The most important key when adding a new offset is to make sure no other offset setting is disturbed. In this example, it means the first diameter ($\varnothing$16) and the last diameter ($\varnothing$24) will both have to be machined with wear offset 03, while the middle diameter only ($\varnothing$35) will be machined using the new wear offset 13.

```
(T03 - FINISH OD TURN)
N26 T0300
N27 G96 S220 M03
N28 G00 G42 X40.0 Z2.5 T0303 M08
N29 X10.0
N30 G01 X16.0 Z-0.5 F0.12
N31 Z-10.0                       (FIRST DIAMETER)
N32 X24.0 C-0.5                  (FIRST SHOULDER)
N33 Z-20.0                       (MIDDLE DIAMETER)
N34 X35.0 C-0.5                  (SECOND SHOULDER)
N35 Z-34.0                       (LAST DIAMETER)
N36 U5.0
N37 G00 G40 X100.0 Z75.0 T0300
N38 M01
```

With the comments it is easier to see that the transition between the first and middle diameter takes place in block N32. That will be the block where the new wear offset T0313 will be added.

In order for the offset to be used for that diameter only, it must be changed back to T0303 before the last diameter is machined. This transition takes place in block N34, and that will be the block where the wear offset T0303 will be reinstated. A word of caution:

> **Under no circumstances** *cancel* **the wear offset, always** *change* **from one number to another while in machining mode**

Here is the new listing with two additions:

```
(T03 - FINISH OD TURN - PART A)
N26 T0300
N27 G96 S220 M03
N28 G00 G42 X40.0 Z2.5 T0303 M08
N29 X10.0
N30 G01 X16.0 Z-0.5 F0.12
N31 Z-10.0                       (FIRST DIAMETER)
N32 X24.0 C-0.5 T0313            (FIRST SHOULDER)
N33 Z-20.0                       (MIDDLE DIAMETER)
N34 X35.0 C-0.5 T0303            (SECOND SHOULDER)
N35 Z-34.0                       (LAST DIAMETER)
N36 U5.0
N37 G00 G40 X100.0 Z75.0 T0300
N38 M01
```

Initial Control Setting

At the control, the two wear offset settings will only apply to the X-axis, because that is the axis that controls all diameters - Z-axis settings must always be the same for both wear offsets 03 and 13 (both on diameter):

WEAR No.	X	Z	R	T
W 001	...	...	...	...
W 002	...	...	...	...
W 003	-0.050	0.000	0.800	3
W 004	...	...	...	...
...	...	...	...	...
W 013	0.050	0.000	0.800	3

Once the offsets are in the program and entered on the offset screen, any additional changes can be made. Keep in mind that the Z-setting can be any required setting but must be the same for both wear offsets. Also, do not forget to enter the same R and T settings for both offsets.

Setting Adjustment

While the explanation so far is correct mathematically and logically, in machining, we also have to think practically and creatively. As every CNC operator knows, the tool wears out as more and more parts are machined. That does not mean the tool is useless, it can still cut many more parts. The tool wear only means a slight change is size of the part - diameter in this example. This is where some creative thinking comes in. Experienced operators want to get as many parts machined from the same insert before adjusting the wear offset.

The solution is to *review* the wear offset settings. Take the two diameters with minus tolerance. Both can be up to 0.120 mm *below* nominal size:

⌀ 16 mm   =>   ⌀ 15.880
⌀ 35 mm   =>   ⌀ 34.880

If we aim at a final diameter just above the minimum, the tool wear will still happen, but it will be much longer before the wear offset will have to be adjusted further.

For the example, let's aim at the diameters ⌀ 15.900 and ⌀ 34.900. As both are controlled by wear offset 03, the new setting of X-axis for both will be negative 0.1 mm (-0.100):

W003 X-0.100

The same consideration has to be given to the middle diameter of ⌀ 24 mm. In this case, we want the tool to cut just slightly above the minimum of 0.030 mm, for example 0.05 mm.

In both cases, as the tool wears out, the diameter machined will increase marginally, but still well within tolerances provided. The control setting will now change:

WEAR				
No.	X	Z	R	T
W 001	...	...	...	...
W 002	...	...	...	...
W 003	-0.100	0.000	0.800	3
W 004	...	...	...	...
...	...	...	...	...
W 013	0.050	0.000	0.800	3

There was no need to change offset 013, as it already had the required amount stored. Here is an example of the results: the tool wears out 0.010 mm, making each diameter larger by that amount:

	New tool - no wear	With 0.010 mm wear
⌀ 16	⌀ 15.900	⌀ 15.910
⌀ 24	⌀ 24.050	⌀ 24.060
⌀ 35	⌀ 34.900	⌀ 34.910

As long as the tool maintains good surface finish, machining can continue on many parts before the insert has to be replaced.

Tolerances on Shoulders - Part B

Drawing tolerances on diameters are far more common than tolerances on shoulder. Tolerance on a single shoulder is more common than tolerance on two or more shoulders. The next illustration shows offsets on shoulders, therefore controlled by the Z-axis:

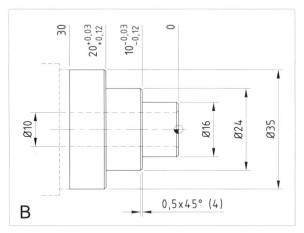

B

The same bi-directional range of tolerances is provided on two shoulders. As one tolerance is in the plus direction and the other in minus direction, two offsets have to be applied. It important to study the previous section that explained in great detail the wear offset setting for diameters. The approach is exactly the same for shoulders. If you understand one, you should have no problems understanding the other. The aim of the settings will also consider the tool wear with emphasis on shoulders only - diameters in this case have no tolerances provided.

As for the sizes desired, the first shoulder at Z-10.0 should be measured as Z-9.950, which requires wear offset 03 to be +0.050 along the Z-axis, while the second shoulder at Z-20.0 should be Z-20.100, which requires wear offset 13 to be -0.100 along the Z-axis:

WEAR				
No.	X	Z	R	T
W 001	...	...	...	...
W 002	...	...	...	...
W 003	0.000	0.050	0.800	3
W 004	...	...	...	...
...	...	...	...	...
W 013	0.000	-0.100	0.800	3

In the original program, the two shoulders have been identified with a comment

```
(T03 - FINISH OD TURN)
N26 T0300
N27 G96 S220 M03
N28 G00 G42 X40.0 Z2.5 T0303 M08
N29 X10.0
N30 G01 X16.0 Z-0.5 F0.12
N31 Z-10.0                   (FIRST DIAMETER)
N32 X24.0 C-0.5              (FIRST SHOULDER)
N33 Z-20.0                   (MIDDLE DIAMETER)
N34 X35.0 C-0.5              (SECOND SHOULDER)
N35 Z-34.0                   (LAST DIAMETER)
N36 U5.0
N37 G00 G40 X100.0 Z75.0 T0300
N38 M01
```

This time, the wear offset change should be applied during the motion towards the shoulder. As tool moves, the length of cut will be reduced or extended:

```
(T03 - FINISH OD TURN - PART B)
N26 T0300
N27 G96 S220 M03
N28 G00 G42 X40.0 Z2.5 T0303 M08
N29 X10.0
N30 G01 X16.0 Z-0.5 F0.12
N31 Z-10.0                   (FIRST DIAMETER)
N32 X24.0 C-0.5              (FIRST SHOULDER)
N33 Z-20.0 T0313             (MIDDLE DIAMETER)
N34 X35.0 C-0.5              (SECOND SHOULDER)
N35 Z-34.0 T0303             (LAST DIAMETER)
N36 U5.0
N37 G00 G40 X100.0 Z75.0 T0300
N38 M01
```

From the two presentations, it is clear that the wear offset setting for the X-axis controls only diameters, whereby the wear offset for the Z-axis controls only shoulders.

*Note - for offsets applied for **grooving**, refer to a separate chapter, starting on page 195.*

Tolerances on Diameters and Shoulders

This third application combines both previous examples. It is far less likely you encounter this rather complex assortment of tolerances, but understanding the general concepts will enable you to apply wear offsets with confidence that the part machined will be within specifications.

If you understand tolerances applied to diameters only and shoulders only, you will find that the combination of both in the drawing will also reflect in combination of the offsets. Additional offsets will be required.

The original program will have to be changed - it must contain two offsets for controlling the diameters (as per part A) and different two offsets controlling the shoulders (as per Part B). Selection of the offset numbers have to be within the range of offsets available at the control.

First, the drawing for *Part C*:

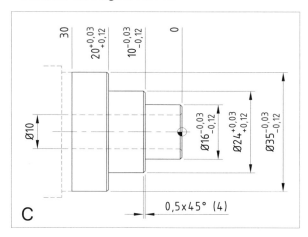

Part C is nothing more than both *Part A* and *Part B* combined - the dimensions and tolerances are all the same. The real challenge is to update the original program to cover *all* tolerances in the drawing. For this combined example, we will need two pairs of offsets:

- **T0303**   ... *controls diameters for the minus tolerance*
- **T0313**   ... *controls diameter for the plus tolerance*
- **T0323**   ... *controls shoulder for the minus tolerance*
- **T0333**   ... *controls shoulder for the plus tolerance*

This distinction does not take into consideration the *combination* of the X-wear and Z-wear offsets. The following illustration shows the expected dimensions when a new tool is correctly setup and part machined with the dimensions established earlier:

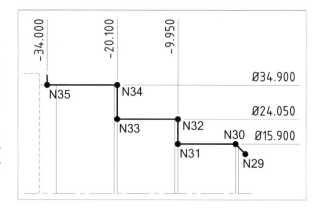

CNC Control Setup for Milling and Turning
LATHE OFFSETS   171

In order to achieve these dimensions, look at how the offsets are distributed over the critical portion of the finish contour:

N29	X-0.100	Z+0.050
N30	X-0.100	Z+0.050
N31	X-0.100	Z+0.050
N32	X+0.050	Z+0.050
N33	X+0.050	Z-0.100
N34	X-0.100	Z-0.100
N35	X-0.100	Z+0.050/Z-0.100

→ W003 X-/Z+
→ W013 X+/Z+
→ W023 X+/Z-
→ W033 X-/Z-

From this chart, the wear offset settings in the program for *Part C* will be:

```
(T03 - FINISH OD TURN - PART C)
N26 T0300
N27 G96 S220 M03
N28 G00 G42 X40.0 Z2.5 T0303 M08
N29 X10.0
N30 G01 X16.0 Z-0.5 F0.12
N31 Z-10.0                  (FIRST DIAMETER)
N32 X24.0 C-0.5 T0313       (FIRST SHOULDER)
N33 Z-20.0 T0323            (MIDDLE DIAMETER)
N34 X35.0 C-0.5 T0333       (SECOND SHOULDER)
N35 Z-34.0                  (LAST DIAMETER)
N36 U5.0
N37 G00 G40 X100.0 Z75.0 T0300
N38 M01
```

The setting of the wear offsets is the last step before actual machining:

WEAR				
No.	X	Z	R	T
W 001	...	...	...	...
W 002	...	...	...	...
W 003	-0.100	0.050	0.800	3
W 004	...	...	...	...
...	...	...	...	...
W 013	0.050	0.050	0.800	3
...	...	...	...	...
W 023	0.050	-0.100	0.800	3
...	...	...	...	...
W 033	-0.100	-0.100	0.800	3

WORKING WITH WEAR OFFSETS

When working with wear offsets, always decide first on the size you want to achieve. Take one offset at a time first, then evaluate its relationship with other offsets, if necessary.

Regardless of how many tolerances are present in the drawing, or even if there are no tolerances, there is only one objective in machining - *to prevent scrap*.

Definition of scrap for *outside* diameter machining:

> For turning - OD cutting :
> ... if the machined diameter is *TOO SMALL*

Definition of scrap for *inside* diameter machining:

> For boring - ID cutting :
> ... if the machined diameter is *TOO LARGE*

Also watch for a possible scrap on faces and shoulders. Always use *Geometry* offset for the initial setting or when a major change in setup is done. Always use *Wear* offsets for dimensional adjustments.

Wear offsets should be rather small. The larger the offset amount, the more likely it is that some other measurement was not done correctly.

Watch for decimal points in the input - older machines may not support decimal points for offsets, and an entry of -0.240 mm will be -240, In imperial units, an entry of 0.006 inches will be 0060 or just 60.

All offsets should always be set before any program testing. It can easily happen that a program may run well when there are no wear offsets (particularly the R and T entries), and when these values are input later, the program may not work.

If you get an alarm that you suspect may be related to offsets, check the offset setting first. After, check your setup, and also check the program.

Common Offset Change Example

A large number of jobs does not require the relative complexity of the examples presented in this chapter. Most offset adjustments will be on the part diameter, and to a lesser extent on shoulders.

In the final section we will look at some common scenarios and solve a problem by adjusting the appropriate wear offset.

Actual diameter, when measured, offers three possible outcomes - within tolerances, above and below the tolerance range. For external diameters, a larger size can be recut (in most cases), smaller size is a scrap. For internal diameters, smaller size can be recut (although poor surface finish may result), larger size is a scrap.

The example will have the objective to machine an nominal external diameter of ⌀ 132 mm:

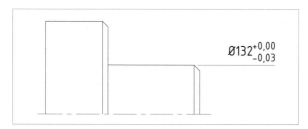

The specified tolerance is within the range of 132.000 down to 131.970 mm. At the time of dimensional check the wear offset (for T02) was **W002 = -0.075**:

WEAR				
No.	X	Z	R	T
W 001	...	...	...	...
W 002	-0.075	0.020	0.800	3
W 003	...	...	...	...

The desired size is ⌀ 131.980 mm or close to it.

◆ 1. Diameter was measured as 131.985
    ... ON SIZE = PART IS OK - no need for adjustment

Correction for the NEXT part ...
... WEAR OFFSET IS NOT CHANGED

◆ 2. Diameter was measured as 132.063
    ... OVERSIZE = CAN BE RECUT TO SIZE (*)

Correction for the NEXT part must include ...
... WEAR OFFSET MADE SMALLER

Calculations:

Required ⌀ :   131.980
Measured ⌀ :   132.063
Difference: 131.980 - 132.063 = -0.083 mm
New X-offset: -0.075 + -0.083 = **-0.158**

(*) *A very small amount of stock for recut may result in poor surface finish*

◆ 3. Diameter was measured as 131.874
    ... UNDERSIZE = SCRAP

Correction for the NEXT part must include ...
... WEAR OFFSET MADE LARGER

Calculations:

Required ⌀ :   131.980
Measured ⌀ :   131.874
Difference: 131.980 - 131.874 = +0.106 mm
New X-offset: -0.075 + +0.106 = **+0.031**

Can you see a pattern in the calculations? Both examples use the *same formula*. Keep in mind that a new offset is always *added* to the old offset, regardless of being positive or negative.

Diameter Offset Adjustment Formula

Offset calculation formula (adjusted offset value):

$$\text{New X-wear offset} = \text{Required } \varnothing - \text{Measured } \varnothing + \text{Current X-wear offset}$$

If dimensions along the Z-axis are in tolerance, there is no need to adjust the Z-wear offset.

WEAR OFFSET PARAMETERS

Two system parameters relate to tool wear offset:

5013   Maximum value of tool wear compensation

This parameter limits the maximum tool wear offset value allowed. If this amount is exceeded, the control will return an alarm. The amount should be very small, preferably not zero.

**5014   Maximum value of incremental input
            for tool wear compensation**

This parameter limits the amount of wear offset adjustment. If this amount is exceeded, the control will return an alarm. The amount should also be very small, preferably not zero.

17 TURNING AND BORING

This lathe oriented chapter is about external and internal stock removal on CNC lathes. It is most likely that every CNC lathe operator will machine parts that use external machining (generally called *turning*) and internal machining (generally called *boring*). On modern CNC lathes, this machining will be programmed using special cutting cycles. Operators do not have to know how to program such cycles, but they should fully understand *how these cycles work*. The main benefit of such knowledge results in the ability to make small changes at the machine that can produce significant improvements in many aspects of machining.

The best way to illustrate the concept is an illustration, a drawing. A very simple drawing shown below has been selected for further evaluation. It contains simple external and internal contours. The part size in the drawing is the initial size, the operations required are part of a *second* operation only. Surfaces within jaws have been pre-machined in the first operation (irrelevant here).

LATHE CYCLES G70 - G71 - G72

These powerful and flexible lathe cycles are called *multiple repetitive cycles* - a kind of parametric data input for machining. Programming these cycles and input of machining data is the job of CNC programmer. Keep in mind that the main purpose of any cycle is not only to make part programming easier and shorter, but also that any parametric cycle can be very easily edited at the machine, by the CNC operator. To understand how these cycles work, you need to know the format, structure, and application of the multiple repetitive cycles. This chapter covers three commonly used cycles for turning and boring - for roughing and finishing operations. The lathe cycles covered in this chapter are:

- **G70**   **Finishing cycle**
- **G71**   **Roughing cycle**   ... **horizontal direction**
- **G72**   **Roughing cycle**   ... **vertical direction**

with a brief mention of **G73** - *Pattern repeating cycle*

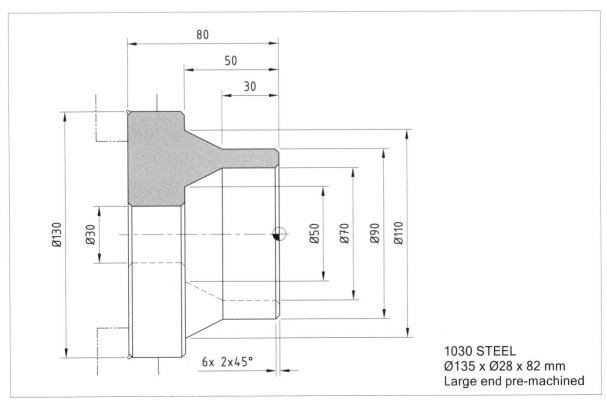

1030 STEEL
Ø135 x Ø28 x 82 mm
Large end pre-machined

Concepts of Turning and Boring Cycles

Multiple repetitive cycles were introduced by Fujitsu Fanuc in the early to mid 1980's, when a true computer based control system became more widespread. This was the first time when a shortcut was introduced to lathe machining. Basic fixed cycles for drilling on mills are a bit older. However, the concept date back another 15+ years, when programming languages were common.

Of the early language based programming methods, most names have been forgotten - Compact II®, Split®, APT®, GeoPath®, Numeridex®, and quite a number of others. In such a programming environment, the basic concept of stock removal was based on the removal from a *defined area*. Once the area had been defined, other parameters were used to fine tune the stock removal process, parameters such as depth, stock allowance, etc.

This concept has been adapted to modern programming, using the 'cycle' method.

Establishing an Area

Any geometrical area can be defined by the **minimum** of three points that do *not* form a single line - like a triangle. The illustration below shows the area of a simple taper to be machined out of a round stock:

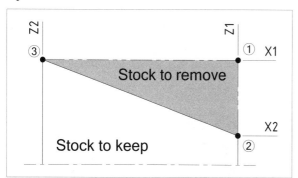

The emphasized word *minimum* in the last paragraph suggests there could be more than three points. In terms of CNC machining, a contour can have many *contour change points* to define the desired shape and sizes for machining. That is where the simple illustration above comes in - how is the contour represented? The part contour is a single line between points *2* and *3* - *a taper*. Point *1*, on the other hand, represents a point to start from - a *starting point*.

Multiple repetitive cycles are based on the same concept of three points minimum. Adapting this concept to the initial drawing (*page 173*) and changing some terminology, a multiple repetitive cycle can be defined.

Contour Points

If the point *2* in the illustration defines the *first* XZ coordinate of the contour, and if the point *3* of the same illustration defines the *last* XZ coordinate of the contour - *and* there are other specified points between the two points, a unique contour for machining can be defined.

Apply the previous generic illustration to the actual drawing. First, the *outside* contour (in scale), as machined with an external turning tool:

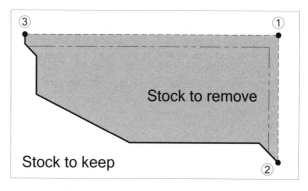

Next is the *inside* contour, for a different toolpath (also in scale, but one with a smaller factor), this time for *internal* machining, using a suitable boring bar:

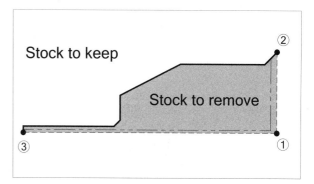

In both illustrations there is a noticeable change from the original description of areas. The change is not in the concept itself, but in its application. The conceptual definition does not take any clearances into effect. Yet, clearances are part of machining, and not including them in the area definition would cause severe machining problems. For that reason, any area that defines actual machining must include any necessary clearances.

During evaluation of a program received for machining, the location of point *1* may *not* correspond to its position in the illustration. This is not a requirement of the cycle itself, but a reflection of programming efficiency.

Program Listing

A CNC operator will be presented with a part program to be used as the source for machining. His or her job is to interpret the program correctly and be able to make some changes, if and when they become necessary.

As the large end of the part is not relevant here, it is important to understand what shape the material will have for the second operation, which *is* important here. As the following illustration shows, the shaded area represents the material that was removed at the large end.

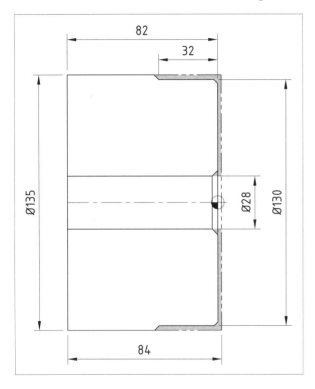

The original stock size of ⌀135 x 84 x ⌀28 mm has been reduced to ⌀135 x 82 x ⌀28 mm, as 2 mm have been faced off. The remaining 2 mm will be faced off in the second operation.

Both the outer and the inner chamfers have also been completed during this operation.

➻ *Note:*

During a two-sided lathe operation that requires facing at both ends, it may be necessary to split the facing stock differently, not always by equal amounts. A smaller amount at one end may require two or more cuts at the other end.

When the part is reversed, the remaining material has to be removed - that will be the subject of further study.

The shaded portion of the next illustration shows the material to be removed during second operation.

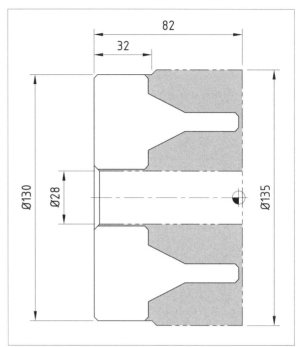

The program will most likely start with a facing cut. Part zero is at the centerline and the front finished face. In the program, the initial clearance will be 1.5 mm above the actual stock diameter, and only one facing cut will be required:

```
N1 G21 T0100        (ROUGHING OD TOOL - R0.8)
N2 G96 S140 M03
N3 G00 X138.0 Z0 T0101 M08
N4 G01 X25.0 F0.33
N5 G00 Z2.5
N6 ..
```

When facing to an existing hole, some programs may not face far enough, and the face will not be flat at the end of cut. At the control, and if necessary, you may have to extend the facing motion along the X-axis. Use the following practical guideline:

> **The *minimum* clearance programmed below an existing diameter must be double the tool nose radius**

In the example, the tool nose radius is R0.8. Double that amount to 1.6, and when subtracted from the existing ⌀28 mm, the X-axis end point should not be greater that **X26.4**. With an additional clearance, the **X25.0** in the program is sufficient.

Contour Shape - External

The shape of the contour to cut (for both external and internal operations) has been already shown on *page 174*. One important observation is the fact that each cut *starts and ends away* from the actual contour. In other words, it must include clearances and tool run-off. The program listing for the external contour continues with block N6:

```
N1 G21 T0100          (ROUGHING OD TOOL - R0.8)
N2 G96 S140 M03
N3 G00 X138.0 Z0 T0101 M08
N4 G01 X25.0 F0.33
N5 G00 Z2.5
N6 G41 X140.0
```

The tool is now at the *point 1* of the external contour. This point will become the **start point** (SP) of the external contour. Program will continue with the cycle:

```
N1 G21 T0100          (ROUGHING OD TOOL - R0.8)
N2 G96 S140 M03
N3 G00 X138.0 Z0 T0101 M08
N4 G01 X25.0 F0.33
N5 G00 Z2.5
N6 G41 X140.0 (X140.0 Z2.5 = START POINT SP)
N7 G71 U3.0 R0.5
N8 G71 P9 Q15 U2.0 W0.0125 F0.5
N9 G00 X81.0          (FIRST BLOCK OF CONTOUR)
N10 G01 X90.0 Z-2.0 F0.13     (FRONT CHAMFER)
N11 Z-30.0 F0.3               (DIAMETER)
N12 X110.0 Z-50.0                (TAPER)
N13 X126.0            (START OF END CHAMFER)
N14 X131.0 Z-52.5 F0.13       (END CHAMFER)
N15 X136.0            (RUN-OFF CLEARANCE)
N16 ..
```

These few blocks will remove all material from the defined area and leave stock for finishing. Finishing can be done with the same tool, or another tool, to control dimensions, surface finish, and tool life more closely. The block N9 contains the tool position of the contour start, and block N15 contains the tool position of the contour end. They are represented by *points 2* and *3* in the initial illustration. When these generic points are replaced with block numbers, the illustration will not change much:

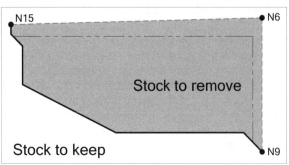

The area to remove by cutting is not a simple triangular area, it is a more complex shape. This complete contour shape is represented by the starting block N9 and ending by the block N15. The main key to this definition is the second block of the **G71** cycle, block N8 to be specific. In this block are the addresses *P* and *Q*. The first address (P-address) refers to the block *number* representing the *first point* of the contour, and the Q-address refers to the block *number* representing the *last point* of the contour.

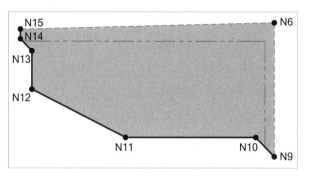

Once the CNC control system is 'aware' of the material area to be removed, the roughing operation is completed, including the stock allowance left for finishing. Then, the finishing cycle **G70** can be applied to the same tool or another tool. The difference is only a tool change in the program. Here is the completed part program for the same tool, finishing cut included:

```
N1 G21 T0100          (ROUGHING OD TOOL - R0.8)
N2 G96 S150 M03
N3 G00 X140.0 Z0 T0101 M08
N4 G01 X25.0 F0.33
N5 G00 Z2.5
N6 G41 X140.0 (X140.0 Z2.5 = START POINT SP)
N7 G71 U3.0 R0.5
N8 G71 P9 Q15 U2.0 W0.125 F0.5
N9 G00 X81.0          (FIRST BLOCK OF CONTOUR)
N10 G01 X90.0 Z-2.0 F0.13     (FRONT CHAMFER)
N11 Z-30.0 F0.3               (EXT DIAMETER)
N12 X110.0 Z-50.0                (TAPER)
N13 X126.0            (START OF END CHAMFER)
N14 X131.0 Z-52.5 F0.13       (END CHAMFER)
N15 U5.0 (X136.0)     (RUN-OFF CLEARANCE)
N16 G70 P9 Q15 S175         (FINISHING CYCLE)
N17 G40 G00 X300.0 Z50.0 T0100   (INDEX POS)
N18 M01                        (OPTIONAL STOP)
```

The finishing cycle **G70** uses the same block numbers represented by *P* and *Q*, since they represent the finished contour of the part. From a program like this, a wealth of information can be found, but only if you understand the cycles **G71** and **G70**, all their parameters, and how they work internally, within the CNC system.

All modern controls, including as the older Fanuc control models 0-T, 16-T, 18-T, 21-T, as well as the much newer *i-series*, such as 16i-T, 18i-T, 21i-T, 30i-T, 31i-T, and 32i-T, etc., require a two-block format for the multiple repetitive cycle **G71**.

G71 Cycle Format for Two Blocks

Modern control systems listed above require a double block entry for the **G71** cycle. Check your control manual for possible format exceptions - the standard programming format is:

G71 U.. R..
G71 P.. Q.. U.. W.. F.. S..

➡ *where ...*

First block:

U = Depth of roughing cut
R = Amount of retract from each cut

Second block:

P = First block number of the finishing contour
Q = Last block number of the finishing contour
U = Stock amount for finishing on the X-axis diameter
W = Stock left for finishing on the Z-axis
F = Cutting feedrate (mm/rev or in/rev) overrides feedrates between the P-block and Q-block
S = Spindle speed (m/min or ft/min) overrides spindle speeds between P-block and Q-block

CNC programmers have to understand the cycle format in order to develop the program in the first place. CNC operators have to understand the cycle format - *and its inner workings* - for making intelligent changes at the machine control. Before getting into that subject, let's see the same **G71** cycle applied to internal machining (boring).

Contour Shape - Internal

The first sixteen blocks of the program example were all related to the external cutting (outside diameter). Block N18 completed the activity of *tool 1* (T0101). The next tool will be *tool 2* (T0202) - a boring bar.

The turning cycle **G71** is used for both external and internal material removal in turning. Watch for some minor differences in the cycle when used for boring, but the programming format is exactly the same.

First, the program itself will be shown, followed by a matching illustration and some follow-up explanation.

```
N19 T0200          (25 MM DIA BORING BAR - R0.8)
N20 G96 S130 M03
N21 G42 G00 X24.0 Z2.5 T0202 M08 (START POINT)
N22 G71 U2.5 R0.5
N23 G71 P24 Q31 U-2.0 W0.125 F0.3
N24 G00 X79.0            (FIRST BLOCK OF CONTOUR)
N25 G01 X70.0 Z-2.0 F0.125     (FRONT CHAMFER)
N26 Z-30.0 F0.25               (INT DIAMETER)
N27 X50.0 Z-50.0                       (TAPER)
N28 X34.0               (START OF END CHAMFER)
N29 X30.0 Z-52.0 F0.125          (END CHAMFER)
N30 Z-81.0 F0.25             (END OF LAST BORE)
N31 U-2.0 (X28.0)           (RUN-OFF CLEARANCE)
N32 G70 P24 Q31 S150
N33 G40 G00 X250.0 Z50.0 T0200     (INDEX POS)
N34 M30                            (PROGRAM END)
%
```

In the next illustration, the block numbers represent the three key points - the SP point, P-point, and Q-point:

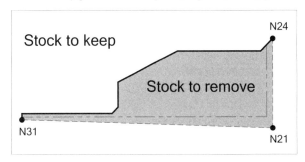

The second illustration also includes the individual contour points between the P-point and the Q-point:

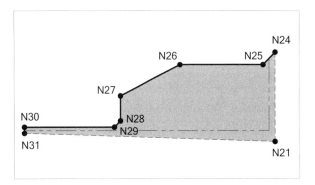

The program as shown so far is the one received in the shop. Once you understand the principle of the **G71** and **G70** cycles, it is easy to evaluate the program and make any changes that improve the part quality and overall cycle time. The question is, what to look for?

PROGRAM FEATURES TO WATCH

When programming or optimizing the **G71** and **G72** lathe cycles, some general comments can be useful:

1. P..is the block number where the contour starts
   ...in second block for a two-block format
2. Q..is the block number where the contour ends
   ...in second block for a two-block format
3. Watch the P and Q numbers as they relate to existing block numbers
4. Change of block numbers, such as re-sequencing, could influence the P and Q numbers
5. Also make sure that no block number is duplicated in the program
6. Feedrates programmed between P and Q blocks are for finishing cycle G70 only
7. Start point for the cycle is defined as the last XZ coordinate before the G71/G72 is called
8. For safety reasons, use the same start point for roughing with G71/G72 and finishing with G70
9. Cycle starts and ends at the initial start point, and the return to this point is automatic
   Do not move to the start point inside of P-Q block range
10. Start point in the X-axis determines the actual depth of the first rough pass in G71 or width of cut in G72
11. For external cutting, increasing the start point diameter will decrease the actual depth of the first cut, decreasing the start point diameter will increase the actual depth of the first cut
12. For internal cutting, increasing the start point diameter will increase the actual depth of the first cut, decreasing the start point diameter will decrease the actual depth of the first cut
13. G41 and G42 tool nose radius offset must be started *before* the cycle is called, and must be cancelled by G40 *after* the cycle is completed
    Do not use G41 or G42 inside of P-Q range
14. Clearances, such as run-off or change in direction, should always be greater than double the tool nose radius
15. Subprogram cannot be called within the P-Q range
16. Understand Type I and Type II cycle for G71 (page 187)

Based on these observations, how do they apply to the part program as received in the shop? The items 1 to 9 have been satisfied, and there is nothing to consider in terms of item 13. Items 10 to 12 are all related and should be looked at in more detail. The same applies to item 14. First, let's evaluate the start point location.

Start Point Location

The location of the start point for turning and boring cycles is more than just a convenient and safe position. If you find that the first depth of cut is too light or too heavy - or the tool does not remove any material at all - look at the start point coordinates. In the example, the start point for the X-axis is programmed as **X140.0**. As the stock is ⌀135 mm, the 2.5 mm (0.1 inches) clearance seems reasonable. However, the result will be that the first cut will be in the air - no material will be removed. Here is why:

When the tool reaches the programmed start point location of **X140.0 Z2.5**, the next block is the cycle call. In the cycle are two parameters that will force the tool to move out of its current position - the first one is the *stock amount*. In the program, the stock on diameter is programmed as 2 mm. The tool immediately moves 2 mm on diameter and now is at **X142.0**. The Z-location will be adjusted as well, but that is not important for the depth. In addition, the tool will also move by the amount of wear offset, which we assume is set to zero.

From this shifted position of **X142.0**, the second cycle parameter comes into effect - the *depth of cut*. In the program, the actual depth is 3 mm (per side), so the tool moves back down by 6 mm, to the new start position of **X136.0**. Now the first cut will start. The problem is that the stock is only ⌀135 mm, so no cutting will take place for the first cut. When the start point in the program is changed - say to **X135.0** - the actual diameter of the first cut will be:

135.0 + 2.0 + 0.0 - (2 x 3) = 131.0

The result is **(135 - 131) / 2 = 2** mm physical depth of cut. Still less than the programmed depth, but at least acceptable. You can optimize the X-start location to control the first depth of cut, to get 'under the skin' of some cast or forged materials.

For the internal cut, the calculation is the same, in the opposite direction. The programmed X-start point is **X24.0**, the hole is ⌀28 mm, depth of cut is 2.5 and stock is 2 mm on diameter. The first cut diameter will be:

24 - 2 + 0.0 + (2 x 2.5) = 27.0

Again, no first cut. Changing the start point to **X28.0** will make the actual depth of first cut 1.5 mm.

Unfortunately, these are common errors in programming. A knowledgeable CNC operator will either catch such errors when studying the program, or will be able to change the program after discovering the error during part setup and program proving.

Clearances

In the example, the clearances are at the start and end of each contour. Both start points in block N9 (**X81.0**) and in block N24 (**X79.0**) have been programmed correctly, based on the **Z2.5** position.

The actual tool motion from block N14 to N15 is programmed as incremental distance of **U5.0**, or 2.5 mm per side. This amount will guarantee that the chamfer will be completed. *Keep in mind that the tool nose radius offset is in effect.*

For the internal cut, there is a major problem in the program - if you run this program as is, you will never bore the part. The control system will issue an error message, informing you about a **CRC** (cutter radius compensation) error. This error means that the tool radius cannot fit into the space provided.

This error is quite common, and even some experienced programmers may make it here and there. It often happens when a large boring bar is used for a hole that is only marginally larger. The reason behind the error is that the programmer is afraid of contact between the boring bar and the part on the negative side of the centerline.

Back to the above example. The existing hole in the part is drilled to ⌀28 mm, the last bore is ⌀30 mm, and the boring bar has ⌀25 mm shank. Also to be considered is the minimum boring diameter for the bar. Let's use this bar to illustrate:

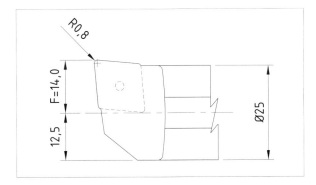

The *'F'* dimension is a standard tool catalogue dimension indicating the distance between boring bar centerline and the tip of a gage insert radius. The gage radius is also defined in the tooling catalogue. Also in the catalogue will be a minimum bore diameter, for example, we will use ⌀27 mm. Also note the tool nose radius - it is a common size of R0.8 mm (R0.0313 inch).

In the program, the last block for boring is block N31:

```
N31  U-2.0
```

This block contains a boring bar motion from the completed diameter of 30 mm, to the clearance diameter of 28 mm. After that, the boring bar returns automatically to the start point. Note the situation carefully - there are three consecutive motions:

- ⌀30 mm cutting   ... Z motion NEGATIVE
- 2 mm run-off   ... X motion NEGATIVE
- Return back to start point ... Z motion POSITIVE

Graphically, it looks like this:

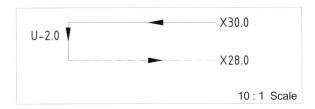

The program would work, *IF* there were no tool nose radius compensation (cutter radius offset). This offset is necessary, and with it in effect, the program will not work and an alarm at the control will be the result.

Consider the same pictorial representation with the tool nose radius added:

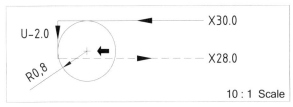

The above image shows the position of the tool, actually its radius, at the end of the bore, *before* the retract clearance motion. The cutter radius offset is in effect, and programmed to the left of the contour, as **G41**. The control always 'looks ahead' and calculates the next tool position. In this case, the radius needs to be to the *left of the return motion* - the motion back to the start point. For that, the new tool location would be as shown next:

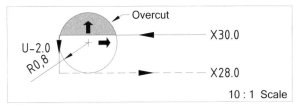

The X-motion will be positive, in order to position the tool properly. That results in overcutting, which the control system does not allow - hence the alarm.

The solution? The clearance retract in block N31 must be *increased* to accommodate the double tool radius plus some clearance. The current tool nose radius is R0.8 mm, the full circle is ⌀1.6 mm per side, or 3.2 mm on diameter. That does not provide any clearance. Changing the U-2.0 to U-4.0, for example, would work well for this example, but there is another consideration.

The nose radius R0.8 (0.0313 inch) is one of three most common radiuses used for turning and boring:

- R0.4 mm = 1/64 = 0.0156 inches
- R0.8 mm = 2/64 = 0.0313 inches
- R1.2 mm = 3/64 = 0.0469 inches

The program will be more flexible, if the clearance is large enough to accommodate the R1.2 (0.0469) nose radius. In that case, if the insert is changed to a larger one, no change in the program will be necessary:

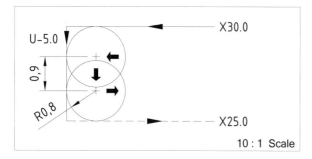

The corrected block N31 will be:

```
N31 U-5.0 (X25.0)
```

With the physical retract motion of only 0.9 mm per side, there is no danger that the boring will collide with the hole surface.

Tool Change Clearance

Another kind of clearance relates to the tool change. When one tool is completed, the turret moves away, and is indexed to another tool station. In the example, the turning tool (T01) is replaced by a boring bar (T02).

The tool motion away from the part is relatively small, to avoid increasing the cycle time. Usually, the programmer may use 25 to 50 mm or 1 to 2 inches as a common clearance. While such clearance is sufficient, you have to watch the clearance when a short tool is followed by a long tool. This is particularly important when these tools are mounted next to each other in the turret. This is the case in the program example. When you evaluate the program as presented, you may discover a flaw.

Here is the tool change clearance for tool T01:

```
N17 G40 G00 X300.0 Z50.0 T0100    (INDEX POS)
```

And here is the tool change clearance for tool T02:

```
N33 G40 G00 X250.0 Z50.0 T0200    (INDEX POS)
```

Both are programmed as 50 mm (2 inches), which seems reasonable. When you consider the tool tip clearance in relationship to the actual overhang of each tool, the clearance for T01 may need to be evaluated. The tool setup will be close to these dimensions:

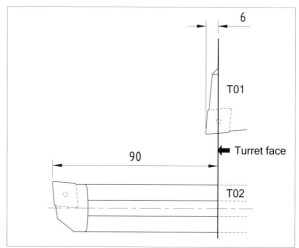

The key concern here is that when the tool T01 indexes at only 50 mm off the front face, the boring bar tip will be **behind the part** by 34 mm (90 - 6 - 50 = 34). In order to move the tool to a position where the boring bar tip will be on the *positive side* of the front face (Z0), the clearance programmed in block N17 should be increased. There is no need for detailed calculation, as long as the clearance is greater than the boring bar overhang. In the example, `Z125.0` will be sufficient:

```
N17 G40 G00 X300.0 Z125.0 T0100    (INDEX POS)
```

There is no problem for indexing from T02 back to T01 (for a new part), and the X-axis clearances are all reasonable - generally above the largest outside diameter of the part.

➪ *Note:*

In most CNC lathe applications, there is enough clearance between tools and the above discussion may not apply. However, due to the fact that safety is always a major concern, the slight increase may be justified.

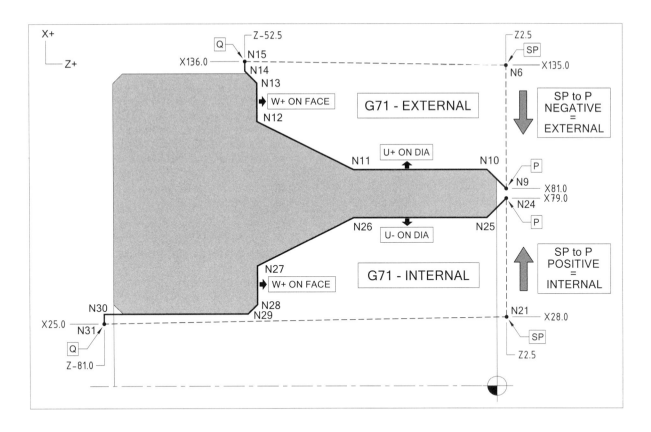

Program after Changes

The final program has all the changes discussed earlier underlined. Other changes that influence machining can be done after a few parts have been completed. They include spindle speeds, cutting feedrates, and depths of cut.

Here is the optimized part program:

```
N1  G21 T0100           (ROUGHING OD TOOL - R0.8)
N2  G96 S150 M03
N3  G00 X140.0 Z0 T0101 M08
N4  G01 X25.0 F0.33
N5  G00 Z2.5
N6  G41 X135.0     (X135.0 Z2.5 = START POINT SP)
N7  G71 U3.0 R0.5
N8  G71 P9 Q15 U2.0 W0.125 F0.5
N9  G00 X81.0           (FIRST BLOCK OF CONTOUR)
N10 G01 X90.0 Z-2.0 F0.13     (FRONT CHAMFER)
N11 Z-30.0 F0.3               (EXT DIAMETER)
N12 X110.0 Z-50.0                     (TAPER)
N13 X126.0            (START OF END CHAMFER)
N14 X131.0 Z-52.5 F0.13         (END CHAMFER)
N15 U5.0 (X136.0)         (RUN-OFF CLEARANCE)
N16 G70 P9 Q15 S175           (FINISHING CYCLE)
N17 G40 G00 X300.0 Z125.0 T0100  (INDEX POS)
N18 M01                         (OPTIONAL STOP)
N19 T0200        (25 MM DIA BORING BAR - R0.08)
N20 G96 S130 M03
N21 G42 G00 X28.0 Z2.5 T0202 M08 (START POINT)
N22 G71 U2.5 R0.5
N23 G71 P24 Q31 U-2.0 W0.125 F0.3
N24 G00 X79.0         (FIRST BLOCK OF CONTOUR)
N25 G01 X70.0 Z-2.0 F0.125     (FRONT CHAMFER)
N26 Z-30.0 F0.25                (INT DIAMETER)
N27 X50.0 Z-50.0                        (TAPER)
N28 X34.0             (START OF END CHAMFER)
N29 X30.0 Z-52.0 F0.125          (END CHAMFER)
N30 Z-81.0 F0.25             (END OF LAST BORE)
N31 U-5.0 (X25.0)         (RUN-OFF CLEARANCE)
N32 G70 P24 Q31 S150
N33 G40 G00 X250.0 Z50.0 T0200     (INDEX POS)
N34 M30                           (PROGRAM END)
%
```

On a simple part, the purpose of the evaluation was to look at what possible errors can happen, what to look for, and how to apply correction at the machine, during setup or over the first few parts.

There are other issues that may have to be resolved, so a short *'watch list'* may come handy.

PROGRAM DATA TO WATCH

The part program delivered to the machine shop for production generally falls into three categories:

- Program is totally useless
- Program is absolutely perfect
- Program is good but adjustments may improve it

It does not take too much imagination - if the program is totally useless, it may happen once or even twice before the incompetence of the CNC programmer is discovered and dealt with. Practically, this is an extreme case.

The second possibility, a program that is *'absolutely perfect'*, is actually nothing more than a program that needs no further modifications by the CNC operator during the part setup and subsequent production run. In many applications, this is also an extreme case, albeit as a positive event. In most cases, such programs are written by programmers with considerable skills and experience. This is the ideal - *the perfect* - part program.

In reality, working with such programs is unusual, but the good programs definitely prevail. Although each item listed below can take a page or two of details, the list itself (with a few comments) should be sufficient enough to provide some basic understanding of what to watch for and for possible changes during setup:

Start Point Too High or Too Low

Start point for a turning/boring cycle influences the depth of the first cut. This has been covered on page 178.

Positive and Negative Stock Allowance

The amount of stock (stock left for finishing) is programmed in the second block of the **G71** or **G72** cycle as address *U* and *W*. This amount can be positive or negative.

The following illustration shows the positive and negative orientation of the U/W stock for turning and boring:

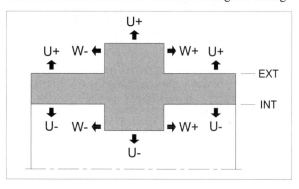

W-Stock Amount Too Heavy

While the positive and negative direction of the stock allowance is always important, equally important is the actual *amount* programmed. In **G71** cycle, the stock left on part diameter (the U-address) is generally close to the tool nose radius per side, but the W-stock requires special attention. For the **G72** cycle, the logic is the same, but the axes are reversed (more on **G72** later in this chapter). The amount of stock allowance (*i.e.*, stock left for finishing) is programmed in the second block of the two-block **G71** or **G72** cycle as address *U* and *W*, for X-axis and Z-axis respectively.

Special attention should be given to the programmed W-stock amount in the **G71** cycle. This is particularly important when facing a 90° shoulder, externally or internally, with an insert whose lead angle is 5° or less. This common situation applies to all three most frequently used cutting inserts - the *C* type (80°), the *D* type (55°), and the *V* type (35°). The 80° insert has a 5° lead angle clearance, whereby both the 55° and 35° inserts have only 3° lead angle clearance.

Why the Special Attention?

Unlike in manual lathe work, where the tool motion is normally changed quite often in various directions, in most CNC lathe work, the tool follows a path that does not change into opposite directions. That means the cutting is always uniform. When it comes to facing a shoulder, the very narrow lead angle of the insert will 'bite' into the stock left on the shoulder. The heavier the stock, the bigger 'bite'. The result can be a severe chatter, stalling the machine, or even pulling the part out of the chuck.

The illustration below shows the situation and also provides a rather lengthy formula to see what the 'bite' depth *D* will be with a programmed W-stock amount.

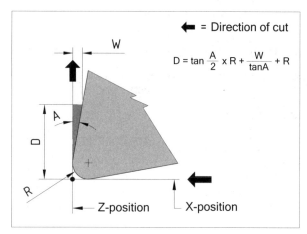

In the formula, the individual entries apply already known data, in order to calculate the actual - *physical* - depth of cut D, when the insert reaches the Z-position, but before making the shoulder cut (**G71** cycle):

➡ where ...

D = Actual depth of cut at the face
A = Lead angle of the insert
R = Radius of the insert
W = Stock left on face for finishing (as programmed)
X-position = Target position for the X-axis
Z-position = Target position for the Z-axis

During setup, the provided formula may not be a great help - it seems to be too complex, pocket calculator is not always available, and more than one try may be necessary to confirm the result even with a calculator. Fortunately, there is a rule-of-thumb that can be applied:

W-stock amount programmed	Millimeters	Inches
	W0.125	W0.005

Without counting on any specific formula, these suggested W-stock amounts should be suitable for virtually all 90° shoulder machining, using inserts with 3° to 5° lead angle.

Rules-of-thumb are no more than suggestions gathered from experience. They work in most cases, but not in all cases. It means you should feel free to adapt the 'rule' from 'as is', and change it to your working preferences.

Return to Start Point in the Cycle

When the last entity of the contour toolpath has been machined, it is followed by a retract motion *away* from the part. This retract is X-positive for external contours, and X-negative for internal contours.

Retract amount is nothing more than a clearance motion that should always be the last motion of the cycle, programmed by the Q-address reference to the matching block number.

When you are checking the program, make sure that the motion back to the start point (generally Z-positive for **G71**) is *not* part of the P to Q range. This is a serious error and the control system will issue an alarm.

P and/or Q Reference Defined Incorrectly

The addresses *P* and *Q* in multiple repetitive lathe cycles are most important. They are pointers to the beginning and the end of contour. *P* represents the block number of the first point of the contour, *Q* represents the block number of last point of the contour.

Even if the program is originally correct, many errors happen during editing, either at the control system or away from it. Adding or removing blocks from the P-Q range may affect the block numbering. Also watch when using a CNC editing software. The vast majority does *not* renumber P's and Q's while changing block numbers. One software that does renumber is NCPlot®.

Also note the general rules for block numbering:

- Block numbers can be in any order
- Block numbers can be duplicated
- Block number can have any increment
- Block numbers can be in strategic blocks only
- Block numbers are not mandatory

Although all four items are correct, that does not make them practical. having block numbers in ascending order makes most sense, having them in descending order is ok but makes less sense. Having block numbers mixed makes no sense at all, and neither does having duplicate block numbers.

Some programmers only use block numbers at tool change. Having block numbers in increments of 5 or 10 is quite common, although it may have an adverse effect on memory capacity. Some programs do not use block numbers at all, although they are required in the turning cycles described in this chapter. Also a common practice is to have only the first and the last block of the finishing contour numbered - for example:

```
...
G71 U.. R..
G71 P100 Q100 U.. W.. F.. S..
N100
...
...
...
...
N200
G70 P100 Q200 S..
...
```

> Make sure that block numbers applied to the contour P-Q are not repeated elsewhere in the program

This reminder is very important, particularly for large programs where all blocks are numbered.

Feedrates for Roughing and Finishing

Look at all the feedrates programmed for each tool and you will often notice that they change even for the same tool. This is not an unusual programming practice, and experienced programmers use it all the time.

In the example used in this chapter, there is a noticeable distinction between feedrates applied to chamfers and those applied to the rest of tool motions (diameters, tapers, shoulders). There is a good reason, and you will understand it if you can answer the following question:

What is the result of cutting a chamfer 0.25x45° with the programmed feedrate of F0.3 (0.3 mm/rev.)? The answer is that the chamfer is a mess.

If the feedrate per revolution is greater than the actual chamfer length, the tool will not have the time to actually cut the chamfer - it will create a small ridge that will lack all qualities of a chamfer. For that reason, do not be surprised to see very small feedrate on chamfers.

When it comes to feedrates for **G71** nad **G72** lathe cycles, look at the final program on *page 181* (T01):

The **G71** cycle call has feedrate programmed as F0.5, which is 0.5 mm/rev. However, in block N10 (chamfer) the feedrate changes to F0.13, immediately after, in block N11 it changes again to F0.3, and there is another change later on. The same approach is taken for T02, in feedrates for boring. How do the feedrates work?

Normally, any feedrate programmed after the current feedrate cancels the previous feedrate. After all, only one feedrate can be in effect at any time. In lathe cycles it is different. In our example, the F0.5 will work for *all motions* of the roughing cycle. Yes, the feedrates programmed for the finishing contour between block N9 and N15 will be ignored in roughing.

When the **G70** finishing cycle is called, then - and only then - the other feedrates will be applied.

Cutter Radius Offset

Cutter radius offset is also called *tool nose radius offset* in turning and boring. In either case, when lathe cycles are used, some rules also apply:

- **Cutter radius G41 or G42 must be applied *before* the cycle is called**
- **Cutter radius offset cancel G40 must be applied *after* the cycle is completed**
- **No G40, G41, or G42 must be programmed within the P-Q range**
- **Cutter radius offset is effective for finishing only, it is disabled for roughing**

Small Inside Radius

Most contours for turning and boring include not only linear motions, but also motions along an arc. These arcs will have various radiuses, but they can be divided into two significant groups:

- **Outside arcs**
- **Inside arcs**

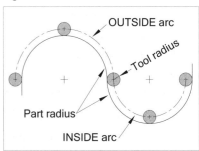

A large number of problems during machining (turning, boring, and milling) can be attributed to one factor:

> The inside arc radius is larger than the part radius

On CNC lathes, it may happen for a number of reasons. One common reason is when you change an insert to another one with a larger corner radius, for example, from R0.8 to R1.2 (R0.0313 to R0.0469).

A simple program illustrates the problem:

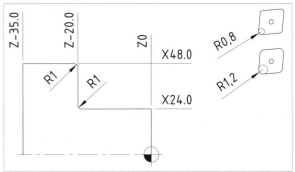

```
...
G00 G41 X24.0 Z2.5 T0505 M08
G01 Z-19.0 F0.2
G02 X26.0 Z-20.0 R1.0 F0.1          (INSIDE ARC)
G01 X46.0 F0.2
G03 X48.0 Z-21.0 F0.1               (OUTSIDE ARC)
G01 Z-...
```

As long as the tool radius is *smaller* than 1.0 mm, the program will work fine. If you change the radius to a larger one, an alarm will be the result. The same alarm will be issued if there is a program or offset entry error.

Regardless of the tool radius, the outside arc will never be affected, just watch for the internal arcs!

G72 AND G73 TURNING CYCLES

The emphasis so far was on the multiple repetitive cycle **G71** (and **G70** for finishing). For tuning or boring with focus on vertical machining, the **G72** cycle is used in the programs.

G72 Lathe Cycle

To illustrate the difference between the **G71** and **G72**, consider the following drawing and the program that machines the radiuses:

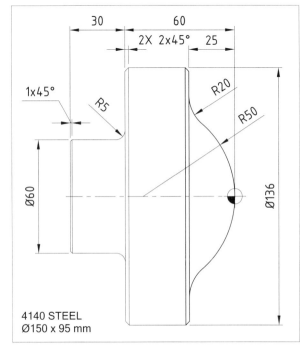

4140 STEEL
Ø150 x 95 mm

```
G21 T0100
G96 S140 M03
G00 G41 X155.0 Z0 T0101 M08
G01 X-1.8 F0.2
G00 Z2.5
X138.0
G72 W2.0 R0.5
G72 P100 Q200 U1.5 W1.0 F0.4
N100 G00 Z-28.0
G01 X132.0 Z-25.0 F0.1
X107.238
G03 X76.599 Z-17.857 R20.0
G02 X0 Z0 R50.0
G01 X-1.8
N200 Z2.5
G70 P100 Q200 S175
G40 G00 X250.0 Z50.0 T0100
M30
%
```

- There are no block numbers, except for the N100 and N200 used for P and Q block in the cycle
- The machining consists of a series of face cuts
- Width of cut is 2.0 mm (W2.0)
- There are no straight diameters, so the *U* and *W* stock are more flexible
- Finishing is done with the same tool

As you see, the logic is identical to the **G71**, but the main motions are perpendicular to the centerline rather then parallel to it. They are, in fact, facing cuts.

G72 Cycle Format for Two Blocks

The programming format for **G72** is almost identical to the **G71** programming format. If the control requires a double block entry for the **G72** cycle, the programming format is:

```
G72 W.. R..
G72 P.. Q.. U.. W.. F.. S..
```

➥ *where ...*

First block:

W = Depth (width) of roughing cut
R = Amount of retract from each cut

Second block:

P = First block number of the finishing contour
Q = Last block number of the finishing contour
U = Stock amount for finishing on the X-axis diameter
W = Stock left for finishing on the Z-axis
F = Cutting feedrate (mm/rev or in/rev) overrides feedrates between P block and Q block
S = Spindle speed (m/min or ft/min) overrides spindle speeds between P block and Q block

G72 turning/ boring cycle is not suitable for parts that require cutting straight diameters. As the tool faces down, the cutting is smooth and the cycle works well. When the tool starts cutting a diameter, it has to move in the Z-positive direction only. Tool design is not suitable for this type of cutting, although it is possible. A major concern is the surface finish. For the tool to cut in the opposite Z-direction than it is designed for, the U-stock must be small, and the finish quality will be difficult to maintain.

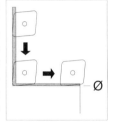

TURNING AND BORING

G73 Pattern Repeating Cycle

This is the last of the three multiple repetitive lathe cycles that is used for turning and boring. **G73** is called the *pattern repeating cycle*, as it repeats the finish contour a specified number of times. Each repetition will bring the tool closer to the part, until the final contour is machined to drawing dimensions.

The **G73** cycle can be used for machining of forged or cast material. It can also be used for parts where the basic contour shape had been pre-machined. Because of its nature, this pattern is also called *closed loop* cycle.

G73 Cycle Format For Two Blocks

Format is similar other cycles, just watch the U/W differences in the first and second block:

```
G73 U.. W.. R..
G73 P.. Q.. U.. W.. F.. S..
```

➼ where ...

First block:

U = X-axis distance and direction of relief - per side
W = Z-axis distance and direction of relief
R = Number of cutting divisions

Second block:

P = First block number of the finishing contour
Q = Last block number of the finishing contour
U = Stock amount for finishing on the X-axis diameter
W = Stock left for finishing on the Z-axis
F = Cutting feedrate (mm/rev or in/rev) overrides feedrates between P block and Q block
S = Spindle speed (m/min or ft/min) overrides spindle speeds between P block and Q block

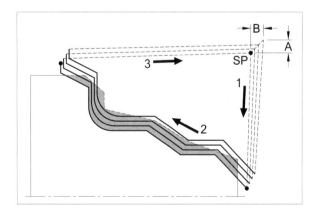

The schematic illustration shows the general concept of the **G73** cycle. The dimension *A* is the amount of the U-relief of the first block *plus* one half of the U-stock allowance from the second block. The *B* dimension is the amount of the W-relief plus the W-stock allowance from the second block.

Note that there is no programmable depth of cut amount available. For this cycle, it is not necessary, as the number of repetitions combined with other data determines the actual depth of cut during machining. Changing the R-amount to more divisions will make the depth of cut smaller, changing the R-amount to a fewer divisions will make the depth of cut larger. That assumes all other settings do not change.

The most ideal use of this cycle is for material where the roughing shape is larger (for OD) or smaller (for ID) than the finished contour but has the same shape. having such an equal amount is rare but not impossible, many investment castings are good candidates for **G73** cycle. With a large difference between the rough shape and the finished shape is significant, expect some 'air cutting'.

ONE-BLOCK FORMAT CYCLES

This section does not bring anything new in terms of cycles, but it does cover format of cycles for older CNC lathe controls that used only a single block of a cycle entry. This was the original cycle programming and many older controls are still used in machine shops. The main reason Fanuc changed the format to two blocks was to add flexibility at the control without changing system parameters. Otherwise, the one-block cycles perform the same machining process.

G71 Cycle Format for One Block

The one-block programming format for the older **G71** cycle is:

```
G71 P.. Q.. (I..) (K..) U.. W.. D.. F.. S..
```

➼ where ...

P = First block number of the finishing contour
Q = Last block number of the finishing contour
I = Distance and direction of rough semi finishing in the X-axis - per side
K = Distance and direction of rough semi finishing in the Z-axis
U = Stock amount for finishing on the X-axis diameter
W = Stock left for finishing on the Z-axis

D = Depth of roughing cut
F = Cutting feedrate (mm/rev or in/rev) overrides feedrates between P block and Q block
S = Spindle speed (m/min or ft/min) overrides spindle speeds between P block and Q block

G72 Cycle Format for One Block

The one-block programming format for the older **G72** cycle is:

```
G72 P.. Q.. (I..) (K..) U.. W.. D.. F.. S..
```

➼ *where ...*

P = First block number of the finishing contour
Q = Last block number of the finishing contour
I = Distance and direction of rough semi finishing in the X-axis - per side
K = Distance and direction of rough semi finishing in the Z-axis
U = Stock amount for finishing on the X-axis diameter
W = Stock left for finishing on the Z-axis
D = Depth of roughing cut
F = Cutting feedrate (mm/rev or in/rev) overrides feedrates between P block and Q block
S = Spindle speed (m/min or ft/min) overrides spindle speeds between P block and Q block

The *I* and *K* entries are not always available. They manage the last continuous cut before final roughing.

G73 Cycle Format for One Block

The one-block programming format for the older **G73** cycle is similar to both the **G71** and **G72** cycles:

```
G73 P.. Q.. I.. K.. U.. W.. D.. F.. S..
```

➼ *where ...*

P = First block number of finishing contour
Q = Last block number of finishing contour
I = X-axis distance and direction of relief - per side
K = Z-axis distance and direction of relief
U = Stock amount for finishing on the X-axis diameter
W = Stock left for finishing on the Z-axis
D = The number of cutting divisions
F = Cutting feedrate (mm/rev or in/rev) overrides feedrates between P block and Q block
S = Spindle speed (m/min or ft/min) overrides spindle speeds between P block and Q block

G71 TYPE I AND TYPE II CYCLES

A special consideration should be given to the most common roughing cycle **G71**. That is how it handles *undercuts* or *recesses, pockets* and *cavities*.

Normally, the direction of cut for outside cutting is in X+ and Z- for rear type lathes. Contour block may include X+ without the Z-axis and Z- without the X-axis. This is normal type of programming and Fanuc calls it *Type I* **G71** cycle. In practice, it means there are no undercuts within the contour.

If the drawing calls for an undercut (pocket), the program will have to specify how the undercut will be handled during roughing. Suitable tool selection also has to be made. Note the difference in programming format - it applies only to the first block of the finish contour, called by the P-address:

Type I

➼ *Format example:*

```
G71 U.. R..
G71 P100 Q200 U.. W.. F.. S..
N100 X..           (Z OR W IS NOT PROGRAMMED)
...
...
N200 ...
...
```

Type II

➼ *Format example:*

```
G71 U.. R..
G71 P100 Q200 U.. W.. F.. S..
N100 X.. Z..       (Z OR W IS PROGRAMMED)
...
...
N200 ...
...
```

From the CNC operator's point of view, it is an easy change to make during setup. The amount of Z is usually the same as the Z-position of the start point. W0 can also be specified. Consider changing to Type II cycle when the tool tries to cut a heavy depth in the undercut.

The finishing **G70** cycle will be performed as a continuous cut between the points *P* and *Q*.

SUMMARY

Removing material on CNC lathes is normally done by using *Multiple Repetitive Cycles*. There are four cycles used for turning and boring:

- **G70** Finishing cycle, used for part shape generated by either G71, G72, or G73

- **G71** Roughing cycle used for turning or boring where the direction of cut is primarily horizontal

- **G72** Roughing cycle used for turning or boring where the direction of cut is primarily vertical

- **G73** Roughing cycle used for turning or boring where the finished shape is cut repeatedly

The **G71** cycle is most commonly used for external or internal roughing. The roughing is mainly in the horizontal direction, while roughing with **G72** is mainly in vertical direction. If you understand the format and usage of this cycle, you will learn the others more easily.

There are two formats for this cycle, one format occupies only one program block, the other occupies two program blocks, depending on the control type.

For the one block entry, the depth of cut is measured per side in the units of the smallest increment, without the decimal point. For example, 2.7 mm depth will be programmed as *D2700*. For imperial units, an example of 0.125 depth will be programmed as *D1250*).

Even if the depth of cut is suitable for machining, you may consider changing it in order to influence the *last diameter depth*. This change may eliminate last cuts that are very small.

18 TOOL NOSE RADIUS OFFSET

The previous chapter addressed the major issues of lathe offsets sufficiently. In this chapter, the previous subject of lathe offsets will continue, with the special focus at the tool nose radius offset, and some issues that both the CNC Programmer and the CNC operator have to deal with, at least once in a while.

BRIEF REVIEW

Tool nose radius offset is nothing more for a lathe than a cutter radius offset for a mill. It's just a different name for the same control feature. Some CNC lathe manufacturers prefer to separate these terms, to suggest that there are some differences between radius offset in milling and turning. This distinction is more superficial than practical, because the internal functions are the same for both machine types. That is not to say that differences do not exist - they do.

In order to understand the difference between the two types of cutter radius offsets, you have to understand the differences in tooling. For both types of machining applications, cutter radius offset is used mostly for semi-finish and finish contouring. For both types of machining, it is important that the physical edge of the cutting tool follows the contour machined. Any contouring tool for milling can be represented by a circle that has the size of the tool diameter.

The previous chapter introduced four major subjects:

- **Geometry offset**
- **Wear offset**
- **Imaginary tool point**
- **Tool tip orientation number**

When it comes to actual setup, lathe tools are generally set as a distance between the cutting edge and part zero. This is a major difference from a milling cutter, where the setup is always measured to the cutter center. For milling applications, work offsets **G54**-**G59** are used for this purpose. On CNC lathes, the equivalent measurement is called *Geometry* offset. Work offsets can be used on CNC lathes as well, but that is not a common practice. Think of the geometry offset as being the lathe version of the work offset.

For reference, the geometry offset for both external and internal tool is set to the imaginary point (virtual point) - not to the nose radius center (TNR):

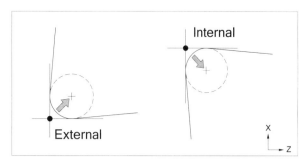

In order to understand how tool nose radius offset works, a simple contour will be evaluated with and without the offset programmed.

TOOLPATH EVALUATION

For the evaluation, an external contour will be used:

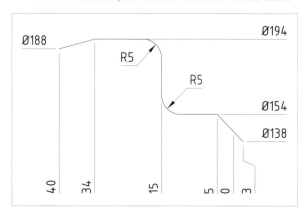

The contour has all geometrical features of a typical contour machined on CNC lathes:

- **Horizontal lines**
- **Vertical line**
- **Angular lines**
- **CW and CCW arcs**

Standard D-style 55° tool will be used, and only finishing toolpath.

189

Program Listing - No Offset

The program listing is correct in terms of the toolpath itself, but it does not contain the cutter radius offset:

```
N1 G21 T0100
N2 G96 S300 M03
N3 G00 X138.0 Z3.0 T0101 M08
N4 G01 X154.0 Z-5.0 F0.2
N5 Z-10.0
N6 G02 X164.0 Z-15.0 R5.0
N7 G01 X184.0
N8 G03 X194.0 Z-20.0 R5.0
N9 G01 Z-34.0
N10 X188.0 Z-40.0
N11 G00 U10.0
N12 X250.0 Z50.0 T0100
N13 M30
%
```

The cutting insert used has the common nose radius of 0.8 mm (0.0313 inch):

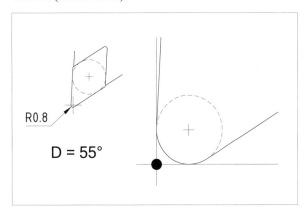

Keep in mind one very important fact - it is the *imaginary tool point* that was the point of reference during geometry offset setup. During toolpath, it will be *this* point that will be located at the endpoint of each contour entity. The next seven illustrations will show the effect in detail, individually for each entity.

For the front chamfer, not enough material has been removed, resulting in insufficient cutting:

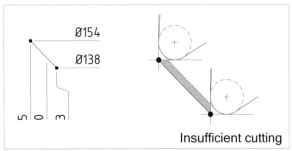

Insufficient (or excessive) cutting is caused by the empty space between the command point of the insert and the physical edge of the insert. The purpose of tool nose radius offset is to adjust the tool position in such a way that the tool nose edge is always tangent to the contour entity being machined

The next entity is a diameter, and since the geometry offset was set on a diameter, every contour diameter will be as per drawing. Note that the six o'clock position of the insert is the same as that of geometry setup:

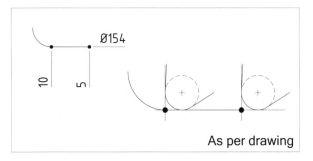

The inside arc will also be wrong, for the same reason as the one described for the front chamfer, resulting in an insufficient cut and much larger radius:

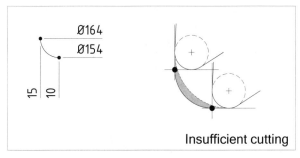

Just like diameters, when machining walls, faces and shoulders that are perpendicular to the lathe centerline, there will be no dimensional error, as the Z-axis geometry offset had been set from a face. Note that the nine o'clock position of the insert is the same as that of geometry setup:

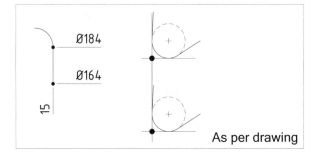

TOOL NOSE RADIUS OFFSET

The outside arc radius will also be larger than the radius specified in the drawing, resulting in an insufficient material removal:

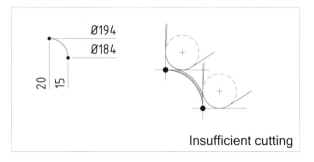

Insufficient cutting

Another diameter of the contour will be machined correctly and to drawing dimension:

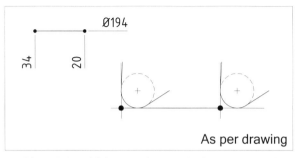

As per drawing

Although insufficient cutting results in a part that does not match drawing dimensions, there is no danger to the cutting tool itself:

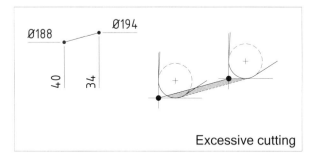

Excessive cutting

Summary

From the detailed evaluation it is clear that without tool nose radius offset the machined part will not be correct and will result in scrap. The only exception will be if the contour does not include angular or round entities at all, which is not likely for virtually all lathe jobs.

Yet, including the **G41** *or* **G42** command is a very simple addition to the program.

Program Listing - With Offset

In the corrected program, only a small but significant change has been made, causing the part to turn on size:

```
N1  G21 T0100
N2  G96 S300 M03
N3  G42 G00 X138.0 Z3.0 T0101 M08
N4  G01 X154.0 Z-5.0 F0.2
N5  Z-10.0
N6  G02 X164.0 Z-15.0 R5.0
N7  G01 X184.0
N8  G03 X194.0 Z-20.0 R5.0
N9  G01 Z-34.0
N10 X188.0 Z-40.0
N11 G00 U10.0
N12 G40 X250.0 Z50.0 T0100
N13 M30
%
```

Two preparatory commands **G42** and **G40** have been added to the program, in blocks N3 and N12 respectively.

Compare the *'before'* and *'after'* insert positions:

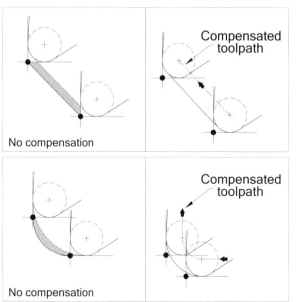

Offset Screen

For the tool nose radius offset to work, **G41** *or* **G42** has to be in the program and the *R* and *T* set in the control:

GEOMETRY			↓	↓
No.	X	Z	R	T
G 001	...	...	0.800	3

FACING TO CENTERLINE

One common machining operation on the CNC lathes is facing a solid part. Regardless of the operation that follows, it is important to have the front face perfectly flat all the way to the centerline. There are two possible reasons why a small tip is left where the face meets the spindle centerline:

- **Tool height set incorrectly** ... *setup error*
- **Incorrect tool position** ... *program error*

Tool Height Error

If the tool height is set incorrectly, the only remedy is to change it. This is usually done by placing a suitable shim under the tool holder, if the height is too low. If the height is too high, grinding off the contact surface of the tool holder will be necessary.

Incorrect Tool Position

If the tool height is correct, the problem is most likely in the program itself. Consider the following program segment:

```
N1 G21 T0100
N2 G50 S2750
N3 G96 S300 M03
N4 G00 X33.0 Z0 T0101 M08
N5 G01 X0 F0.2
N6 ...
```

In the program, the tool moves to the spindle centerline, which is X0, so the program looks correct. The problem is that there is no provision in the program for the *tool nose radius*. This will be the result of running the above program:

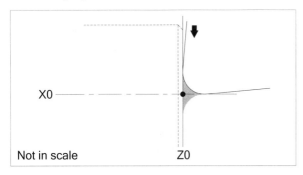

The undesired tip can cause problems for subsequent operations, for example, accuracy of a drill position. It can also create problems during assembly, if the face is not flat. The problem can also be esthetic.

A subject earlier in this chapter described the effect of geometry offset on a diameter of the part. With or without a tool nose radius offset, the part diameter will be machined as per drawing - and that is the exact situation in this case. The solution is simple - change the X0 in the program to an X-position *below centerline*, which will be a negative position. This corrected position should be *'just right'* - after all, there will be a certain amount of drag, since the spindle rotation below centerline is opposite to the one used for cutting.

What is *'just right'*? The minimum amount is the tool nose radius (TNR), doubled for diameter programming:

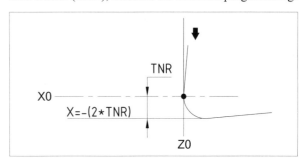

The next two tables show the minimum X-position and *suggested* X-position for the three most common corner radiuses:

TNR - Metric	Minimum X	Suggested X
R0.4	X-0.8	X-0.9
R0.8	X-1.6	X-1.7
R1.2	X-2.4	X-2.5

TNR - Imperial	Minimum X	Suggested X
R0.0156	X-0.0313	X-0.0350
R0.0313	X-0.0625	X-0.0660
R0.0469	X-0.0938	X-0.0980

The updated program will be the same as the last one, except the change of X-position in block N5:

```
N1 G21 T0100
N2 G50 S2750
N3 G96 S300 M03
N4 G00 X33.0 Z0 T0101 M08
N5 G01 X-1.7 F0.2
N6 ...
```

G41 - Yes or No?

There is no radius offset in the last program example. Does its absence have something to do with the tool tip problem? The answer is *no*. Many CNC programmers *do* include the **G41** command for facing, just to be consistent with the use of tool nose radius offset throughout the whole program.

The inclusion of the offset **G41** for facing will become very important for the next topic.

OFFSET CHANGE

In the previous chapter, the wear offset was changed from one to a different one for the purpose of maintaining dimensional tolerances. In those instances, the direction of cutting had not been changed.

Changing offset because the tool has changed cutting direction is much more common on lathes than on machining centers. A simple drawing example and a short program will illustrate a possible problem that may occur during the program check stage of the part setup.

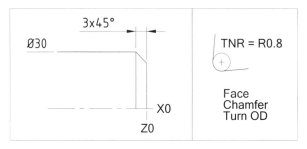

The program starts with a face cut, then the chamfer and continues diameter turning. **G41** is in effect for the face, **G42** is in effect for the chamfer and the diameter:

```
N1 G21 T0100
N2 G50 S2750
N3 G96 S300 M03
N4 G41 G00 X33.0 Z0 T0101 M08
N5 G01 X-1.7 F0.2
N6 G42 G00 X25.0 Z2.0
N7 G01 X30.0 Z-3.0 F0.15
N8 Z-..
...
```

When you evaluate the program, all data appear to be correct. In the following illustration, the enlarged tool motions can be viewed as individual motions (blocks N4 to N7). The control system only interprets the XZ point definitions and defines tool motions based on the preparatory G-codes.

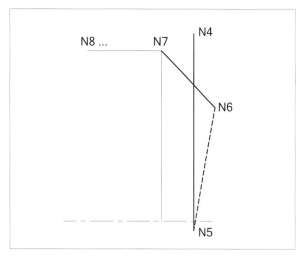

This program will not work as presented and has to be changed. During the program check, you will find that the facing cut has started well, but *it will not be finished*.

The CNC system will follow all program instructions, including the **G41** - *Left* and **G42** - *Right* commands. The program calls the facing cut to start in block N4 and end at X-1.7 in block N5, in the **G41** (left) mode. This facing motion is followed by a rapid motion in the opposite direction, creating a wedge. In order to be to the left, the tool nose radius also has to be to the left of the motion from block N5 to block N6. Since the control does not allow overcutting, it just follows instructions and remains in the **G41** (left) mode. The result is that the face cut will *not* be completed:

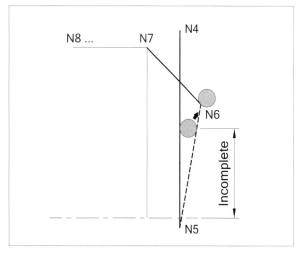

Just like cutter radius offset for milling, the tool nose radius offset is also the 'look ahead' type that prevents overcutting - and it is the *overcutting* error message that will be the result of the above program.

To correct this problem, the program itself has to be changed. The change in the program has to eliminate the wedge between the two tool motions. Just changing the Z-clearance alone will not do it, the angular motion has to be replaced by a rectangular motion:

```
N1 G21 T0100
N2 G50 S2750
N3 G96 S300 M03
N4 G41 G00 X33.0 Z0 T0101 M08
N5 G01 X-1.7 F0.2
N6 G00 Z2.0                 (*** BLOCK ADDED)
N7 G42 G00 X25.0       (*** CLEARANCE REMOVED)
N8 G01 X30.0 Z-3.0 F0.15
N9 Z-..
...
```

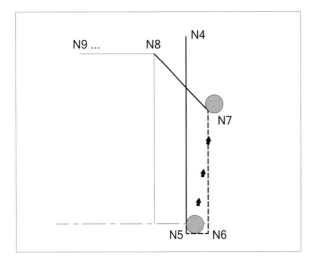

As long as the rectangular motion can contain the full circle of the nose radius, the face cut will be completed.

CLEARANCE AMOUNT

The previous program used a 2 mm clearance for the tool to move away from the front face. The last illustration shows that this amount was sufficient.

The rule of clearance is simple:

Clearance must be greater than double nose radius

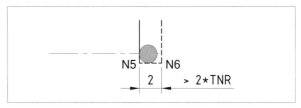

During machining, it may often become necessary to change the insert and - *along with it* - its nose radius. When such radius is smaller, for example, changed from R0.8 to R0.4, the program does not need to change.

On the other hand, a change from R0.8 to R1.2 will require an increase of the 2 mm clearance, otherwise an overcutting alarm will occur:

Considering the three most common nose radiuses available on lathe inserts (shown in both metric and imperial sizes), it is easy to establish that a clearance that is greater than the largest nose radius of the three will always be sufficient for majority of work:

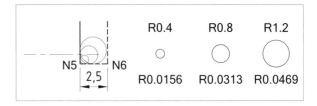

19 GROOVING ON LATHES

One of the common lathe operations performed on cylindrical parts is *grooving*. Grooving is a machining operation that will produce relatively narrow and shallow channel - called *a groove* - between two walls that are normally parallel to each other. This channel - *this new groove* - will always reflect the physical shape of the grooving tool used for the machining operation:

- A *square* grooving tool with *sharp* corners ...
  ... *will make a groove that has sharp internal corners*

- A *round* grooving tool ...
  ... *will make a groove with fully rounded bottom*

- A *square* grooving tool with *rounded* corners ...
  ... *will make a groove with internal corner radiuses*

The list can continue - an angular grooving tool will leave groove walls that match the angle of the grooving tool, etc. There are many types of grooves that can be machined on a CNC lathe. In spite of the many possibilities, the basic principles for both programming and machining remain the same.

GROOVE MACHINING

As with any other CNC toolpath, groove machining on CNC lathes always follows the CNC part program. Part programmers can use many methods of machining a groove, from a very simple one to those where dimensional accuracy and surface quality are of extreme importance. This applies to both manual and computer generated toolpath

The Simplest Groove

Regardless of the groove type, the machining process in its absolutely simplest form has only three motions:

- Rapid to clearance position XZ
- Feed in to the groove depth
- Rapid out of the groove

While certainly simple, this very primitive groove cutting will produce a groove that is very seldom practical and it certainly does *not* belong to the category of precision machining.

The width of the groove will be determined by the grooving tool insert, and its surface will be very poor. In addition, there will be no corner breaks, and no way to control its size or surface finish.

In spite of all these disadvantages, there is one application of this simplest groove, and that is when it is used to rough-out a groove and prepare it for finishing, generally with another grooving tool.

Standard Grooves

The two keys to a successful groove cutting is the tool insert, combined with suitable toolpath. Always think of a grooving tool as a specialized tool that does normal cutting (grooving), together with forming its own shape into the material. Both cutting and forming will be done at the same time.

Most standard grooves have walls that are perpendicular to one of the machine axes but may have different orientation:

- **Standard grooves machined on *diameter* have walls perpendicular to the X-axis**

  ... these are so called ***diameter grooves***

- **Standard grooves machined on *faces* have walls perpendicular to the Z-axis**

  ... these are so called ***face grooves***

- **Standard grooves machined on *angles* (usually 45°) have walls at an angle to both axes**

  ... these are so called ***corner grooves*** or ***neck grooves***

There are many grooves that will not have their walls parallel to each other but machined at an certain angle. The typical group in this special category is a common *O-ring* groove.

The vast majority of grooves machined on CNC lathes are oriented at 90° to the spindle center line (perpendicular grooves). There are also grooves that are parallel to the spindle center line, but they are used far less often. These grooves are commonly called the *face grooves*.

Typical Groove

It may be a bit of a stretch to define any groove as being typical, but most grooves are not the rough-type grooves but grooves with various degree of precision that share many characteristics. *Part-off* or *cut-off* operations resemble grooving, but are never considered precision operations and are not covered in this chapter.

As machining operations go, most lathe grooving operations are not too difficult or complicated in terms of tool setup. A high quality groove always starts with a high quality part program and a suitable grooving tool.

At the time of programming, both grooving tool size and its shape are selected by the programmer. The design of the tool that matters most is generally an interchangeable carbide insert. For common square grooves, this is typically an insert that is narrower than the groove width, leaving stock for finishing. This approach produces a groove of a better quality, as the insert follows a groove contour rather than just plunges into the material. For utility type grooves, the tolerances and finish quality are more open than for precision grooves.

Grooving Tool

Many designs of grooving tools are offered by many tool companies, but their basic design is the same. The carbide insert is mounted in a holder, either for right hand or left hand orientation. The insert itself has to match the holder and is generally defined by its width and maximum depth of cut. Its corners have a very small radius to add strength and are not considered during programming. The illustration below shows typical tools.

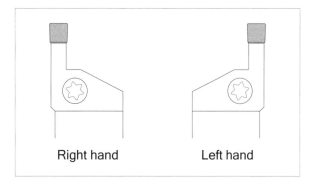

Right hand   Left hand

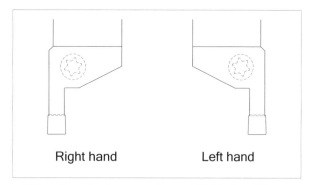

Right hand   Left hand

When these tools are used on a CNC lathe of the rear type (most common design), they are mounted in such a way that the insert and the clamp face away from the operator (see above illustration).

Although the right hand orientation is far more common than the left hand orientation, there are times, when the left hand tool may be a better choice to make the required groove. When working with a grooving tool, the most important part is the carbide insert.

Grooving Insert

Even the most common square grooving inserts come in several designs, particularly in the area of chip breaking. Square grooving inserts also have a very small corner radius, generally ignored in most programs. Another very important aspect of insert selection is its *maximum cutting depth*. This is also a consideration during the programming process. The CNC programmer not only selects the insert size and shape, but also the *command point* or *reference point* of the grooving insert.

Insert Reference Point

All lathe tools require a reference point. This subject was described in the chapter *Lathe Offsets*, starting on *page 161*. Like any other tool type, this is a point that is used for exact positioning of the tool relative to the part.

Unlike a single typical turning or boring tools, for grooving tools, the reference point can be in three general locations of the insert (rear lathe example used, viewed from the operator's position):

- **Lower left corner**   ... *most common*
- **Lower right corner**   ... *less common*
- **Lower middle position**   ... *least common*

The following illustration shows the reference point of the insert in relationship to the reference point of the part (part zero = program zero).

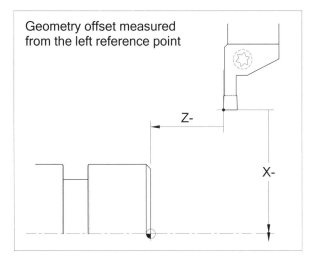

Geometry offset measured from the left reference point

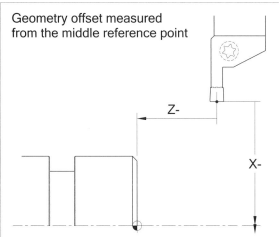

Geometry offset measured from the middle reference point

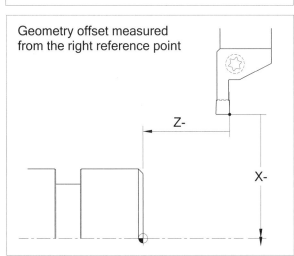

Geometry offset measured from the right reference point

The middle reference point is seldom used.

Control Settings

The measured amounts in both X-axis and Z-axis will be entered into the **GEOMETRY** offset number specified in the program. For example, program block

```
G00 X60.0 Z-23.1 T0404 M08
```

Geometry offset 04 is used in the program. During actual setup, the measured X-diameter was *-634.570* and the Z-length was *-241.760*.

The geometry offset entry will be:

GEOMETRY

Offset No.	X-axis	Z-axis	R	T
01				
02				
03				
04	-634.570	-241.760	0	0
05				
...				

A similar display will be shown for the WEAR offset.

Program Evaluation

In order for the CNC lathe operator to cut a precision groove with proper control settings, it is important to study and understand the provided part program. The following five program examples will present different objectives for the final groove:

- **Tolerances are not specified**
- **Depth tolerances specified**
- **Position tolerances only**
- **Width tolerances only**
- **Both position *and* width tolerances**

As each of the five examples is equally important, they will be discussed separately in detail. In all examples, the focus will be on how to handle each requirement at the control panel. This applies to both initial setup and adjustments during a production run. In all examples, the program will assume all pre-machining (turning or boring) had been completed and the program itself is correct in all aspects. The same basic external groove will be used for all examples, with the necessary changes for each example. The rules that apply to an external groove are similar to internal grooves and even face grooves.

Basic Groove Dimensions

The following four general drawings show the most common methods of how groove dimensions are applied in the engineering drawing.

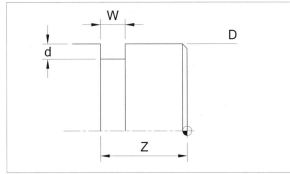

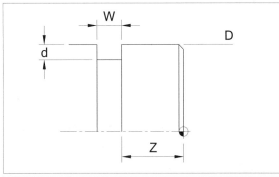

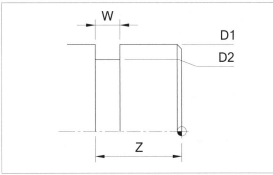

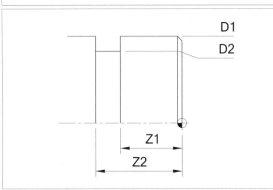

Keep in mind that the method of dimensioning is a design engineering feature with the purpose of one part interacting with another within the completed design. Other dimensioning methods also exist, usually variations on those shown here.

Groove Corner Breaks

A typical square groove will most likely have a corner break at the two outside corners. The corner break is usually a very small chamfer or a radius. Many drawings do not specify the corner size, so the CNC programmer makes a corner size suitable for handling and cosmetic purposes.

Drawing Used for Examples

A basic drawing will be used as an example of various control settings, specifically for the **GEOMETRY** and **WEAR** offsets:

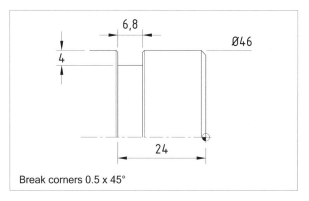

Break corners 0.5 x 45°

This is a simple groove with all relevant dimensions provided. No special requirements are necessary. This is a groove where no tolerances are specified.

TOLERANCES NOT SPECIFIED

The drawing example is a typical square groove, with no special tolerances required, although surface finish is somewhat important. Dimensions typical to many drawings will be used:

- Diameter   D = 46 mm
- Left groove wall location   Z-24.0
- Groove width   W = 6.8 mm
- Depth   d = 4 mm

   (Groove diameter   38 mm)

The programmer has selected a 5 mm wide grooving tool as tool number 4. Both geometry and wear offsets are the same - T0404. Program listing follows the illustration of important calculations:

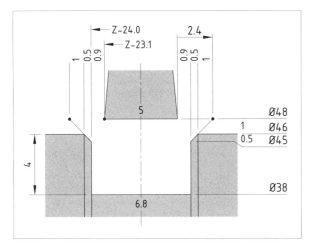

```
(T0404 - 5 MM GROOVING TOOL)
N41 T0400
N42 G96 S150 M03
N43 G00 X60.0 Z-23.1 T0404 M08
N44 X48.0               (AT CENTER OF GROOVE)
N45 G01 X38.1 F0.25 (CENTER CUT + CLEARANCE)
N46 G00 X48.0           (RETRACT TO START POINT)
N47 W-2.4               (START OF LEFT CHAMFER)
N48 G01 X45.0 W1.5 F0.1     (LEFT CHAMFER)
N49 X38.1 F0.2                  (LEFT WALL)
N50 G00 X48.0 Z-23.1    (BACK TO START POINT)
N51 W2.4                (START OF RIGHT CHAMFER)
N52 G01 X45.0 W-1.5 F0.1    (RIGHT CHAMFER)
N53 X38.0 F0.2                  (RIGHT WALL)
N54 Z-24.0                      (SWEEP BOTTOM)
N55 G00 X48.0 Z-23.1    (BACK TO START POINT)
N56 X100.0 Z75.0 T0400  (TOOL CHANGE POSITION)
N57 M01
```

The program represents a classic approach to machining a precision groove - a single plunge in the middle, retract, followed by left chamfer and left wall cutting. After retract, the tool cuts the right chamfer, the right wall and sweeps the bottom before final retract. It will be used in the upcoming sections with modifications for specific drawing requirements. The drawing - as presented above - does not have any dimensional tolerances specified. In this case, maintaining dimensions should not present any problems. You still want to maintain dimensions as required, but the range is a bit larger than if tolerances were specified.

Let's evaluate what can go wrong, than add some scenarios with common tolerances that may be required in precision grooving.

Dimensional Problems

Groove dimensional problems occur in several areas:

- Grooving depth
- Grooving position
- Grooving width
- Grooving width *and* position

If the program has been written correctly, all adjustments can be done at the control, using various offset adjustments. For the explanation, the **WEAR** offset screen will be used, not the **GEOMETRY** offset screen described on *page 197*. Wear offset looks identical to the geometry offset. Initial setting for offset 04 will be X0.000 and Z0.000. If the tool and related settings are all correct, the wear offset setting can remain at zero, as no adjustment are necessary.

WEAR				
Offset No.	X-axis	Z-axis	R	T
...				
04	0.000	0.000	0	0
...				
...				
...				

The *R* and *T* settings are not applicable, as grooving of this type does not required cutter radius offset.

DEPTH CONTROL

➡ *NOTE - This section specifically addresses offset adjustments to control the groove DEPTH only*

On vertical grooves, the groove depth is controlled by the X-axis settings, for horizontal grooves the depth is controlled by the Z-axis settings. The vertical groove orientation is much more common and will be covered in more detail.

With or without any specific tolerances of the groove bottom diameter, to fine tune the groove depth means adjusting the groove bottom diameter.

When measured, the groove diameter can be on size, oversize or undersize. During setup, an experienced operator will adjust the offset to produce a slightly larger diameter for external grooves and slightly smaller diameter for internal grooves. The idea is that the diameter can be measured, offset adjusted, and groove recut.

Keep in mind one important factor:

> **Offset adjustment of a groove depth in X-axis will always be on diameter**

Although the actual groove diameter is not directly on the drawing, it is easy to calculate that **46 - 2 x 4 = 38**.

External and Internal Grooves

Some grooves are dimensioned in the drawing as the outside diameter *D1* and groove diameter *D2*, as described on *page 198*. On the same page, some grooves are dimensioned by the diameter *D1* and groove depth *d*.

Regardless of how the drawing defines the groove dimensions, the part program will almost always use the groove *diameter*. Diameter can be measured by a vernier or a micrometer, groove depth by a depth gage. Offset adjustments are always applied to the groove diameter.

External Grooves

Let's look at two examples of offset adjustment for an oversize (O/S) and an undersize (U/S) groove diameter of an *external* groove. Expected diameter of the groove is 38 mm, as per drawing:

● **EXT Example 1 - Measured groove is ⌀ 38.8 mm**

For an *external* groove, a **larger** diameter means the groove is not deep enough (U/S). To adjust the offset, the groove diameter has to be made smaller, and a *negative* offset will be needed to force the tool to cut closer to the centerline. For the 38.8 mm oversize diameter, the offset adjustment will be:

    38 - 38.8 = -0.800   *is the offset adjustment*

● **EXT Example 2 - Measured groove is ⌀ 37.7 mm**

For an *external* groove, a **smaller** diameter means the groove is too deep (O/S). To adjust the offset, the groove diameter has to be made larger, and a *positive* offset will be needed to force the tool to cut further away from the centerline:

    38 - 37.7 = +0.300   *is the offset adjustment*

Now for the *internal* grooves - drawing is not necessary, the approach to offset adjustment is the same as for external grooves.

The example groove is expected to be ⌀ 65 mm.

Internal Grooves

As the expected *internal* groove diameter is ⌀ 65 mm, any **larger** diameter will be oversize (O/S), and any **smaller** diameter will be undersize (U/S).

● **INT Example 1 - Measured groove is ⌀ 65.2 mm**

For an *internal* groove, a **larger** diameter is a groove too deep (O/S). To adjust the offset, the groove diameter has to be made smaller, and a *negative* offset will be needed to force the tool to cut closer to the centerline:

    65 - 65.2 = -0.200   *is the offset adjustment*

● **INT Example 2 - Measured groove is ⌀ 64.75 mm**

For an *internal* groove, a **smaller** diameter means the groove is not deep enough (U/S). To adjust the offset, the groove diameter has to be made larger, and a *positive* offset will be needed to force the tool to cut further away from the centerline:

    65 - 64.75 = +0.250   *is the offset adjustment*

Do you see a pattern here? Do the calculations suggest something? There is a certain pattern and the calculation for all four examples have one feature in common - they all use the same formula:

> **Offset adjustment for X-axis grooves is ... :**
>
> # EXPECTED DIA - MEASURED DIA

An illustration may be useful for better understanding:

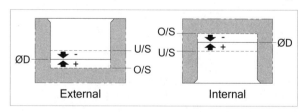

External   Internal

On the **WEAR** offset screen, offset adjustment of -0.8 is input - original offset 04 in X-axis was 0.000:

WEAR				
Offset No.	X-axis	Z-axis	R	T
...				
04	-0.800	0.000	0	0
...				

POSITION CONTROL

➡ *NOTE - This section specifically addresses offset adjustments to control the groove POSITION only*

Groove position is the exact distance from the Z-axis part zero to the groove wall as dimensioned in the engineering drawing. This location is generally the left or the right wall of the groove. Although it may seem irrelevant whether the left side or the right side of the groove is dimensioned, keep in mind one general rule:

> Drawing dimensions and tolerances are part of the engineering design and must always be maintained

In order to adhere to the rule, to control the groove position in its relationship to Z0, there is one more element that has to be considered - *insert reference point* (see *page 196*).

Unlike a typical insert for turning or boring, where only one reference point of the insert is used, grooving tool has two options:

- Reference point at the insert bottom and the LEFT edge
- Reference point at the insert bottom and the RIGHT edge

During program development, the programmer has to make a selection - the left edge or the right edge of the insert? While reasons for selecting one over the other are often a matter of personal preference, there are also sound reasons in therms of practicality and maintenance of dimensional accuracy.

Left Side Reference Point

Regardless of how the groove is dimensioned, setting the reference point of the grooving insert at the bottom and the *left edge* is far more common than any alternative. The reason is simple and, for the CNC operator, quite welcome. Setting the reference point to the left side edge is easier for part touch-off during setup and is consistent with setting all other tools.

The geometry measurement is to the intersection of the left side and bottom edge of the grooving insert. Tolerances related to the position and/or width of the groove can easily be handled by adjusting various offset settings.

Right Side Reference Point

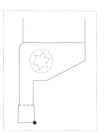

Setting the reference point of the insert to the right side edge may have some advantages, when the right wall of the groove is dimensioned on the drawing. In this case, the dimension will correspond with the Z-axis value in the program, when the tool cuts that wall.

In spite of this minor advantage, the disadvantages are far greater. The insert edge at right is harder to set at the machine and a face touch-off is not possible.

Reference Point Set in the Middle

By mentioning the left edge versus right edge pluses and minuses, what about setting the insert reference point to the middle of the insert width? Some programmers like to do that, as it makes groove programming symmetrical, therefore easier in some ways.

The problem with this method is that while the programming may be easier, it is much more difficult to set such reference point at the machine.

Offset Adjustment

Before you adjust the position of the groove along the Z-axis, keep in mind one rule:

> Position change of the groove does not change its width or depth

The above statement indicates that the offset change takes place only along the Z-axis. The example drawing shows a 24 mm dimension to the left wall of the groove:

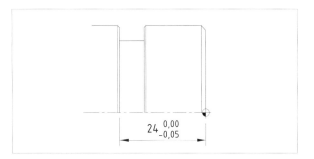

For the example a tolerance has been added. Measuring the distance from the front face (Z0) to the left wall two situations will require offset adjustment:

- Left wall too much to the left ... more than 24.000
- Left wall too much to the right ... less than 23.950

For the last example, no tolerance was given. In this example, the tolerance has to be carefully evaluated.

Programmed Target Dimension

Programmers normally use the *nominal* dimensions and ignore the tolerances - the initial example is no exception. You may find that some programmers may use a middle of the tolerance range in the program. In many ways it is a matter of opinion. The opinion promoted here is to use nominal dimension of *Z-24.0* and leave the offset adjustment to the CNC operator. That also means the target dimension has to be established at the control first. What is the best Z-target at the control? Z-24.000? Z-23.950? Somewhere in the middle? Middle makes sense - but not necessarily the *exact* middle.

As the same tool cuts many grooves, it gets slightly smaller by wear. Its cutting capabilities are still intact, but the tolerance of 50 microns (~0.002 inch) has to be taken into consideration. Do we aim the offset adjustment more to the left or to the right? What is better to aim for - Z-23.960 or Z-23.990, for example? Since the groove cannot be more than Z-24.000, it is better to aim further from zero, so Z-23.990 is a better choice. As the grooving tool becomes even slightly smaller, the actual dimensional shift will be further from the nominal size, but still well within tolerances. The following two examples will aim for the target dimension for the groove position of **Z-23.990**.

➡ *Example 1 - Measured distance is Z-23.930*

This dimension does *not* result in a scrap, because the grooving tool is *smaller that the groove width*. Had the tool width been the same as the groove width, the part would be a scrap. The offset has to be adjusted for the next part, only to conform to the target of Z-23.990. As the measured dimension is too far from the groove left face (closer to Z0), the adjustment must be done in the direction of the left face, which is Z-negative direction:

Offset adjustment	=	Target - Measurement	
Offset adjustment	=	-23.990 - -23.930	
	=	-0.060 mm	

The wear offset *04* must be adjusted by the negative amount of 0.060 mm (-0.060).

➡ *Example 2 - Measured distance is Z-23.990*

This dimension is right on. There is no need of adjustment, just monitoring the size for subsequent grooves. The offset calculation follows the same formula:

Offset adjustment	=	Target - Measurement
Offset adjustment	=	-23.990 - -23.990
	=	0.000 mm

The wear offset *04* does not require adjustment.

Note that in both calculations, the numbers were *negative*, subtracted from each other.

> Offset adjustment for Z-position is ... :
> # EXPECTED Z - MEASURED Z

There are other ways to find the offset adjustment amount - the method presented used the same formula for both examples.

> Use caution
> when adding or subtracting negative numbers

One final note that relates to the groove Z-position. Sometimes, it is necessary to change the tool itself. For example, the programmed grooving tool of 5 mm is not available, but a smaller insert of 4 mm or a larger insert of 6 mm is available. In this case, *the program itself would have to be changed*. Look at the changes required for a 6 mm grooving tool (underlined). The 1 mm difference between tools has been distributed by ±0.5 mm:

```
(T0404 - 6 MM GROOVING TOOL)
N41 T0400
N42 G96 S150 M03
N43 G00 X60.0 Z-23.6 T0404 M08
N44 X48.0                (AT CENTER OF GROOVE)
N45 G01 X38.1 F0.25 (CENTER CUT + CLEARANCE)
N46 G00 X48.0         (RETRACT TO START POINT)
N47 W-1.9              (START OF LEFT CHAMFER)
N48 G01 X45.0 W1.5 F0.1     (LEFT CHAMFER)
N49 X38.1 F0.2                    (LEFT WALL)
N50 G00 X48.0 Z-23.6  (BACK TO START POINT)
N51 W1.9              (START OF RIGHT CHAMFER)
N52 G01 X45.0 W-1.5 F0.1    (RIGHT CHAMFER)
N53 X38.0 F0.2                   (RIGHT WALL)
N54 Z-24.0                     (SWEEP BOTTOM)
N55 G00 X48.0 Z-23.6  (BACK TO START POINT)
N56 X100.0 Z75.0 T0400 (TOOL CHANGE POSITION)
N57 M01
```

WIDTH CONTROL

> **NOTE** - *This section specifically addresses offset adjustments to control the groove WIDTH only*

Many groove designs require the groove width to be within certain tolerances, usually to allow for a physical match with another part during assembly. For standard horizontal groove used for examples in this chapter, that means offset adjustments in the Z-axis, and Z-axis only. The drawing change will also require other changes:

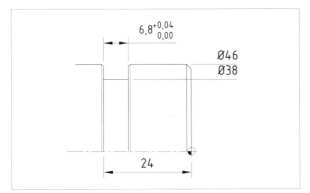

All original dimensions have been preserved, but the grooving width of 6.8 mm has to be revised and a tolerance added. The tolerance is to the plus side, which means the groove can be wider up to 0.04 mm (~0.0015 inch), but cannot be narrower.

Program Requirements

The program as presented originally (see page 199), has to be changed as well. Ask yourself - *why?* After all, the wear offset *04* has been programmed and is active throughout the whole program. If the program remains as is, yes, you can manipulate the wear offset *04* in the Z-axis as much as you want, but that will *not* change the groove width, only its position. That was the subject of the previous section.

The major change in the part program for the groove is to use of *TWO OFFSETS* for the same tool. While experienced CNC operators generally have no problems working with one geometry or wear offset per tool, they may not have the opportunity to use two (or more) offsets for the same tool.

After all, the position is good, so why change it? It is the *width* that needs to be controlled and that can only be achieved by one offset controlling the position, such as wear offset *04*, and *another* offset controlling the width, such as offset *14*, for example.

Program Change

Experienced programmers will anticipate the need for groove width and apply the necessary code to the program. Unfortunately, often it is left at the CNC operator to make the change.

The program using two offsets should contain a message that two offsets have been programmed and even identify them. In the program, you should check for:

- Blocks containing offsets
- Numbering of offsets
- Offset cancellation

The original program has been adjusted to control groove width separately from groove position, using two wear offsets - *04* and *14*. Keep in mind that each offset has its own special purpose:

- Wear offset 04 controls the groove precise location
- Wear offset 14 controls the groove precise width

What about the depth? That is a separate issue altogether and is not a factor here. The conclusion is:

> **Settings for offsets controlling the groove position and groove width separately MUST have the identical X-amount**

In other words, if you want to control the groove depth *in addition* to position and width, X-offset setting for both offsets *04* and *14* must be the same.

Here is the original example, modified for two offsets:

```
(T0404 - 5 MM GROOVING TOOL)
N41 T0400
N42 G96 S150 M03
N43 G00 X60.0 Z-23.1 T0404 M08
N44 X48.0                   (AT CENTER OF GROOVE)
N45 G01 X38.1 F0.25 (CENTER CUT + CLEARANCE)
N46 G00 X48.0           (RETRACT TO START POINT)
N47 W-2.4               (START OF LEFT CHAMFER)
N48 G01 X45.0 W1.5 F0.1       (LEFT CHAMFER)
N49 X38.1 F0.2                    (LEFT WALL)
N50 G00 X48.0 Z-23.1    (BACK TO START POINT)
N51 W2.4 T0414          (START OF RIGHT CHAMFER)
N52 G01 X45.0 W-1.5 F0.1      (RIGHT CHAMFER)
N53 X38.0 F0.2                   (RIGHT WALL)
N54 Z-24.0 T0404              (SWEEP BOTTOM)
N55 G00 X48.0 Z-23.1    (BACK TO START POINT)
N56 X100.0 Z75.0 T0400  (TOOL CHANGE POSITION)
N57 M01
```

Study the underlined changes - cutting on the left uses wear offset *04*, cutting on the right uses wear offset *14*. Offsets will take effect during the actual tool motion.

The example will show the offset change in its simplest form - wear offset *04* is set to zero.

→ **Example 1 - Groove width needs additional 0.015 mm:**

The offsets are set as expected:

WEAR

Offset No.	X-axis	Z-axis	R	T
...				
04	0.000	0.000	0	0
...				
14	0.000	0.015	0	0
...				

In the next section, both groove position *and* groove width will be combined.

POSITION AND WIDTH CONTROL

→ **NOTE - This section specifically addresses offset adjustments to control both the groove POSITION and its WIDTH**

This particular section does not really bring anything new except a couple of examples. If you understand the wear offset adjustments from the last two sections, you will understand this section as well. Take the last example requirement:

→ **Example 1 - Groove width needs additional 0.015 mm:**

Although the change requirement is the same, this time the wear offset *04* contains settings related to the groove position - these settings have to be respected:

WEAR

Offset No.	X-axis	Z-axis	R	T
...				
04	0.035	-0.026	0	0
...				
14	0.035	-0.011	0	0
...				

This is an area where changing wear offsets needs good understanding of the basic concepts and concentration while making the change. It's all in the numbers.

From all the dimensional possibilities and various examples, this is the rarest but also the toughest one in two areas. First, the programmer has to do a really good job to apply the idea, then the CNC operator has to apply the idea in practice. One without the other will present only problems, not solutions.

For a skilled CNC programmer it is a relatively simple task to program the necessary features. The burden of success is on the CNC operator, providing the program itself is correct.

GEOMETRY OFFSET

All the discussion here was about the wear offset and how to change it for a particular objective. What about the *GEOMETRY* offset?

In most work, once the tool is set and its geometry offset established and entered into the control system, there is no need to make any changes to it. CNC lathes have settings separate for the *GEOMETRY* and *WEAR* offsets, and all adjustments should always be done on the *WEAR* offset screen.

Change of Tool Width

This subject has already been raised - see the program example on *page 202*. The only remaining question is *how* the change of tool width affects the geometry offset.

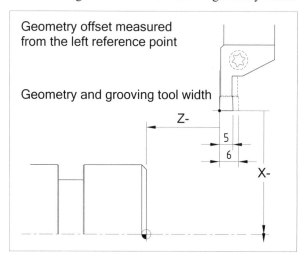

Looking at the illustration, it should not affect it at all, because of the tool design, although minor adjustments may be necessary because of tolerances. This is also one major advantage of setting the command point to the *left side* of the grooving tool, but check the tool anyway.

Program Requirements - Summary

When the programmer encounters a drawing that has a groove dimensioned with tolerances, he or she will evaluate the supplied dimensions and evaluate the specified tolerances. The result of this initial evaluation is how the available offsets are utilized. Understanding the program intentions is always a must on the part of the CNC operator.

A skilled CNC programmer will realize very soon in the program development that to control the groove position only can be done with a single offset. The same skilled programmer will also realize that to control the groove width only can also be realized with a single, but different offset. As the wear offset for groove position is always required, the challenge for both programmers and operators is to combine them together. Combining the offset that controls the groove position and the offset that controls the groove width cannot be done with a single offset.

The solution is to incorporate TWO unique offsets for the single grooving tool in the program.

GROOVING CYCLES

Under the multiple repetitive cycles in Fanuc manual you will find two that relate to grooving, although they are described as *drilling* cycles:

- G74   End face peck drilling cycle
- G75   Outer / Inner diameter peck drilling cycle

By Fanuc definitions only, there are no grooving cycles available for CNC lathes. On the other hand, if you look carefully what parameters each cycle allows, either cycle can be used for grooving as well, not just for drilling. For details, consult Fanuc documentation. This section only covers basic descriptions, as these are not primary cycles for majority of grooving. These descriptions are important to the CNC operators, as they allow correct interpretation of each cycle in the program, and allow informed changes, if necessary.

In either cycle description, there is the word *'peck'*. That means the cut can be an interrupted cut. Keep in mind that these cycles are basically drilling cycles, *not* intended for precision grooving, but may be very useful for roughing grooves, particularly those are fairly wide and require a number of steps the grooving tool has to make.

G74 Cycle

Of the two cycles, this one is used rarely for grooving. It is useful for roughing wide grooves oriented along the Z-axis. The format for a two-block cycle is the same for drilling and grooving - the explanation is adapted to grooving:

G74 R..
G74 X.. Z.. P.. Q.. R.. F..

➡ where ...

R = Return amount (retract clearance between cuts) (first block only)
X = Start diameter of the groove
Z = Final groove depth
P = Movement amount in X direction (without sign) (equivalent to the width of cut)
Q = Depth of cut in Z direction (without sign) (depth of each cut)
R = Relief amount of the tool at the cutting bottom (X-axis clearance after each peck)
F = Cutting feedrate

In essence, this drilling cycle can be used to rough out a groove located on the face of the part. Subsequent finishing will be required. Both applications are the responsibility of the CNC programmer.

G75 Cycle

Fanuc lathe controls offer a similar cycle listed under the multiple repetitive cycles section in the manuals - the **G75** cycle. This cycle is designed for grooves that also require an interrupted cut, for example, those machined in tough materials, deep grooves, or in any other type of machining where the chip breaking will improve the machining process. The **G75** cycle can also be used for cutting multiple grooves that share the same characteristics. While this cycle is not very suitable for grooves of high precision and surface finish, it can be used very successfully for wide groove roughing in X-direction. There is no stock allowance in this cycle, so the roughing parameters must account for that. This cycle is the direct opposite of the **G74** cycle, described in the previous section for a two-block application only.

The **G75** cycle also can be programmed in two formats, depending on the control unit. As this cycle is more practical than the **G74** cycle, both one-block and two-block program formats will be presented.

For the older 10/11/15 Fanuc controls, use the one-block format:

```
G75 X.. Z.. I.. K.. D.. F..
```

➥ where ...

X	=	Final groove diameter
Z	=	Z-position of the groove (last groove for multiple grooves)
I	=	Depth of each cut (positive value)
K	=	Distance between grooves (positive value) (for multiple grooves only)
D	=	Relief amount at the end of cut
F	=	Cutting feedrate in selected units

Spindle speed can also be programmed in the same block.

A two-block format is used for recent controls, such as 0i/16i/18i/20i/21i/30i/31i/32i Fanuc systems:

```
G75 R..
G75 X.. Z.. P.. Q.. R.. F..
```

➥ where ...

R	=	Clearance for each cut (return amount) (first block only)
X	=	Final groove diameter
Z	=	Z-position of the groove (last groove for multiple grooves)
P	=	Depth of each cut (positive value)
Q	=	Distance between grooves (positive value) (for multiple grooves only)
R	=	Relief amount at the end of cut (second block)
F	=	Cutting feedrate in selected units

A program sample for a one-block input **G75** cycle - drawing is not necessary, just the format and the meaning of individual data. This is a sample for a single groove. As the example represents only a single groove, note that there is no Z-location and there is no distance between grooves.

```
...
N11 T0100
N12 G96 S175 M03
N13 G00 X28.0 Z-11.0 T0202 M08
N14 G75 X20.0 I1.0 F0.15        (1 MM DEPTH)
N15 G00 X100.0 Z200.0 T0200
N16 M01
...
```

For a two-block format, the program will change only slightly. This time a basic drawing will be used - input data are also important:

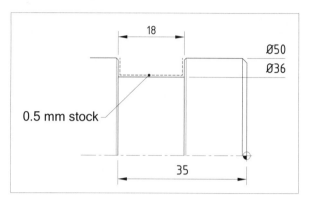

```
...
N31 T0500                       (5 MM GROOVING TOOL)
N32 G96 S175 M03
N33 G00 X52.0 Z-34.5 T0505 M08
N34 G75 R0.25
N35 G75 X37.0 Z17.5 P4000 Q1000 R0.1 F0.15
N36 G00 X100.0 Z200.0 T0500
N37 M01
...
```

Also note that the **G75** command has to be repeated in both blocks. This is consistent with other multiple repetitive cycles. Addresses *P* and *Q* do not take a decimal point and must be input in accordance of minimum increment rule for the selected units (0.001 mm or 0.0001 inch). In the example *P4000* is 4.000 mm and *Q1000* is 1.000 mm. The example groove will leave 0.5 mm stock on both walls and the bottom. Groove finishing is required and should be included in the program.

The main difference between the two formats is that the relief amount for each cut can only be programmed in the two-block format. In the one-block format, this amount is set by a system parameter.

Both formats do support programming of the relief at the bottom of the groove.

20

THREAD CUTTING

There is a common feeling amongst CNC programmers and operators that when it comes to threading, it either goes smoothly or it creates nothing but problems. Although such a feeling is not always supported in reality, there is no doubt that threading operations on a CNC lathe can - *and do* - present various problems.

Modern threading almost always uses a threading cycle, so it is important for the operator to understand not only the general structure of the cycle and meaning of individual data, it is equally important to understand the threading process itself.

THREADING PROCESS

On a CNC lathe, the method of machining a thread falls into two categories:

- Thread cutting
- Tapping

The majority of threads are cut rather than tapped. Cutting a thread is actually a combination of material removal by cutting and forming simultaneously. Another method of making a thread is by forming only (no cutting), which produces a rolled thread. This method is not part of general CNC operations.

CNC threading process requires the following items are selected carefully:

- Threading tool
- Threading insert
- Spindle speed and feedrate

All three have to work together. Although all three items are selected by the programmer, it is really only the spindle speed and feedrate that appears in the program directly. Threading tool and insert may be described in comments and it is up to the operator to make sure they are suitable for the given job.

The threading process itself is also a part of the program and is called the *single point thread cutting*. Many threads on manual lathes can be done by using threading dies - this is not the case of thread cutting on CNC lathes. The threading tool uses a single insert that has to cut the thread at multiple depths to reach the final depth.

THREADING TOOLS

The threading tool and the threading insert are separate entities. The threading tool is a tool holder mounted in the turret and is made of mild steel. A threading insert is inserted into the tool holder and secured, usually by a clamp. A single tool holder can accept several different inserts, not only for threading, but also for grooving. Such a tool holder can remain in the turret for extended period of time, only the insert will change.

Threading tool holders are divided into two groups:

- External and Internal
- Right Hand and Left hand

Most CNC lathes accept both groups, so it is important to mount the holder according to the program.

Threading Tool Holders and Inserts

The most common threading tool holder has a special pocket milled at the tip that accommodates the insert. In the illustration, a right hand and left hand tool holders are shown, as viewed with the insert positioned up.

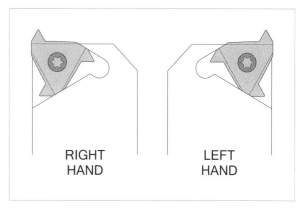

The insert shown is only one of several designs available. Some inserts have only two cutting tips and often require different holder for right hand than for left hand orientation. Check the tooling catalogue for inserts that a particular holder can accept. Also keep in mind that the actual position in the turret will be different for each hand of the holder.

Internal holders (threading bars) are also available in right hand and left hand orientation. The illustration shows a right hand and left hand threading bars, as viewed with the insert positioned up.

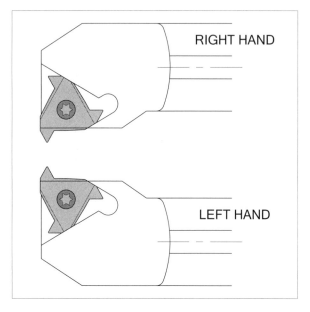

As for external threading tools, the bar orientation will be different for right hand and left hand bars.

Hand of Tool Selection

Whether to select a right hand threading tool or a left hand tool for setup is not up to the CNC operator. Although CNC lathes can generally accept both types, it is the programmer who selects the tool, the spindle rotation and the direction of cut. All three selections are important factors in threading, external or internal.

In most applications, threading is not a heavy operation, so the hand of tool makes no difference. Right hand tools are generally the preferred selection in most machine shops, because of their availability and cost. For heavy duty turning and boring on slant bed lathes, left hand tools provide forces towards the solid part of the machine, therefore they are more rigid. Of course the concept is the same for threading, but threading is hardly a heavy duty operation, in most cases anyway.

Right hand tools on a rear type CNC lathe (slant bed) are mounted *'upside down'*, meaning they face away from the operator. During setup the insert orientation does not present any problems, as the turret is oriented to show the tool with the insert facing towards the CNC machine operator.

The first indication of the hand of thread is in the program itself - in the form of how the spindle is programmed. Consider the following program blocks:

```
N43 T0600                        (OD THREADING)
N44 G97 S700 M03
N45 G00 X.. Z.. T0606 M08
...
```

Block N44 is the spindle speed command, applied for tool T06, a threading tool. The common **G97** is typical to threading - spindle speed *not* cutting speed. S700 specified the spindle rotation in revolutions per minute. The real clue is the rotation function:

- M03   Spindle rotation clockwise (CW)
- M04   Spindle rotation counter clockwise (CCW)

Most threads are right hand, and programmed in the Z-negative direction, towards the chuck, collet, or a similar work holding device. For these threads, the *M03* function will be used in the program. To compare the tool setup with the program, the next eight illustrations should provide a complete reference. All apply to a slant bed, rear type of CNC lathes, the most common design.

First, the two options for the *right hand external tool:*

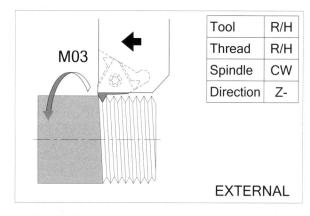

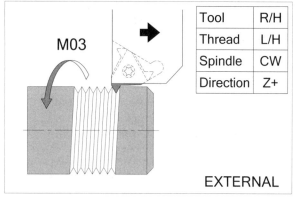

Two other options are for the *left hand external tool:*

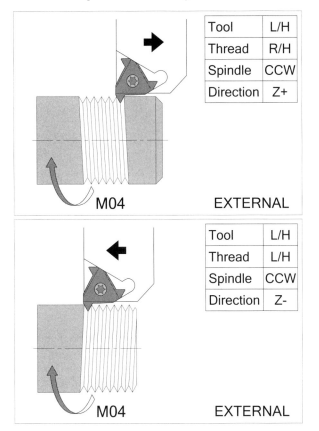

Tool	L/H
Thread	R/H
Spindle	CCW
Direction	Z+

EXTERNAL

Tool	L/H
Thread	L/H
Spindle	CCW
Direction	Z-

EXTERNAL

Both external and internal threading follows the same definitions. The method described for external threads is also applied to internal threads.

First, the two options for the *right hand internal thread:*

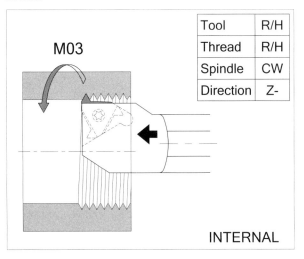

Tool	R/H
Thread	R/H
Spindle	CW
Direction	Z-

INTERNAL

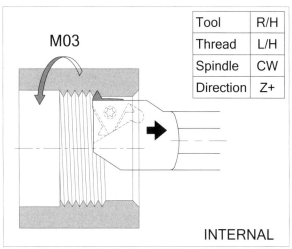

Tool	R/H
Thread	L/H
Spindle	CW
Direction	Z+

INTERNAL

The other options are for the *left hand internal thread:*

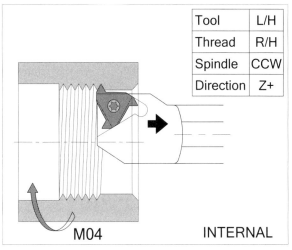

Tool	L/H
Thread	R/H
Spindle	CCW
Direction	Z+

INTERNAL

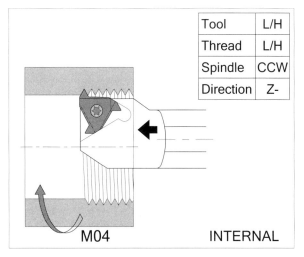

Tool	L/H
Thread	L/H
Spindle	CCW
Direction	Z-

INTERNAL

Use the previous illustrations to compare your understanding of the tool setup against the programmer's intent. Regardless of which threading method is used, the tool selection, spindle rotation and the direction of cut will always be critical to successful threading.

THREADING INFEED

The **G76** cycle, presented shortly, contains quite a number of settings that can be changed at the CNC machine, generally for the purpose of improving cutting efficiency or quality of the thread. One of the important changes is to control the *infeed of the thread*.

Threading infeed is the direction of the threading tool when in enters each threading pass. There are four common infeed methods (60° V-thread will be used):

- Radial infeed (straight)
- Flank infeed
- Modified flank infeed
- Alternating flank infeed

As most threads are located on a diameter, the infeed direction is relative to the spindle centerline - either as *perpendicular* or at an *angle*. All infeed methods have advantages and disadvantages.

Radial Infeed

Radial infeed is perpendicular to the machine axis. As the threading insert enters each cut, both insert edges are engaged equally.

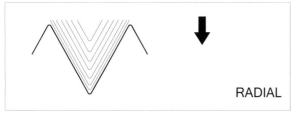

The equal cutting pressure prevents chipping of the insert edge, which is the main advantage of this method. This relative advantage for all roughing passes turns into a disadvantage for the last pass, as it may introduce tool chatter that was not present for previous cuts. Another disadvantage applies to cutting threads in tough materials, particularly high-strength steel, such as 4340. As the chips from both edges concentrate in the same limited space, a small channel can be developed and the threading tip can get damaged. Generally not recommended.

For some threads (*ie.,* square), this is the infeed to use.

Flank Infeed

Flank infeed is a common description for a threading tool entering each pass at an angle.

The angle of entry is at 30°, one half of the standard thread form. When the insert enter each cut, it is only one edge that does all the cutting. This method usually minimizes burr buildup, but also causes certain amount of rubbing caused by the opposite edge, the non-cutting edge. The surface finish deteriorates even more in certain soft materials, such as copper or some types of aluminum. Marginally better than radial infeed.

Modified Flank Infeed

For the majority of threading jobs, using the *modified flank infeed* is the most preferred method. Similar to flank infeed, modified flank infeed enters each threading pass at an angle that is not 30° (for standard V-threads, metric or imperial). Instead, the angle is 29.5° or less, depending on program settings.

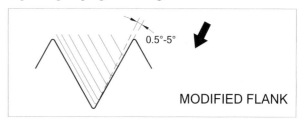

The angular difference from 30° only provides clearance for each threading pass, it does *not* change the actual angle of the thread. The cutting will be done by both edges of the insert.

> Note - Keep in mind that the threading tool itself has a 60° included angle, so the final thread will still be 60°.

Using this method of thread infeed provides the most satisfactory results for most threads. A certain amount of experimentation may be necessary, but often it is worth doing. The positive results can be transferred to other threads. Changing infeed is only one of the methods of improving thread quality - don't forget depth of cut and the number of threading passes.

Alternating Flank Infeed

As the illustration shows, both edges of the threading insert cut equally, due to the left-right and right-left tool shift for each threading pass. That feature alone extends the insert life and improves surface finish.

Is there a disadvantage in this method? Unfortunately, yes. The disadvantage is that the control system has to support this feature. Using the long hand threading method **G32** (see *page 213*) can eliminate the problem, as it is a truly manual method. On the other hand, the **G32** method, as powerful as it is, does not provide for convenient adjustments at the control. **G76** cycle offers this feature, but with some restrictions.

Alternating flank infeed method has particular benefits for large threads and threads that are generally designed to transfer motion (metric trapezoidal or *Acme* forms of thread).

THREADING INSERT

CNC part program seldom limits the operator to a particular insert. Even if the insert is specified, it is usually described in a program comment, and often only suggested. In selection of threading inserts (or any inserts), the CNC operator has a certain choice. For threading, the inserts come in two types:

- Full profile
- Partial profile

As usually, any choice has its pluses and its minuses.

Full Profile

Threading with full profile can be expensive, but the benefits may be worth the extra investment, particularly for large volume of parts, where the same thread is machined. *The same thread* are the key words when it comes to full profile. The cost of insert itself is not as much of a problem as the cost of keeping inventory for full profile inserts of various pitches. As you see from the illustration, full profile insert can only make *one thread pitch size*.

The next illustration shows a typical full profile of a threading insert. Note that the insert contacts both the *crest* (top) and the *root* (bottom) of the thread:

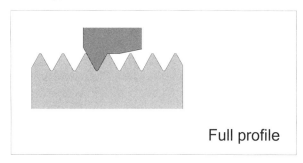

Full profile insert is also called a *'full topping'* insert. Its main benefit is that the final thread will have full depth with cleaned-up crest (top) and root (bottom). Another benefit of full profile threads is that because of their strength, they require fewer threading passes than when standard inserts are used. This feature somewhat offsets their higher initial costs.

Partial Profile

Partial profile insert is also called a *'non-topping'* insert. Compare the above illustration of a full profile with a similar illustration of a partial profile:

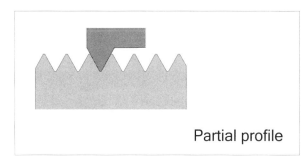

A fairly large range of threads can be cut by partial profile inserts, as there is no 'topping'. In theory, a fine thread and a coarse thread can be cut with the same insert. In practice, however, size of the tip radius will determine the actual range of threads suitable to cut.

For small threads, the tip radius is also small, but for coarse threads the insert requires additional strength and the radius must be larger. Generally, a partial profile inserts are manufactured with a smaller tip radius, which means they have to cut deeper for medium size threads.

Although it is a practice by some CNC operators to adjust the cutting tip radius manually by grinding it, keep in mind that any coated insert will lose some of its coating.

Clearances

Carbide insert is supported by a special shim, called an *anvil*. Anvil provides clearance for the insert and selecting the right anvil results in better utilization of the insert itself. For some threading tools, the anvil can be adjusted to match the thread helix angle. This adjustment improves cutting conditions and extends insert life.

Multitooth Inserts

Normal threading insert has a single cutting tip. When an insert is designed with more than one cutting tip, it is called *multitooth insert*. The concept of threading with multitooth inserts is progressive metal removal. Each threading tip of such an insert is designed to cut deeper into the material. The result is major savings in time, as fewer passes are required. About 60-70% cycle time can be saved when compared with a single point threading. Since the distribution of metal removal is spread over several tips, the tool life is also increased.

All these positive comments are offset by some disadvantages. The CNC lathe requires suitable power to match the higher cutting forces of these inserts. A typical multitooth insert can be quite wide, which means the last tip of the series has to have enough clearance to get out of the thread at full depth. Also, some chatter may be expected when threading tubular parts with thin wall.

THREAD RELATED DEFINITIONS

Threading is a special method of machining. As such, it uses many words that are exclusive to this method. Because of their common use, it is important to understand at least the most basic terms and definitions. This brief overview identifies those that are most commonly used in drawings, programs, and during actual machining.

Pitch

Pitch is the distance between two corresponding points of adjacent threads, measured parallel to the axis:

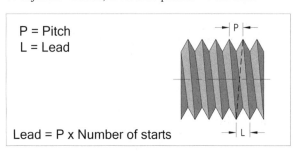

Lead

Lead is the amount of travel along an axis, measured in one full revolution (360°):

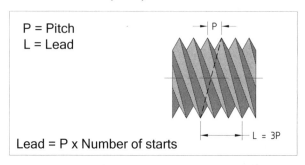

Note that in both illustrations, the lead has been defined the same. Lead is always the pitch multiplied by the number of threads. For single start threads, which make the majority of all thread cutting operations, the lead is multiplied by one, resulting in the same amount as the pitch. For example, a single start thread with 1.25 mm pitch will have the lead of 1.25 mm. A three start thread will have a pitch *P* but the lead will be **3 x P**. For example, if the pitch is 1.25 mm, the lead will be three times that amount, or 3.75 mm.

Threading Feedrate

Lead is very important for the threading feedrate - in fact, the lead *is* the feedrate:

> **Feedrate for single point threading is always the thread LEAD**

Feedrate is programmed in units per revolution.

Other Definitions

Amongst other common thread related terms, *Pitch diameter*, *Crest*, *Root*, *Major* and *Minor diameter* are used most frequently - a simple illustration identifies these terms (external thread shown):

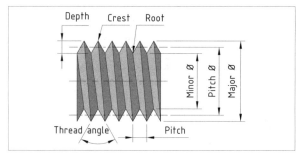

Pitch Diameter

When measured on a straight thread, **pitch diameter** is an imaginary diameter where the width of the thread and the width of adjacent space are the same. It is one of the most important dimensions for thread measurement but not normally needed for programming.

Crest

Crest is the top of the thread. For external diameters, crest is part of the outside diameter, for internal threads, crest is part of the inside diameter (diameter of the bore).

Root

Root is part of the thread bottom. For external diameters, root is at the bottom of the thread, for internal threads, root is at the largest diameter of the thread.

Major Diameter

Major Diameter is a cylinder formed by the crests of the thread - it is the *largest* diameter of the thread.

Minor Diameter

Minor Diameter is a cylinder formed by the roots of the thread - it is the *smallest* diameter of the thread

TPI - Threads Per Inch

Unique to imperial threads, **TPI** is an abbreviation for *Threads Per Inch*. This designation follows the nominal size of the thread - for example:

3/4-16   3/4 = 0.75 thread diameter - 16 threads per inch

16 threads per inch results in 0.0625" pitch, based on the formula:

Pitch = 1 / TPI

Hand of Thread

Hand of thread defines the thread rotation:

Right Hand Thread - Thread that makes a screw to rotate clockwise (majority of threads)

Left Hand Thread - Thread that makes a screw to rotate counter clockwise (minority of threads)

Tool orientation, spindle rotation and direction of cut are all important - see *page 208*.

METHODS OF PROGRAMMING

From the previous subjects, you can see that single point threading is a multi-pass operation. Considering the threading insert with a very small nose radius and the high feedrates it requires, it is easy to see the adverse comparison with turning.

Threading, or more accurately - *single point threading* - is process that uses a single pointed tool in the shape of the thread, and through a series of passes cuts the thread to the desired size. There are three methods used by CNC programmers:

- Longhand threading   G32
- Box threading   G92
- Repetitive cycle threading G76

Longhand Threading - G32

This method is the original method of all thread cutting. It uses a preparatory command **G32** (or **G33**). Each motion is programmed as an individual block, which means there are at least four blocks required for each threading pass. Typical threading process is shown here:

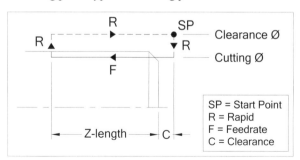

This method can result in very long programs that are virtually impossible to edit and may be too long to store in the control memory. **G32** is not a threading cycle.

Box Threading - G92

The so called *box threading* was the first attempt at a threading cycle, with the objective to make programming simpler and shorter. Most controls use preparatory command **G92** for this type of threading. The reason for the name 'box threading' is because the four basic threading motions are always parallel to either the X or the Z axis, resulting in a rectangle - or a *box*.

This is a cycle with very limited options and is seldom used in modern CNC threading.

Repetitive Cycle Threading - G76

Of the three methods listed, this is the most common method of programming threads. It is a very advanced cycle that has a number of features every CNC operator should understand thoroughly. Like any complex application, this cycle has a few limitations, and it may not be suitable for some special threads.

Over the years, the **G76** threading cycle has evolved from a single block input into a two-block input. The reason for the change was to provide the CNC operator with a cycle containing as many settings as possible that can be easily edited at the machine, if necessary. The earlier one-block cycle offered only a limited editing.

For editing at the machine, it is important to know the basic thread data in order to apply them to the cycle.

THREADING DATA

Several threading terms have been covered earlier in this chapter. These and a few others will have a direct connection to the **G76** cycle:

- **Thread form** ... V-thread, Trapezoidal, etc.
- **Thread depth** ... per side, between crest and root
- **Infeed method** ... straight, flank, modified flank
- **First pass depth** ... to control number of passes
- **Cutting feedrate** ... always the thread lead

Depth of Thread

Depth of thread is always measured per side, as actual depth. Two formulas are used in CNC thread programming, one is for external threads, another for internal threads:

External Threads

Metric:	Thread depth = 0.613 x pitch
Imperial:	Thread depth = 0.613 / TPI *or* 0.613 x pitch

Internal Threads

Metric:	Thread depth = 0.541 x pitch
Imperial:	Thread depth = 0.541 / TPI *or* 0.541 x pitch

When selecting or changing thread depth, consider the above formulas as a starting point only. Threads have different tolerances and fits, and fine-tuning the depth at the machine is often one way to meet the thread dimensional requirements.

Depth of the First Pass

Full depth of a thread cannot be machined in a single pass, and several passes are required. The actual number of passes is not normally programmed, the control system will calculate it. Threads made in hard materials and deep threads will need more passes than shallow threads or threads made in softer materials.

The **G76** cycle calculates the number of threading passes automatically, based on the total *thread depth* and the *first pass depth*. For a given depth, the greater the first pass depth, the fewer passes will be generated, and vice versa. The first pass depth is normally selected for the thread type, insert used, and the stock material.

Threading Feedrate

Only one rule exists for the thread cutting feedrate:

> **Threading feedrate = Lead of the thread**

The majority of threads are single start threads. For those threads, both lead and pitch of the thread are identical. For multi start threads, the feedrate is:

Feedrate = Pitch x Number of thread starts

Multi start threads are covered later in this chapter (see *page 218*).

Angle of Thread

Angle of thread is the included angle between two adjacent flanks. V-threads have a 60° angle, Whitworth has 55°, Acme has 29°, Metric Trapezoidal thread is 30°, and special German type pipe thread PG is 80°. All these thread angles are supported by the **G76** cycle - their main purpose is to define the infeed angle for the threading motions generated by the cycle.

Feel free to experiment with different angles for any given thread. The thread shape and size are result of the insert form, not the infeed angle.

Spring Pass

Although the term *'spring pass'* is not used in the cycle parameters, it is a common threading term. Spring pass is always performed at the end of the threading and refers to one or more threading passes that are made at the same diameter. The main purpose of spring pass(es) is to release tool pressure and remove the last remnants of the material. Tool chatter may be a side effect.

G76 THREAD CUTTING CYCLE

The **G76** multiple repetitive threading cycle is a true achievement in software engineering. With only one or two blocks of program, many different thread forms can be easily programmed and machined. The vast majority of CNC threading operations uses this cycle.

One-Block Format

Older Fanuc controls that are still used in production used the **G76** cycle with all data confined to one block. These controls include Fanuc models 10T, 11T, and 15T, as well as the early version of the 0T model. In the program, the cycle call is preceded with a data that define the start point of the cycle, and is followed by one block of the **G76** cycle:

```
<... XZ starting point ...>
G76 X.. Z.. I.. K.. D.. A.. F.. (P..)
```

➥ where ...

- X = *(a)* **Last diameter of the thread**
  (absolute diameter)
  ... or
  (b) **Distance from the start point to the last thread diameter**
  (incremental distance and direction)
- Z = **End of thread along the Z-axis**
  (can also be an incremental distance W)
- I = **Radial difference between start and end positions of the thread at the final pass**
  (I0 used for straight thread can be omitted)
- K = **Height of the thread**
  (positive radial value - decimal point is allowed)
- D = **Depth of the first threading pass**
  (positive radial value - no decimal point)
- A = **Angle of thread**
  (only A0, A29, A30, A55, A60, and A80 are allowed)
- F = **Feedrate of the thread**
  (same as the thread lead)
- *P = **Method of infeed (P1 - P4)** P1 = default
  (may not be available on all controls)

> *)* Optional reference:
> *CNC Programming Handbook*, Ed. 3, pages 363-366, and page 368 for details on the P-address

Programming the thread angle other than zero in the **G76** cycle, will result in a flank infeed, which provides longer tool life and better thread finish. Alternating flank infeed is also possible in the **G76** cycle, by programming the *P1-P4* option in the cycle (check your control system for availability).

Two-Block Format

All latest Fanuc lathe controls, in this case the Fanuc T-series *0i / 16i / 18i / 21i / 30i / 31i / 32i* (as well as related legacy controls without the *i-suffix*) support the two-block **G76** format. The purpose of the **G76** two-block format and function remain the same, only the actual data input has changed. If your control supports the two-block programming, its format is:

```
<... XZ starting point ...>
G76 P.. Q.. R..
G76 X.. Z.. R.. P.. Q.. F..
```

➥ where ...

First block:

- P = ... **is a six-digit data entry in three pairs:**
  Digits 1 and 2
  (number of finishing cut repeats - range is 01-99)
  Digits 3 and 4 - number of leads for gradual pull-out
  (0.0-9.9 times lead, no decimal point (00-99))
  Digits 5 and 6 - Angle of thread
  (only 00, 29, 30, 55, 60, 80 degrees allowed)
- Q = **Minimum cutting depth**
  (positive radial value - no decimal point)
- R = **Fixed amount for finish allowance**
  (decimal point is allowed)

Second block:

- X = *(a)* **Last diameter of the thread**
  (absolute diameter)
  ... or
  (b) **Distance from the start point to the last thread diameter**
  (incremental distance and direction)
- Z = **End of thread along the Z-axis**
  (can also be an incremental distance W)
- R = **Radial difference between start and end positions of the thread at the final pass**
  (R0 used for straight thread can be omitted)
- P = **Height of the thread**
  (positive radial value - no decimal point)
- Q = **Depth of the first threading pass**
  (positive radial value - no decimal point)
- F = **Feedrate of the thread**
  (same as the thread lead)

The P and Q addresses are used twice in the two blocks - do not confuse the P/Q addresses of the first block with the P/Q addresses of the second block. They have their own meaning that applies within the particular block only!

Some features of the two-block **G76** cycle are:

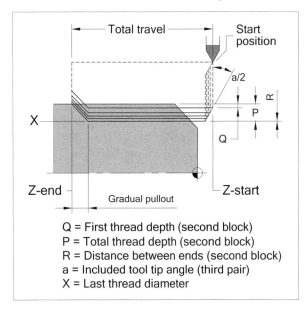

Q = First thread depth (second block)
P = Total thread depth (second block)
R = Distance between ends (second block)
a = Included tool tip angle (third pair)
X = Last thread diameter

The Z-position at the start and end of threading are clearly shown. The gradual pull-out is programmed as the second pair of the P-address in the first block (digits 3 and 4 - `P..01..`). Remaining parameters are described in the cycle definition.

The *Gradual Pullout* setting may need some explanation. During threading, the end of thread can be either in an open space (such as recess) or in the solid material. For open space, there is no need to pull the tool out of the thread gradually, so the setting of the `P--00--` will be zero, meaning no gradual pull-out.

For threads ending in solid material, at least one pitch length pull-out will be required `P--10--`:

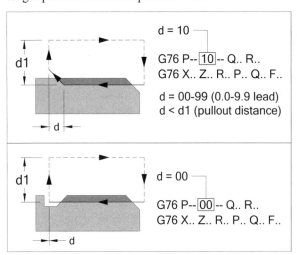

G76 Cycle Examples

The following examples (threading blocks only) are presented with the focus on possible program changes at the CNC lathe, primarily during setup. These changes may also be necessary to optimize threading operations.

➡ **Example 1 - Metric external thread - M36x4:**

This thread ends in a recess.

```
...
N11 G97 S600 M03
N12 G00 X45.0 Z10.0 T0101 M08     (START POINT)
N13 G76 P010060 Q050 R0.04
N14 G76 X31.096 Z-50.0 P2452 Q420 F4.0
...
```

Spindle speed is programmed in **G97** mode, *revolutions per minute*. Spindle rotation **M03** is for right hand tool. For the example, the spindle will rotate at 600 r/min.

Start Point is 4.5 mm above the 36 mm diameter. For best results, it is recommended to have the X-clearance greater than the thread lead (lead = pitch in this case). For the Z-axis, the front clearance has to absorb the acceleration of the tool from zero feedrate to 4 mm/rev. Two to three times the lead is the general rule of thumb.

First block of **G76**, the P-address. The three pairs are:

`P01....`   01 = one finishing pass

`P..00..`   00 = No gradual pullout

`P....60`   60 = included angle of the insert

The Q-address is the **minimum cutting depth**. Q050 represents 0.050 mm or 50 microns. As the insert cuts deeper it also cuts less with each pass. The Q-amount determines the minimum cutting depth that will not be any smaller for any remaining passes.

The R-address is the fixed amount for **finishing allowance**. R0.04 is 0.040 mm or 40 microns. This is the last depth of cut.

Second block of the **G76** cycle contains other cutting data required for the threading:

X-address is the last diameter cut - for V-threads, the formula to calculate the depth is:

0.613 x pitch = 0.613 x 4 = 2.452 mm

For external threads, the X-diameter is calculated by subtracting double the depth from the nominal diameter:

36 - 2 x 2.452 = 31.096 mm = X31.096

Z-address is the end of thread dimension as per drawing (not shown). In the example, it is 50 mm from the front face of the part.

The P-address in the second block is the **depth of thread**, which is always measured per side. *P2452* is equivalent to 2.452 mm. Note there is no decimal point - it is implied if you count *three* numbers from the back.

The Q-address is the **depth of the first pass**. *Q420* is equivalent to 0.420 mm (no decimal point is written). This is the deepest amount the tool will cut - all subsequent passes will be progressively smaller.

Finally, the F-address us the **cutting feedrate** in *mm/rev* and must be equivalent to the thread lead. For single start threads, the lead amount is equivalent to the pitch.

Editing - Metric Example

Assuming the program is generally correct, for the purposes of optimization you should *not* change:

- X, Z, and F in the second block

Other data can be changed, if necessary. The angle in the first block can be changed to achieve a different in-feed angle, for example, from 60° to 55°. Always watch for changing the P-depth (second block), and always make one change at a time.

➡ *Example 1 - Imperial external thread - 1-1/12 - 6:*

This thread ends in solid material.

```
...
N21 G97 S750 M03
N22 G00 X1.9 Z0.35 T0202 M08      (START POINT)
N23 G76 P021060 Q005 R0.003
N24 G76 X1.2957 Z-2.0 P1022 Q0160 F0.166667
...
```

The explanations for the imperial thread example are similar to those shown for the metric thread *(Example 1)*. Imperial threads are defined by the number of threads per inch (TPI). 6 TPI is six threads per inch and it means the thread pitch is **1/6 = 0.1666666**.

Spindle speed is programmed in **G97** mode, *revolutions per minute*. Spindle rotation **M03** is for right hand tool. For the example, the spindle will rotate at 750 r/min.

Start Point is 0.2 inches above the 1.5 diameter. For best results, it is recommended to have the X-clearance greater than the thread lead (lead = pitch in this case). For the Z-axis, the front clearance has to absorb the acceleration of the tool from zero feedrate to 0.16666 in/rev. 2x - 3x the lead is the general rule of thumb.

First block of **G76**, the P-address. The three pairs are:

P02....   02 = two finishing passes

P..10..   10 = One lead gradual pullout

P....60   60 = included angle of the insert

The Q-address is the **minimum cutting depth**. Q005 represents 0.005 inches. As the insert cuts deeper it also cuts less with each pass. The Q-amount determines the minimum cutting depth that will not be any smaller for any remaining passes.

The R-address is the fixed amount for **finishing allowance**. R0.003 is 0.003 inches. This is the last depth of cut.

Second block of the **G76** cycle contains other cutting data required for the threading:

X-address is the last diameter cut - for V-threads, the formula to calculate the depth is:

0.613 x pitch = 0.613 x 0.166667 = 0.1022 inches

For external threads, the X-diameter is calculated by subtracting double the depth from the nominal diameter:

1.5 - 2 x 0.1022 = 1.2957 inches = X1.2957

Z-address is the end of thread dimension as per drawing (not shown). In the example, it is 2 inches from the front face of the part.

The P-address in the second block is the **depth of thread**, which is always measured per side. *P1022* is equivalent to 0.1022 inches. Note there is no decimal point - it is implied if you count *four* numbers from the back.

The Q-address is the **depth of the first pass**. *Q0160* is equivalent to 0.0160 inches (no decimal point is written). This is the deepest amount the tool will cut - all subsequent passes will be progressively smaller.

Finally, the F-address is the **cutting feedrate** in *in/rev* and must be equivalent to the thread lead. For single start threads, the lead amount is equivalent to the pitch. For leads that do not round to four decimal places, the F-address allows up to six decimal place entry to increase accuracy, particularly for long threads.

Editing - Imperial Example

The same rules apply here as for the metric example.

Also keep in mind that factors outside of the program also influence overall thread quality - they include setup rigidity, tooling selection, part support, etc.

MULTI START THREADS

Threading operations are generally focused on cutting the so called standard threads, which are - *in great majority* - threads with a single start. For various special purposes, threads may also be defined as double-start, triple-start, quadruple-start, etc. Any thread that has more than one start is a multiple start thread.

Multi start threads are mostly used to transfer motion. For example, over the same length, a double start thread with 2 mm pitch will transfer motion twice as fast as a single start thread with 2 mm pitch.

From the CNC operator's perspective, the four most important program features are the *spindle speed*, the *threading feedrate*, the *Z-axis front clearance* and the amount of *shift* between threads. Here is an example.

➡ *Example - M75x2.5 - 3 Start Thread*

```
...
N31 G97 S150 M03
N32 G00 X91.0 Z18.0 T0303 M08
N33 G76 P010060 Q060 R0.05
N34 G76 X71.935 Z-70.0 P1533 Q300 F7.5
N35 W2.5
N33 G76 P010060 Q060 R0.05
N34 G76 X71.935 Z-70.0 P1533 Q300
N35 W2.5
N33 G76 P010060 Q060 R0.05
N34 G76 X71.935 Z-70.0 P1533 Q300
...
```

Both X and Z clearances are more than sufficient in the example. Note the rather very slow spindle speed (more on the subject in the next section). The feedrate is the thread lead and the shift *W* is always the thread pitch. The number of shifts is always one less than the number of starts. Also watch for the overall clearance in front of the thread, particularly if the tailstock is used or nearby.

SPINDLE SPEED LIMITATION

It may seem the issue of spindle speed for single point thread cutting is not worth much consideration. For most modern CNC controls that is true, except for some extreme cases. The situation is different for older systems.

Just like for turning and boring, the CNC programmer selects the spindle speed based on the material, tool used, and the setup conditions. While these selections are equally important in threading, there is one additional situation to be considered. It is a situation unique to threading and it is described as *spindle speed limitation*.

The question of *'why limit the spindle speed at all?'* can only be answered once you consider the fact that there is a direct relationship between the spindle speed and the cutting feedrate (lead of thread).

Machine Specifications

In an earlier chapter starting on *page 5*, various machine specifications have been described. One of them is listed in the lathe section, on *page 7 - Maximum cutting feedrate*. The particular lathe described there is a modern lathe that has maximum cutting feedrate of 1265 mm/min or 50 in/min.

In the majority of threading applications, you will not have to worry about the spindle speed at all, as long as the thread quality is good. There is at least one exception where you should pay attention to the spindle speed - *multi start threading*. In fact, the more starts, the more important it is to watch the spindle speed.

Consider the previous example - spindle speed was set to 150 r/min and feedrate at 7.5 mm/rev. Converted to mm/min, it amounts to **150 x 7.5 = 1125 mm/min**. This is just under the maximum feedrate allowed. The basic rule is simple - the heavier the thread lead (feedrate), the slower the spindle speed has to be. Maximum cutting feedrate is given in *units per minute*, **not** in *units per revolution*. The formulas are:

Units/min = r/min x Units/rev

Maximum r/min = Maximum cutting feedrate / Thread lead

Older machines have much lower maximum cutting feedrate and calculating the spindle speed is important.

THREADING DEFAULTS

Various system parameters relate to threading in general and **G76** cycle in particular. See *page 251* for more details. Just keep in mind that incorrect settings of system parameters may result in dangerous situations.

Optional reference:

For more details on the programming aspect of threading, refer to **CNC Programming Handbook**, *3rd Edition, Chapter 38 - Single Point Threading, pages 349-382.*

21 WORKING WITH BLOCK SKIP

By design, CNC machining centers and lathes offer unparalleled possibilities for machining flexibility. However, this flexibility is not achieved automatically - it requires that all elements of the CNC development process work together without fail. Flexibility in this sense may have several meanings, generally those that relate to the program flow or to required dimensions in general and dimensional tolerances in particular. In either case, both the part program and the part setup have to work together - a single error in either area will cause a problem during machining.

Several programming methods can be used to control the program flow or dimensional accuracy. One such tool CNC programmers have at their disposal is a function called *Block Skip* or - incorrectly - *Block Delete*. This simple function is programmed by using a special symbol - the forward slash /.

BLOCK SKIP FUNCTION

As the name of the function indicates, block skip is used to bypass (skip) one or more blocks during machining. There are several practical uses for this function. They include:

- Optional machine zero return
- Variable stock allowance
- Thread sizing
- Trial cuts
- Similar parts
- One part - two materials
- ... other applications

In order to work with block skip at the machine, you need to know how the program is tied to the actual machining. Block skip function has to be present in the part program in order to be used at the machine. Take, for example, one common application of block skip - return to machine zero (home position). Normally, on vertical machining centers, only the Z-axis has to be returned to the home position, as this is also the tool change position. Some programmers like to include the home position return for the X and Y axes as well, either at the beginning or the end of the program. This is for the convenience of the operator, in case such return is needed.

If the XY-home return is required for every part, it will be programmed normally. On the other hand, if the XY-home return is needed only occasionally, it will be programmed with the slash symbol. Then, it is up to the CNC operator to send the XY axes home or not.

Block Skip Switch

The programmed slash symbol is the first prerequisite of using the block skip. The other requirement is the status of the switch located on the operation panel of the machine.

The *Block Skip* switch can be a toggle switch or a button switch. In either case, it has two positions, ON and OFF. A push button will be illuminated when ON.

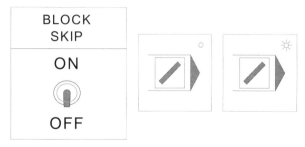

Using the switch is simple:

> To process ALL blocks in the program,
> block skip switch must be OFF

> To bypass blocks WITH SLASH,
> block skip switch must be ON

In the program, the optional return to XY home position will be preceded by the slash symbol:

```
...
N43 G80 Z25.0 M09
N44 G91 G28 Z0 M05
/ N45 G28 X0 Y0         (OPTIONAL RETURN IN XY)
N46 M30
%
```

There are some rules when working with block skip, and any operator should be aware of them.

Using Block Skip

Normally the slash symbol must be located in front of the block, even before the block number N... The *Block Skip* switch must be activated or deactivated *before* program processing reaches the blocks with slash. Also, all blocks with slash should be processed before the switch position is changed.

Block Skip within a Block

Many controls allow the block skip symbol to be placed inside of the block rather than at its beginning. One practical application of this method is two different materials for the same part. For example, the following program will face a part for material that is soft:

```
N1 G21
N2 G17 G40 G80 T01
N3 M06
N4 S1800 M03
N5 G90 G54 G00 X150.0 Y40.0 T02
N6 G43 Z3.0 H01 M08
N7 G01 Z0 F200.0
N8 X-65.0 F300.0
N9 G00 Z3.0 M09
N10 G28 Z3.0 M05
N11 M01
```

The spindle speed and the cutting feedrate have been underlined - these will be the only changes necessary to modify the program to cut the same part by from material that is harder:

- Soft material: S1800   F300.0
- Hard material: S700   F200.0

Another program can be written, but a single program can also be used, providing the block skip can be used within a block. The technique takes advantage of two conflicting commands in the same block:

```
N1 G21
N2 G17 G40 G80 T01
N3 M06
N4 S1800 M03 / S700
N5 G90 G54 G00 X150.0 Y40.0 T02
N6 G43 Z3.0 H01 M08
N7 G01 Z0 F200.0
N8 X-65.0 F300.0 / F200.0
N9 G00 Z3.0 M09
N10 G28 Z3.0 M05
N11 M01
```

With block skip OFF, the hard material will be machined, with block skip ON, the soft material will be machined. The **M03** function *must be before* the slash!

Variable Stock

In CNC lathe work, you will often work with material cut from a bar to smaller pieces that has to be faced off. Ideally, the length of these pieces, often called *billets*, should be the same across the batch. This is not always the case, and some billets will be longer than others. As long as the differences are small, the program may only use one cut for facing. If the differences are significant, two or more facing cuts will be required. Programmers will usually include the proper code into the program, as long as they are aware of the situation.

The following two programs illustrate the required changes from a single face cut method to two face cuts. If the amount of stock exceeds 2 mm width, for example, another face cut will be necessary:

```
N1 G21 T0100
N2 G96 S250 M03
N3 G41 G00 X44.0 Z0 T0101 M08
N4 G01 X-1.7 F0.3
N5 G00 W2.5
N6 G42 X31.0
N7 G01 X40.0 Z-2.0 F0.25
N8 Z-...
...
```

Compare the above program with the modified one:

```
N1 G21 T0100
N2 G96 S250 M03
N3 G41 G00 X44.0 Z2.0 T0101 M08
/ N4 G01 X-1.7 F0.3
/ N5 G00 W2.5
/ N6 X44.0
N7 Z0
N8 G01 X-1.7 F0.3
N9 G00 W2.5
N10 G42 X31.0
N11 G01 X40.0 Z-2.0 F0.25
N12 Z-...
...
```

Note the Z2.0 position for the extra cut, as well as the repeated feedrate. Both are very important. Similar application of block skip an be made for milling face cuts.

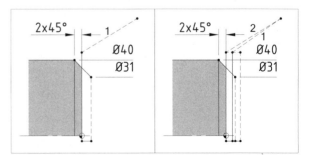

TRIAL CUTS

To maintain precision of the part, trial cuts can be used in certain situations where the part may be difficult to measure or when maintaining dimensional accuracy is important. Trial cut is nothing more than a test cut. There are rules that govern the programming, and these rules influence the machining.

Maintaining Tolerances

There are two key elements to maintain dimensional precision during CNC machining. One is the part program itself, the other covers the control settings and their active maintenance.

As for the first item - *part program* - there is hardly anything the CNC operator can do once the program is available in the shop. Apart from the common speeds and feeds adjustments and similar minor changes, the operator cannot count on making program changes without some degree of authorization (and skill, of course).

The second item - *control settings* - offers many more possibilities, as it is done at the control system itself. Keeping the part dimensions within tolerances is generally achieved by adjustments to the works offset, tool length offset and the cutter radius offset, as needed.

While the first and second items were independent of each other - being related to either the program or the control - there is a third option that should also be well understood. This item combines both program and control settings together. A typical example is a situation where a special program code is developed that allows changes at the control level *before* the actual part is finished. The most characteristic application of this method is a *trial cut*.

Trial Cut Concepts

Trial cut (test cut) is always a cooperative effort between the CNC programmer and the CNC operator. Trial cut toolpath is often added to the CNC program for one very specific reason - *to minimize the possibility of scrap*. Trial cut in a part program can never guarantee total elimination of scrap, but if done well, it can severely minimize such possibility from happening.

> Trial cut is a special series of tool motions that take place in the area of part that will be removed by subsequent machining without leaving any traces

Its main purpose is to catch most setup or even program errors in a safe area, not on the part itself. One of the main benefits of a trial cut is the opportunity to set various tool offsets before actual machining.

Trial Cut Methods

The basic principles relating to trial cuts apply equally for CNC machining centers and CNC lathes. Trial cut is always an *addition* to the part program, provided at a programmer's discretion or can also be added to the program by a skilled operator. Typically, trial cut includes these programming features:

- Temporary cut in an area of part that can be measured
- Program stop command M00 to allow measurement
- Block skip function (/) to use or not to use trial motions
- Restoring of modal commands and functions

Temporary Cut

Trial cut, by definition, is a temporary cut. Usually, the trial cut includes a series of tool motions that have only one thing in common - to allow dimensional inspection *before* actual cutting takes place. A complete application example will be presented later in this chapter.

Program Stop

Including the *Program Stop* function **M00** in the program itself is mandatory. Without it, there would be no way to perform the check of trial cut results. Along with the **M00** function, a short comment should be used:

```
/ N78 M00 (DIAMETER SHOULD BE 34.500 MM)
```

Block Skip

By its own nature, any trial cut is optional. As such, it has be turned ON or turned OFF. All tool toolpath motions relating to the trial cut are included in the part program itself. In order to separate the trial cut (which is optional) from the normal cutting (which is mandatory), the control system uses the *Block Skip* command, defined at the beginning.

Modal Commands Restoration

Modal commands are those that can be specified only once in the program, and keep their mode until a cancellation or a change of mode occurs. Watch for modal commands that are within the block skip range - some may need to be repeated.

MILLING APPLICATIONS

The following drawing represents a part with a tight tolerance on the width - only 25 microns (0.025), or about 0.0010 inches, on the plus side. Undersize is not an option as it will produce scrap. What options do you have as the CNC operator when machining this part?

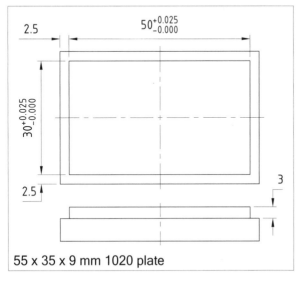

55 x 35 x 9 mm 1020 plate

If the part program does not provide any trial cut, the only way is to carefully control the cutter radius offset. This method is explained in the chapter on the subject, starting from *page 135*.

If the programmer does include a trial cut, it is important to study the program, so the programmer's intentions are interpreted correctly.

Program Sample

For the above drawing, the following program listing uses two tools - ⌀ 16 mm end mill for roughing, and ⌀ 12 mm end mill for finishing. The roughing tool will leave sufficient stock for the second tool, using cutter radius offset setting. The smaller tool will make two contour cuts around the part, the first one being the trial cut, the second one is the finishing contour that has to produce the part width within specified tolerances.

The part used for the program example is intentionally very simple to focus your attention at the method of programming. Study the program carefully, see where the block skip starts and ends. Also look at repeated values. The **G43** in block N28 has to be repeated, as the **G28** command cancels tool length offset.

```
(T01 = 16 MM END MILL FOR ROUGHING   => D51)
(T02 = 12 MM END MILL FOR FINISHING  => D52)

N1  G21
N2  G17 G40 G80 T01
N3  M06         (T01 = 16 MM END MILL - ROUGH)
N4  G90 G54 G00 X-10.0 Y-6.0 S1800 M03 T02
N5  G43 Z2.0 H01 M08
N6  Z-2.8       (0.2 STOCK LEFT AT THE BOTTOM)
N7  G41 X2.5 D51  (==> SUGGESTED D51 = 8.500)
N8  G01 Y32.5 F250.0
N9  X52.5
N10 Y2.5
N11 X-10.0
N12 G40 G00 Y-6.0 M09
N13 G28 Z-2.8 M05
N14 M01     (PART WIDTH SHOULD BE 51.000 MM)

N15 T02
N16 M06         (T02 = 12 MM END MILL - FINISH)
N17 G90 G54 G00 X-8.0 Y-4.0 S2200 M03 T01
N18 G43 Z2.0 H02 M08
/ N19 Z-2.8     (0.2 STOCK LEFT AT THE BOTTOM)
/ N20 G41 X2.5 D52 (==> SUGGESTED D52 = 6.250)
/ N21 G01 Y32.5 F180.0
/ N22 X52.5
/ N23 Y2.5
/ N24 X-10.0
/ N25 G40 G00 Y-6.0 M09
/ N26 G28 Z-2.8 M05
/ N27 M00    (** EXPECTED WIDTH IS 50.500 **)
/ (**** ADJUST D52 OFFSET FOR FINISH SIZE**)
/ N28 G43 Z2.0 H02 M08 (REINSTATE TOOL OFFSET)
/ N29 S2200 M03    (REINSTATE SPINDLE SPEED)
N30 Z-3.0                       (FULL DEPTH)
N31 G41 X2.5 D52    (USES UPDATED OFFSET D52)
N32 G01 Y32.5 F180.0
N33 X52.5
N34 Y2.5
N35 X-10.0
N36 G40 G00 Y-6.0 M09
N37 G28 Z-3.0 M05
N38 M30
%
```

Trial cuts belong to the more advanced category of CNC programming and machining. Applying them correctly requires a certain amount of experience and full understanding of what is actually happening at the control system.

Program Evaluation

In the example, there are several parts of the program that should be considered critical. This is a program for experienced CNC operators who can interpret the program from its listing. Let's look at the details.

Blocks N1 to N14 should present no difficulty in interpretation. They represent the roughing toolpath, using ⌀ 16 mm end mill.

Tool T01 is a roughing tool using drawing dimensions in the program while leaving stock allowance for finishing!

In order to leave stock for finishing, the 'normal' cutter radius of 8.000 will be entered in the radius offset registry as 8.500. This 0.5 mm difference will result in the part with being 0.5 mm larger at each contour side. Ideally, the part width will be 51.000 (0.5 mm added per side, 1 mm per width, since the full width is measured).

Considering the 51 mm width left after rough machining, let's look at the finishing operation - there will be 1.0 mm left for finishing over the part width *or* 0.5 mm per side. There is no room for error - the part that comes off the machine must be within tolerance, and the tolerance is very narrow. At the same time, we have the 1 mm over width to 'play with'.

Tool T02 is covered by blocks N15 to N38. However, these are not ordinary blocks you may be used to from the majority of other programs. This block range includes all expected blocks for the finishing contour, but it also includes blocks that are exclusively dedicated to the trial cut alone. From the program, they are easy to detect - all blocks for the trial cut start with the forward slash symbol - the ***block skip*** function.

Regardless of the actual naming convention, any program block preceded by the forward slash symbol (for example, **/ N22 X52.5**) will be skipped (bypassed) *only* if the *Block Skip* switch is turned on at the machine control panel.

> **No blocks will be physically deleted while the *Block Skip* function is active**

Finishing Tool - Program Details

The roughing end mill (T01) program section (blocks N1 to N14) provided only one benefit - it left 0.5 mm per side stock on all faces of the machined contour. This was achieved by setting the cutter radius offset D51 to 8.500 form the 'normal' 8.000.

For contour finishing, the program listing covers blocks N15 through N38, using tool T02 (⌀ 12 mm end mill). This is the range where the tight drawing tolerance should be controlled. It is very important to understand that the part program alone cannot guarantee good results, and even the best setup at the machine alone cannot guarantee good results either. Part program provides the means, setup at the machine provides the ends.

Conclusion

Keep in mind that although this sample program is correct as presented, it does not necessarily represent the only viable solution. Some programmers may choose to work with advanced datum shift **G10** command, some may prefer two different offsets for finishing, and others may choose a different approach altogether.

Regardless of the programming method used, from the simple milling example presentation, three items become prominent:

- Stock allowance left by roughing
- Stock allowance left by trial cut
- Final size control

➥ *The main key is to measure the part in block N27, and adjust the radius offset against the final width of 50.000 mm, based on the measurement of 50.500 mm*

Keep in mind that while the measurement is done on the width (two sides), the offset adjustment is per side!

TURNING APPLICATIONS

Although the lathe trial cut example is quite different from the milling application, you can still learn a lot as to how trial cuts work, regardless of the machine type. Many applications can be done for both types.

> **To understand the application of trial cuts, evaluate both *milling* and *turning* applications - - they both share the same concepts**

The following lathe example will show many characteristics that are common with those applied to milling. Evaluate the drawing and think how to measure the arcs.

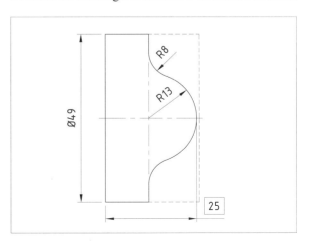

Test Cut Addition

The part will be held in soft jaws on the previously completed flat diameter. Machining the front end will require roughing and finishing toolpath. The front consists of two blend arcs and a small flat shoulder. Parts like this are relatively easy to program, easy to machine - but difficult to measure. In this case, using a trial cut will be more like using a test cut.

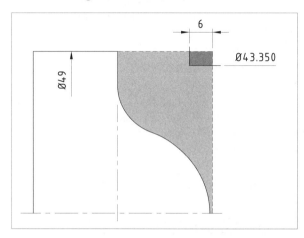

Adding a test cut will have one purpose only - to provide a temporary measurable surface for the finishing tool. Any wear offset adjustment will be done during the test cut. Normally, the test cut is part of the program, but it is still very important to interpret it during setup.

Initial considerations must include:

- **Location** ... close to corner
- **Test cut diameter** ... 43.350 mm
- **Test cut length** ... 6 mm
- **Applied to the finishing tool** ... T02 - stock 1 mm

Location has been selected in a place that does not require too much material removal. Location also must not interfere with the finishing toolpath. Test cut diameter was intentionally selected with three decimal places, to emphasize the need for precision. The length of the test cut is not critical, as long as it is long enough for a measurement with micrometer. A very important part of the test cut is to use the same tool that will do the finishing, including depth of cut and cutting speeds and feeds.

The following program includes all considerations:

```
N1  G21 T0100 (ROUGHING TOOL)
N2  G96 S275 M03
N3  G41 G00 X52.0 Z0 T0101 M08
N4  G01 X-1.7 F0.35
N5  G00 W2.5
/ N6  G42 X45.35      (LEAVES 1 MM STOCK FOR T02)
/ N7  G01 Z-6.0
/ N8  X50.0
/ N9  G40 G00 X100.0 Z50.0 T0100
/ N10 M01

/ N11 T0200 (FINISHING TOOL)
/ N12 G96 S300 M03
/ N13 G42 G00 X43.35 Z2.5 T0202 M08
/ N14 G01 Z-6.0 F0.25
/ N15 X50.0
/ N16 G40 G00 X200.0 Z150.0 T0200
/ N17 M00    (** CHECK DIA - MUST BE 43.350 MM)

/ N18 T0100 (ROUGHING TOOL)
/ N19 G96 S275 M03
/ N20 T0101 M08
N21 G42 G00 X52.0 Z2.5
N22 G71 U2.5 R0.5
N23 G71 P24 Q31 U2.0 W0.125 F0.35
N24 G00 X..
N25 ..
...
N31 X52.0
N32 G40 G00 X100.0 Z50.0 T0100
N33 M01

N34 T0200 (FINISHING TOOL)
N35 G96 S300 M03
N36 G42 X52.0 Z2.5 T0202 M08
N37 G70 P24 Q31
N38 G40 G00 X100.0 Z50.0 T0200
N39 G28 U0
N40 M30
%
```

Using block skip in a program has many applications. In the hands of a CNC operator who has some programming knowledge, block skip may make a big difference in many areas of machining.

22 OFFSET CHANGE BY PROGRAM

Even a small mistake in setup data can have serious consequences in actual machining that follows. A job that worked flawlessly before may be compromised if only a single offset is entered incorrectly. This chapter shows how the supplied part program example *influences* - and in many cases actually sets - all three types of offsets found on CNC machining centers:

- **Work offset**
- **Tool length offset**
- **Cutter radius offsets**

For CNC lathes, the program can set or change both offsets as well:

- **Geometry offset**
- **Wear offset**

All examples of various offset settings so far have been done by the CNC operator, at the machine. There are times when the programmer has a reason to set and/or change various offsets through the program. This method can also be used for MDI operations at the machine, if necessary. Before describing this method, let's look at a fairly common method programmers use, so called *local coordinate system*, using the command **G52**.

LOCAL COORDINATE SYSTEM

Work offsets **G54-G59** (as well as the extended set), have been described in a separate chapter. These offsets are used by the programmer to establish part zero. For certain jobs, programmers often face a decision between a part zero convenient for setup and part zero convenient for calculations. Take the simple drawing shown here. For ease of part setup at the machine, the lower left corner will be the best choice. For ease of programming, the linear pattern will also benefit from part zero at the lower left corner, but for the circular pattern part zero at the circle center would be most suitable for calculations. Using the local coordinate system, the programmer can take advantage of temporarily adjusting the active work offset as required.

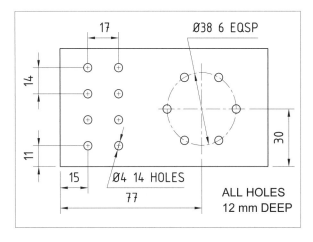

Local coordinate system is a programming feature, not a setup feature. It is important that the operator understands how it works, so the program is interpreted correctly.

Here is a sample program for a spot drill:

```
N1 G21
N2 G17 G40 G80 T01
N3 M06
N4 G90 G54 G00 X15.0 Y11.0 S1200 M03 T02
N5 G43 Z10.0 H01 M08
N6 G99 G82 R2.0 Z-2.3 P200 F175.0
N7 G91 Y14.0 L3 (K3)
N8 X17.0
N9 Y-14.0 L3 (K3)
N10 G90 G52 X77.0 Y30.0 (TEMPORARY ZERO SHIFT)
N11 X19.0 Y0
N12 X9.5 Y16.454
N13 X-9.5
N14 X-19.0 Y0
N15 X-9.5 Y-16.454
N16 X9.5
N17 G80 Z10.0 M09
N18 G52 X0 Y0           (ZERO SHIFT CANCELED)
N19 G28 Z10.0 M05
N20 M01
```

Blocks N11 to N16 use a temporary zero at the center of the bolt circle.

> Work offset must be in effect before G52 can be used in the program

Work Offset Change

As the last caution note indicates, any one of the available work offsets must be in effect for **G52** to work. Keep in mind that **G52** will shift the current offset mathematically - it does not change the actual settings. in the block N18 of the example, the temporary zero shift is canceled:

`G52 X0 Y0`

It simply means that the current work offset will be shifted by the amount of zero in both axes.

Also apparent from the example is that **G52** can only be used of the shift amount is *known*. If the amount is not known, that is where the operator has to set each part zero by using a different work offset. This is a common situation when two or more fixtures are used for setup and by the same program.

Many setup jobs use multiple fixtures, tombstone or a similar arrangement, where the distance between individual part origins is not known. As expected, each part zero will be stored in a different work offset registry. Now, what happens if such a setup is exactly the same when the part is machined in the future? Starting from scratch is always possible, but that is unproductive time that can be avoided. Consider the setup of three vices on a sub plate. The first time the job is run, the operator registers works offsets **G54**, **G55**, and **G56**. Next time the same job is run, all offsets are the same as the last time. In cases like this, the previous offsets can be reused by a datum shift command - programmed with preparatory command **G10**.

DATUM SHIFT

As a preparatory command, datum shift is a programming feature. However, because of its nature, it is often used by CNC operators to reload offsets that have not changed from the previous setup. This method greatly reduces the setup time.

The schematic drawing example above shows a setup for three vises. Although the three vises are permanently located on a sub plate and appear to be in a row, it does not mean their Y-axis setting will be identical. It is necessary to set all three work offsets in both X and Y axes, to guarantee precision of the setup. If the future setup is identical, these setting can be imported though MDI or the program itself.

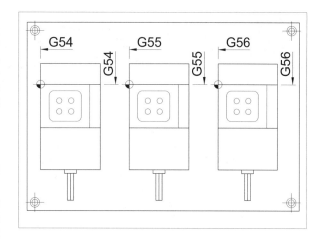

Let's use these initial settings for the three work offsets in the example:

- ➡ *G54 X-752.450 Y-186.540 Z0.0*
- ➡ *G55 X-505.730 Y-186.920 Z0.0*
- ➡ *G56 X-254.910 Y-186.880 Z0.0*

Note the minor changes in the Y-setting - it is only an example, but it attempts to be realistic. Also the Z-setting may be different for each vise if the height is not exactly the same.

Once these measurements have been made, they can be used as program data, together with the **G10**.

G10 Format - Milling Applications

Preparatory command **G10** is described as the *Data setting* command or *Programmable Data Input*. For milling applications, it can be used to set or change the following offsets:

■ **Standard works offsets**	G54 *to* G59
■ **Additional work offsets**	G54 P1 *to* G54 P48
	or G54.1 P1 *to* G54.1 P48
■ **External offset**	non-programmable
■ **Tool length offset**	G43 H..
■ **Cutter radius offset**	G41 D..
	or G42 D..

In order to distinguish which offset the **G10** refers to, the command itself is followed by the address *L* that is associated with a number. The actual setting follows the address *L*. Explanation will be given in individual sections, including an example.

Before getting into details, remember one important rule relating to the dimensional mode - it can be an *absolute* mode or *incremental* mode:

> G10 in ABSOLUTE mode G90
> will always *REPLACE* the current settings

> G10 in INCREMENTAL mode G91
> will always *ADD* to the current settings

Standard Work Offsets

The three work offsets for the example will be associated with the **G10** command in this format:

`G10 L2 P.. X.. Y.. Z..`

The *L2* is an arbitrary address that exclusively identifies the setting as *work offset*. Address *P* is the offset group number, based on the following order:

WORK OFFSET		
Offset Number	In the program:	Sets offset:
1	P1	G54
2	P2	G55
3	P3	G56
4	P4	G57
5	P5	G58
6	P6	G59

Settings from the example can be imported to the control system by the following program:

```
G90                    (ABSOLUTE MODE REPLACES DATA)
G10 L2 P1 X-752.450 Y-186.540 Z0.0 (SETS G54)
G10 L2 P2 X-505.730 Y-186.920 Z0.0 (SETS G55)
G10 L2 P3 X-254.910 Y-186.880 Z0.0 (SETS G56)
```

These settings should be done in absolute mode, so they replace any current settings. That is the reason for including **G90** at the beginning.

- *Example of work offset adjustment:*

Suppose that during setup, you notice the middle vise needs a small adjustment of positive 0.36 mm in the Y-axis. You can handle the problem in two ways:

Replace the **G55** setting (middle vise) for the Y-axis by changing the offset itself from **Y-186.920** to its new amount of **Y-186.560**. There is nothing wrong with this method, but it is more likely that it can lead to errors.

A better way is to use the **G10** command in MDI mode:

`G91 G10 L2 P2 Y0.360`

There is no need to include other data.

Additional Work Offsets

For many complex machining jobs, the six standard work offsets may not be enough. Fanuc control systems offer a set of additional work offsets, usually another 48 of them, for the total of 54 work offsets. In order to distinguish between the six standard offsets and the forty eight additional work offsets, the **G10** format uses a different L-address:

- L2   selects *standard* 6 offsets G54 to G59
- L20   selects *additional* 48 offsets G54 P1 to G54 P48

The program format is similar to the standard offsets:

`G10 L20 P.. X.. Y.. Z..`

For example,

`G10 L20 P3 X-497.726`

will set *additional* offset number three in X-axis only to the amount of -497.726 in absolute mode.

> Additional work offsets
> never interfere with the standard work offsets

External Offset

On modern CNC controls, the external offset is identified as *EXT*, on older controls it is identified as *COM* (for *COMmon* offset). It had been already been described earlier, in the chapter on work offsets - here is only a brief description:

00	EXT
X	0.000
Y	0.000
Z	0.000

> EXTernal offset is used to globally change
> all work offsets by the specified amount

It can also be set or changed by the **G10** command, using the standard offset group *L2* but with **P0**:

```
G10 L2 P0 X.. Y.. Z..
```

➡️ *Example of external work offset setting:*

Using the three vises as an example, suppose that the whole setup is 1.5 mm off in positive X-axis and 0.75 mm negative in Y-axis. of course, you can change each offset separately, but using **G10** is a better choice.

The next question is - should the adjustment be in *absolute* or *incremental* mode? Incremental sounds logical, but some operators may think that absolute is fine, as the most common setting of the **EXT** offset is zero for all axes. In this case, adjustment in absolute mode can be made, but incremental is much safer.

Just because now the **EXT** offset has all axes set to zero, does not mean it will have the same zero settings another day. Using incremental adjustment will update whatever settings there are in the **EXT** offset, including all zeros:

```
G91 G10 L2 P0 X1.500 Y-0.750
```

Tool Length and Cutter Radius Offsets - Shared

Many CNC operators use the **G10** command for tool length offset more often than for the work offset. The benefit is that many tools are used repeatedly for different jobs, therefore with different setups, while the tool length does not change.

Setting tool length offsets with **G10** is more complicated overall, due to the fact that there are three different types of memory offsets:

Offset memory	H-offset Geometry	D-offset Geometry	H-offset Wear	D-offset Wear
Type A	Shared - one column			
Type B	Shared - one column		Shared - one column	
Type C	Separate column	Separate column	Separate column	Separate column

In order to distinguish between the three types of offset memory, **G10** uses different L-address (L10, L11, L12, and L13). Make sure you understand which group number belongs to each offset type.

> Error in offset type can have serious consequences

Offset memory	H-offset Geometry	D-offset Geometry	H-offset Wear	D-offset Wear
Type A	G10 L11 P.. R..			
Type B	G10 L10 P.. R..		G10 L11 P.. R..	
Type C	G10 L10 P.. R..	G10 L12 P.. R..	G10 L11 P.. R..	G10 L13 P.. R..

In all cases, the G10 and its parameters are written in a single block. While the L-address identifies the offset type, the remaining parameters identify:

- P   ... is the offset number
- R   ... is the offset amount

Note - older models of Fanuc controls did not have the wear feature, so there was no reason for distinction between Geometry and Wear offsets. These controls used L1 instead of the L11. This format is obsolete on modern controls, but still supported for compatibility.

It is important to understand that any CNC machine has only one offset memory type, and that is the offset CNC operators focus on. In order to find out which offset memory your machine has, open the offset screen of the control and look at how many columns are available:

- Type A   ... one column
- Type B   ... two columns
- Type C   ... four columns

The following series of examples should give you good understanding how the **G10** offset works for tool length and cutter radius offsets and offset memory *Type A* and *Type B*.

➡️ *Examples of usage - Type A:*

Tool 5 (T05) uses tool length offset 5 (H05) and cutter radius offset 55 (D55). Program contains this block:

```
G91 G10 L11 P5 R0.6
```

This block will change the tool length offset 5 by adding 0.6 to its current setting. If the offset 5 setting was, for example, -312.730, the new setting will be -312.130.

The same tool may also be in need to adjust the cutter radius by -0.025 mm:

```
G91 G10 L11 P55 R-0.025
```

The same tool - and adjustments - will be used for the *Type B* offset memory as well.

◆◆ *Examples of usage - Type B:*

Tool 5 (T05) uses tool length offset 5 (H05) and cutter radius offset 55 (D55). Program contains this block:

```
G91 G10 L11 P5 R0.6
```

This block will change the tool length offset 5 by adding 0.6 to its current setting. If the offset 5 setting was, for example, -312.730, the new setting will be -312.130.

The same tool may also be in need to adjust the cutter radius by -0.025 mm:

```
G91 G10 L11 P55 R-0.025
```

Note that both the geometry and wear offsets have the same format as for *Type A*. That is only because they represent a change of an existing offset, which means using a wear offset. The situation would be different for replacing the offset - for example:

Geometry offset 5 has to be replaced by a new offset amount of -260.492 and the cutter radius offset will be the radius of a 8 mm cutter:

```
G90 G10 L10 P5 R-260.492
G90 G10 L11 P55 R4.000
```

Always watch for the **G90** or **G91** mode.

Tool Length and Cutter Radius Offsets - Separate

Only offset memory *Type C* allows individual adjustment of the tool length offset and cutter radius offset, using the **G10** command. For example:

```
G90 G10 L10 P8 R-341.567
```

will set the tool length offset *H08* to -341.567. If the tool length needs to be adjusted, for example by negative 0.6 mm, the incremental mode can be used:

```
G91 G10 L11 P8 R-0.600
```

The application is similar for the cutter radius offset. In the example, the same tool 8 (T08) is used. First - the initial setting:

```
G90 G10 L12 P8 R7.500
```

This block was used to input radius of a ⌀15 mm end mill into offset number 8, programmed as

```
G41 X.. Y.. D08
```

After a while, if the tool wears out and has to be adjusted by 0.03 mm, the offset has to be changed. In this case, use **G10** with incremental wear offset:

```
G91 G10 L13 P8 R-0.03
```

The incremental method does not require any calculations - the control system will make them for you. Just make sure to enter the amount correctly as a positive of negative value. The equivalent entry in absolute mode requires manual calculation:

```
G90 G10 L13 P8 R7.470
```

G10 For Roughing And Finishing

The example below will use a simple contour to be roughed and finished with two tools. The purpose is to interpret the **G10** format that will be used.

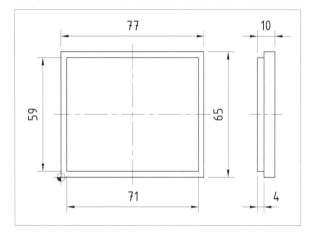

Normally, the programmer will select the cutting tools for roughing and finishing, make a subprogram and assign the tool length and cutter radius offset numbers. For the example, *Type B* offset memory will be used, and the cutting tool - *T01* - will be an ⌀18 mm end mill; for finishing, the *same tool* will be used.

The subprogram will be common to both operations:

```
O6801
N101 G90 G41 G01 X3.0
N102 Y61.0
N103 X74.0
N104 Y3.0
N105 X-14.0
N106 G40 G00 Y-10.0
N107 M99
%
```

◆◆ *Note the absence of D-offset and F-feedrate*

Now the main program - the way it would normally be developed:

```
(T01 = 18 MM END MILL)
N1 G21                 (TOOL 1 IS IN THE SPINDLE)
N2 G17 G40 G80
N4 G90 G54 G00 X-14.0 Y-10.0 S1250 M03
N5 G43 Z10.0 H01 M08
N6 Z-3.75 (LEAVE 0.25 MM STOCK FOR FINISHING)
N7 M98 P6801 D61 F200.0    (*** ROUGH CUT ***)
N8 G01 Z-4.0             (FULL DEPTH CUTTING)
N9 M98 P6801 D51 F175.0 (*** FINISH CUT ***)
N10 G00 Z10.0 M09
N11 G28 Z10.0 M05
N12 M30
%
```

A few items of importance should be mentioned:

- Stock of 0.25 mm left on the bottom for finishing
- Cutter radius offset D61 is used for roughing
- Cutter radius offset D51 is used for finishing
- Both offsets and feedrates are passed to the subprogram

At the machine, the CNC operator will determine the tool length offset, the amount of stock left for contour finish, for example, one half of a millimeter (0.5 mm). Finishing offset will be one half of the tool diameter, or 9 mm. All three offsets will be inserted into the offset register *(Type B)*:

OFFSET		
No.	Geometry	Wear
001	-341.562	0.000
...	...	...
...	...	...
051	9.000	0.000
...	...	...
061	9.000	0.500
...	...	...

For offset memory *Type A*, there would be only a single column, and offset 061 would contain the amount of 9.500. This type of manual calculation can lead to errors, particularly with amounts that have three or four decimal point precision.

The same program can be handled by the **G10** command. In this case, it is the programmer - not the operator - who will decide on the amount of stock. The main program will also change:

```
(T01 = 18 MM END MILL)
N1 G21                 (TOOL 1 IS IN THE SPINDLE)
N2 G17 G40 G80
N3 G90 G54 G00 X-14.0 Y-10.0 S1250 M03
N4 G43 Z10.0 H01 M08
N5 G91 G10 L11 P1 R0.25
N6 G90 Z-4.0
N7 G91 G10 L11 P51 R0.5
N8 M98 P6801 D51 F200.0    (*** ROUGH CUT ***)
N9 G91 G10 L11 P1 R-0.025
N10 G90 Z-4.0
N11 G91 G10 L11 P51 R-0.5
N12 M98 P6801 D51 F175.0 (*** FINISH CUT ***)
N13 G00 Z10.0 M09
N14 G28 Z10.0 M05
N15 M30
%
```

Study the underlined entries carefully. Blocks N5 and N7 apply wear offset to the roughing cut. Block N5 will add 0.25 mm to the current tool length, making the tool motion shorter - the result is a 0.25 mm stock left on the bottom for finishing. Block N7 sets the wear offset to 0.5 mm, making the part width 1 mm longer (0.5 per side).

Once the roughing cut is completed, both wear allowances are removed and original settings are restored. That happens in blocks N9 and N11.

The benefit of this method is that the programmer controls both offsets and the operator concentrates on normal tool length and radius offset settings.

G10 Format - Lathe Applications

Standard two-axis CNC lathe has only two offsets:

- Geometry offset
- Wear offset

On the other hand, each offset has five columns:

GEOMETRY				
No.	X	Z	R	T
G 001	0.000	0.000	0.000	0
G 002	0.000	0.000	0.000	0
G 003	0.000	0.000	0.000	0
G 004	0.000	0.000	0.000	0
G 005	0.000	0.000	0.000	0

The *Wear* offset is almost identical:

WEAR				
No.	X	Z	R	T
W 001	0.000	0.000	0.000	0
W 002	0.000	0.000	0.000	0
W 003	0.000	0.000	0.000	0
W 004	0.000	0.000	0.000	0
W 005	0.000	0.000	0.000	0

On some control models the offset number prefix *G* or *W* is not available.

G10 command for lathe applications has the following format used in the program or MDI mode:

```
G10 P.. X.. Z.. R.. Q..
G10 P.. U.. W.. C.. Q..
```

For multi-axis lathes, additional axes will also be present on the screen and in the **G10** format. As there is no **G90** and **G91** to select the mode of dimensioning, *U* and *W* are incremental motions associated with absolute motions of *X* and *Z* respectively.

Addresses X-Z-U-W

- X = X-axis offset amount   *Absolute*
- Z = Z-axis offset amount   *Absolute*
- U = X-axis offset amount   *Incremental*
- W = Z-axis offset amount   *Incremental*

Absolute and incremental addresses can be combined in the **G10** block.

For CNC lathes that also have the Y-axis, the input method is the same as for X/U and Z/W - the Y-address specifies absolute offset amount, while the V-address specifies incremental offset amount.

Address P

In the format, the P-address is the offset number.

- P0   = Work offset shift
- P1 to P64   = Tool wear offset number
- P10000+(1 to 64) = Tool geometry offset number

Address *P0* in the lathe application of **G10** command is equivalent to the *EXT* work offset described for CNC machining centers.

Addresses R-C-Q

- R = Tool nose radius offset amount   *Absolute*
- C = Tool nose radius offset amount   *Incremental*
- Q = Imaginary tool nose number

▸ *Example of usage - 1:*

To shift current work offset by 0.8 mm in Z-axis:

```
G10 P0 W0.8
```

To reset work offset to zero in both axes:

```
G10 P0 X0 Z0
```

▸ *Example of usage - 2:*

To adjust wear offset number 7 by -0.04 mm in X-axis:

```
G10 P7 U-0.04
```

To set geometry offset 5 to **X-432.671 Z-340.780**:

```
G10 P10005 X-432.671 Z-340.780
```

Often, it is necessary to include the tool nose radius (R or C) and the imaginary tool tip number (Q) when setting or changing geometry offset. Radius of the tool tip appears on the control screen in the R-column. The tool tip numbers are arbitrary and they appear on the control screen in the T-column. The illustration shows tip numbers for a typical CNC lathe of the *rear* orientation.

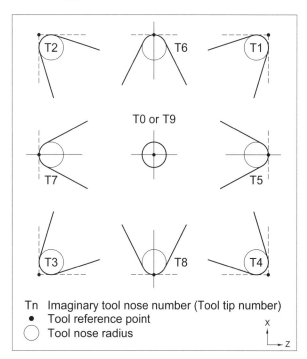

Tn Imaginary tool nose number (Tool tip number)
●   Tool reference point
○   Tool nose radius

Example of usage - 3:

The following example shows setting of three tools as complete geometry offset entry:

Tool 2 = Drill
Tool 4 = Boring bar with 0.4 mm (0.0156") radius
Tool 12 = Turning tool with 0.8 mm (0.0313") radius

The XZ settings are only examples, but they do reflect the type of tool used:

... for the drill:

```
G10 P10002 X-474.395 Z-67.26 R0 Q0
```

... for the boring bar:

```
G10 P10004 X-451.122 Z-83.954 R0.4 Q2
```

... for the turning tool:

```
G10 P10012 X-475.73 Z-177.375 R0.8 Q3
```

Note that all XZ amounts are in absolute mode - their purpose is to set new geometry offsets and replace any old settings that may be in the control.

SETTING PARAMETERS BY G10

The command **G10** can also be used to set various parameters of the control system. A separate chapter focused on system parameters covers this very important subject and also includes parameters set by the program or MDI, including several examples.

This section only covers the format for comparison with other format introduced in this chapter:

```
G10 L50        (PARAMETER SETTING MODE ON)
N.. R..        (FOR NON-AXIS TYPE SETTINGS)
N.. P.. R..    (FOR AXIS TYPE SETTING ONLY)
G11            (PARAMETER SETTING MODE OFF)
```

- N = Parameter number
- R = Parameter setting value
- P = Axis number

Parameters are often input in batches - it is not unusual that if you change one parameter, you may have to change another, associated, parameter as well.

For more details on system parameters, see a separate chapter starting on *page 233*.

23 SYSTEM PARAMETERS

For many shop managers, supervisors, programmers and machine operators, the words *'parameters'*, *'system parameters'*, *'control parameters'*, and particularly *'changing parameters'* cause a little discomfort when faced with the subject. The problem is generally lack of knowledge, therefore some resistance to get deeper into the subject.

The main reason for the skeptics is fear - general fear of the unknown, fear that something can go wrong. Control system parameters are the definite 'unknowns' in the vast area that makes the CNC machine work as it should.

The main reason for a certain amount of apprehension is that setting any parameters the wrong way may cause the whole CNC machine inoperational. The remedy is usually quite expensive. Understanding system parameters in depth is mandatory for service technicians, but even some knowledge from the side of CNC programmers and CNC operators is also very useful.

WHAT ARE PARAMETERS?

System parameters - or just *parameters* - are part of a large group of many internal settings of the control system. Their purpose is to make a connection between the computer part of the machine (control) and its physical part (applications). Virtually any activity of a CNC machine is controlled by those settings. A typical control unit has hundreds and hundreds of parameters belonging to different groups. For common operations of CNC machines, in most cases, there is no need to change parameters at all, as long as the machine functions properly.

While CNC programmers have very little to do with setting system parameters, supervisors, setup personnel, and qualified operators may find that a parameter change here and there can make an improvement in part quality, tool life, surface finish, etc.

Parameters for Operators

Out of the many hundreds of various parameter settings, there is a handful of parameters that every CNC operator should be aware of. Many of these parameters can be changed either permanently or for a specific job. They are divided into several groups:

- Initial settings (defaults)
- Fixed cycles
- Multiple repetitive cycles
- Work offsets
- Geometry and Wear offsets
- Radius offsets
- Communications (DNC)
- Macros
- *... several others*

This is by no means a comprehensive list and a number of other parameters may also be important for CNC machining or programming. User manuals supplied for machines with Fanuc controls also include *Parameter Manual* for individual control models. In this chapter, *Fanuc 16i Model B* and *Fanuc 18i Model B* series for milling (M-series) and turing (T-series) will be used for examples. These are very popular control systems, used with a large variety of CNC machine tools. Other popular series, such as *20i+* or *30i+* use different manuals. Older versions of the same controls (without the *i* suffix) may or may not contain parameters with the same meaning. Referring to the proper documentation is important.

Fanuc Parameter Classification

At the beginning of every Fanuc *Parameter Manual*, there is a listing by function of all parameters a particular control has available, including many for special options. Several of them are useful to CNC programmers and operators, for example, under the listing

Parameters of Canned Cycles

you will find parameters related to fixed cycles (for drilling and other hole operations), multiple repetitive cycles for turning centers, including threading cycles. Others may also be included. For the complete listing, see the *Reference* section on *page 263*.

Access to Parameters

It is easy to imagine the problems that could occur if there were an unrestricted access to system parameters. In order to only *view* current settings of various parameters, there are no restrictions. They can be viewed by selecting function keys and soft keys.

233

The first rule of changing parameters is easy:

> Only QUALIFIED and AUTHORIZED personnel should be allowed to handle system parameters

Note that both *qualified* **and** *authorized* personal characteristics are required.

Viewing Parameters

Current settings of parameters can be viewed on the screen of the control by pressing the **SYSTEM** function key, followed by pressing the **PARAM** soft key:

There are many pages (screen displays) available, and navigating through them is done in two ways:

- Using the PAGE UP and PAGE DOWN function keys
- Using the parameter number, then search for it

Your control manual will describe the required steps in exact detail.

Parameter Protection

Except for a very few parameters that can be changed quickly (so called *'handy'* settings), parameters are protected. Protection can be as simple as a setting of the access to parameters **ON** or **OFF**, but it could also be set by a password on many controls.

Keep in mind that when the protection is removed, parameters can be changed. Also keep in mind that while the parameters are *not* protected, the CNC machine is temporarily inoperable. In fact, when the protection is OFF, the control system will be in *alarm mode*, and alarm message:

P/S100 PARAMETER WRITE ENABLE

will be flashing on the screen. This is normal condition when changing parameters. It prevents any changes that cannot be done while the alarm is active.

Handy Setting - MDI Mode

To enable parameter setting, the control system should be in MDI mode (*Manual Data Input* mode). In this mode, when you press the **OFFSET/SETTING** function key, followed by the **[SETTING]** soft key. The following screen will appear:

```
SETTING (HANDY)          O1234  N0001

PARAMETER WRITE = 0   (0:DISABLE 1:ENABLE)
```

There are several additional settings available on this screen, but the one related to parameters is right at the top. The same screen also shows current program number and block number for reference, both irrelevant for the actual parameter settings. The screen - as illustrated - shows that access to parameters for the purpose of changing them is *disabled*.

Two modes are available for the settings:

- Mode 0 = OFF   [OFF : 0]
- Mode 1 = ON   [ON : 1]

At the bottom of the same screen are two soft keys; they are used to **ENABLE** or **DISABLE** the access to parameter settings:

[ON : 1]   [OFF : 0]

To view parameters, use the **SYSTEM** function key as described earlier.

Once a single parameter or a series of parameters have been changed and the process is completed, it is important to turn the **PARAMETER WRITE** mode back to zero (making it disabled).

> Pressing the RESET button will not change the mode to DISABLED
>
> Set the ENABLE mode first, then press the RESET button

The alarm P/S100 will be cancelled only after setting the *parameter write* to OFF:0 mode.

Backing Up Parameters

Considering how important parameters are for efficient running of a CNC machine, it is surprising how many machine shops and even large companies do not have a backup of parameters for each machine.

Backup Methods

In the early days of CNC, parameters were backed up to tape and so-called *Executive Tape* was created. Modern backup methods also use external devices, but those that are much more efficient and reliable. Backing up current settings of all parameters is always done to an external device, such as a flash card, disk, or personal computer (PC).

Using PC is the most common method, as the same PC is commonly used for other functions, such as DNC operation, program editing and verification, even for part programming, etc. Regardless of which device is used, the process is quite simple.

Interestingly enough, when parameters are input or output using an external device, the control system uses a special interface called **READER/PUNCHER**. Yes, even in this day of super powerful ultra modern controls, the old tape terminology is still alive. Think of the words **READER** and **READ** as *input* of data, and think of the words **PUNCHER** and **PUNCH** as *output* of data.

There are two initial requirements:

- *The first requirement is that the external device (PC used for examples) is connected to the CNC machine*
- *The second requirement is that data settings for communication between the devices are also set*

Output to External Device

Saving parameters as a backup is done through the **EDIT** mode selected from the operation panel of the control system. Display the parameter screen using the **SYSTEM > PARAM** procedure described earlier. Pressing the soft key marked **[(OPRT)]** will display two other soft keys:

This is followed by pressing the soft keys in order, typically:

[PUNCH] > [ALL] > [EXEC]

Only when the **[EXEC]** key is pressed the output will begin. During the output, the word **OUTPUT** will be flashing at the bottom of the screen, indicating that parameters are being copied to the PC (or other device). Copying can be cancelled if required.

Input from External Device

Restoring parameters from a backup device requires that access to parameter writing is enabled:

ON : 1

In all other respects, the input procedure is virtual reversal of the output method. Again, using the **SYSTEM > PARAM** procedure, followed by **[(OPRT)]** key selection, it will be the order of:

[READ] > [ALL] > [EXEC]

to input parameters from the external device.

These are only general procedures described for the purpose of understanding how parameters can be saved and restored. The best way to learn the exact series of steps is to have the manual in hand and follow the steps described there.

PARAMETER TYPES

Fanuc controls divide system parameters into four basic groups:

- **BIT**   and   **BIT AXIS**
- **BYTE**   and   **BYTE AXIS**
- **WORD**   and   **WORD AXIS**
- **2-WORD**   and   **2-WORD AXIS**

The *bit* type parameter is the most common to be changed at the machine. The word *bit* is an abbreviation of two words **B**inary dig**IT**. It is the smallest amount of data computers work with and can have one of only two settings *zero* or *one* (0 or 1). Typically, this setting selects one of only two modes available, such as ON or OFF, PRESENT or ABSENT, YES or NO, etc.

Bits are input in parameters as individual data within an 8-bit format. That also means up to eight different and independent bit settings can be input under a single parameter number. The following method of bit numbering is very important to understand.

Bit Type Description

By a way of introduction, consider the following scenario. It is customary for machine service technicians to talk to the customer on the phone or correspond via e-mails before sending a service person over to solve a problem. The question the technician may ask goes something like this:

'Can you check bit #6 of parameter 3003?'

Do you know how to answer that question? You don't have to know what the technician is actually asking you, but your answer must be accurate. Following the process of displaying parameters on the control screen, you will see this display of parameter **3003**:

3003	MVG	MVX	DEC	DAU	DIT	ITX		ITL

The letters in the above display are for definitions only and will be shown at the control as either the digit *zero* or the digit *one*. Ignore the fact that this parameter has something to do with input and output and check the correct answer - which one is it?

- Bit #6 is   ... ITX
- Bit #6 is   ... DEC
- Bit #6 is   ... other

The correct answer is ... *Bit #6 is other*. In fact, the correct answer is **MVX**.

The reason that **MVX** is the correct answer is based on the rule of bit numbering - how the bits are numbered in the series of eight bits. As a rule, computers start numbering from zero, *not* one, and from right to left, *not* left to right. The following illustration show the numbering of bits in an 8-bit format:

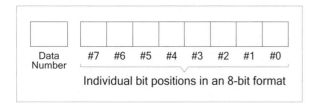

Individual bit positions in an 8-bit format

A single data entry row in the *Parameter Manual* applies to both milling (M-series) an turning (T-series) control systems.

Some 8-bit data may have two rows listed in the *Parameter Manual*:

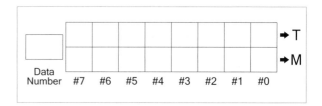

In this case, the upper row indicates settings exclusively for the *T-series* (turning centers), and the bottom row contains setting exclusively for the *M-series* (milling). It is not unusual to find many bits identical to each other, but with two rows, at least one bit will be unique to either control type.

Practical Interpretation

As an example, a particular bit-type parameter will be used to illustrate how it is structured in the *Parameter Manual* and how it can be interpreted by the user.

For the example using *Fanuc 16i -18i* control system, parameter number **3402** will be used. Parameter number is just another meaning of the *Data Number*. This parameter was selected for explanation for three reasons:

- It is practical
- It is easier to understand than others
- It is applicable to both T-series and the M-series

Actual Parameter Example - 3402

The actual entry in the *Parameter Manual* is a small table, followed by options for each bit, starting with the *Data type*:

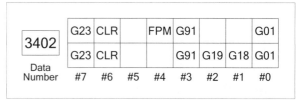

[Data type]   Bit

G01   Mode entered when the power is turned on or when the control is cleared

  0 : G00 mode (positioning)
  1 : G01 mode (linear interpolation)

G18 and G19   Plane selected when power is turned on or when the control is cleared

G19	G18	G17, G18 or G19 mode
0	0	G17 mode (plane XY)
0	1	G18 mode (plane ZX)
1	0	G19 mode (plane YZ)

G91   When the power is turned on or when the control is cleared

0 : G90 mode (absolute command)
1 : G91 mode (incremental command)

FPM   When the power is turned on

0 : Feed per revolution on
1 : Feed per minute mode

CLR RESET   key on the MDI panel, external reset signal, reset and rewind signal, and emergency stop signal

0 : Cause reset state
1 : Cause clear state
For the reset and clear states, refer to Appendix in the Operator's Manual

G23   When the power is turned on

0 : G22 mode (stored stroke check on)
1 : G23 mode (stored stroke check off)

Now, the interpretation of the settings. In many cases, the identifiers are designed in such a way that they indicate some relationship to their meaning. This is not always the case, but anything helps.

Example Details

First, keep in mind that the identifiers of each bit are only *labels*, not actual commands. Second, even if the labels are the same for milling and turning, the difference is often in the actual meaning. Here is the first line:

[Data type]   Bit

This line identifies the *type of the parameter*, in this case, as a bit parameter. For various options, see details on page 235.

The next entry is:

G01   Mode entered when the power is turned on or when the control is cleared

0 : G00 mode (positioning)
1 : G01 mode (linear interpolation)

Although the interpretation of the bit #0 is pretty straightforward, let's look at it in more detail. It should be obvious that if bit #0 is set to 0, the control will default to **G00** mode, which is rapid positioning. If the same bit is set to 1, the control defaults to **G01**, which is linear interpolation. Both are applicable at power ON or when the control is cleared. What are the implications of either setting?

The best way to understand is to see what will happen when the program starts with an XY, XYZ or XZ motion, and there is no **G00** or **G01** assigned to the block. After power ON, such program will interpret the motion as either **G00** or **G01**, based on the parameter setting. Typically, the first *intended* motion of the program is a rapid motion (**G00**). If the motion command is missing and #0 is set to 0, **G00** will take place. However, an interesting situation will arise, if the bit #0 is set to 1. In this case, the control system will issue an alarm, indicating that a *feed rate* is missing. As **G01** is the default, and there is no feed rate in memory after power ON, the control *will* require a feed rate. No feed rate has been programmed, as none was intended.

Author's Opinion:

For safety reasons, it makes more sense to set bit #0 to 1 and make sure G00 is programmed with the motion. Some programmers count on these defaults, which is not a professional approach.

G00 or **G01** will always override the parameter setting, as the program input has higher priority.

The next label is associated with bit #1 and bit #2:

G18 and G19   Plane selected when power is turned on or when the control is cleared

G19	G18	G17, G18 or G19 mode
0	0	G17 mode (plane XY)
0	1	G18 mode (plane ZX)
1	0	G19 mode (plane YZ)

In this case, again, the label refers to a setting when the power to the machine is turned ON or control cleared. Note that these two bits (#1 and #2) in the upper row are empty. That is because the plane selection does not apply to turning controls. Comparing this setting with the previous one, there is a major change. The **G00/G01** selection was a selection from two possibilities. In this case, the selection if from *three* possibilities. As the nature of a bit type parameter is a selection of one of two options, two bits have to be used.

For the three possible modes, the *combination* of settings in bit #1 and bit #2 will be the final selection:

	Label G18	Label G19
	Bit #1	Bit #2
For G17 default	0	0
For G18 default	1	0
For G19 default	0	1

Normally, both bits should be set to 0, as **G17** (XY plane) is the most standard setting.

The bit #3 uses label **G91** for both turning and milling controls. The vast majority of CNC lathes with Fanuc control do not use **G90** or **G91** for absolute or incremental mode, they use **U** instead of **G91**, **W** instead of **G90** (and **H** instead of **C**, if available). This is a good example why you should understand that the label is not always the programming command or function.

Normally, the bit #3 should be set to 0, as virtually all programs start in the absolute **G90** mode.

The bit #4 has a label *FPM*, listed only for the T-series controls. The setting should be 0, selecting feed rate per revolution as the default feed rate mode for turning operations. Feed rate mode in *mm/min* or *in/min* is rare, and is only used if a feed rate is required when the spindle is not rotating. Such per/time feed rate should always be selected in the program by a specific G-code:

- **G98**   Feed rate per minute in mm/min or in/min
- **G99**   Feed rate per revolution in mm/rev or in/rev

In this area, many programmers do *not* program **G99** at all and count on the default setting. Although this approach breaks the rules of consistency and not counting on defaults, it is an acceptable exception. In cases when **G98** and **G99** are used frequently, they both should always be included in the program.

The following bit #5 is not used by either type of control system.

Bit #6 is identified by the label *CLR*. This label suggests the word *'clear'*. It controls whether the RESET key will actually resets specified conditions clears their status. This is a bit that is best left alone at its default setting.

The last bit is #7, labeled as **G23**. While G23 is an actual program command, it is only one of a related pair:

- **G22**   Stored stroke check ON
- **G23**   Stored stroke check OFF

Stored stroke check is a rather fancy name for a three dimensional boundary that can be defined by these commands in the program. This boundary defines space where the tool is not allowed to enter. The inhibited space can be defined as either inside or outside, and the **G22** command defines two diagonal points and makes checking active. **G23** cancels the check. This parameter is used at the power ON, to either check or not to check the stored stroke.

INPUT OF REAL NUMBERS

Real numbers are numbers with decimal point, as opposed to integer numbers, which do not use a decimal point. Parameters generally do not accept a decimal point and serious errors can happen if the specified numeric value is entered incorrectly.

Single bit parameters use only digits 0 and 1, both integers. Parameters relating to offsets, for example, will use the *2-word axis* type. Take this simple parameter:

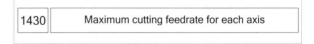

This is a good example of *2-word axis* type of label. Labels of other types (except the bit-type) will have similar labels. Parameter **1430** is quite specific and does not require further explanation. It sets the maximum cutting feed rate for each axis.

Simple enough? It seems so on the surface, but there is more. First, we have to consider the *increment system* and its related *units of input*, then we have to consider the input of actual data *within a particular range*.

These features are quite common for settings of many other parameters, so this example and following explanations are also useful for other data entries.

Increment System

Fanuc controls recognize three increment systems:

- Metric system ... millimeters
- Imperial system ... inches
- Angular system ... degrees

Metric system uses millimeters as units, imperial system uses inches as units, and angular system uses degrees as units.

Units of Input

Units of individual input data are dependent on the increment system, but all use the *units per minute* mode:

- Metric system ... 1 mm/min
- Imperial system ... 0.1 inch/min
- Angular system ... 1 deg/min

These units are used for setting of parameter #1430, and they all have to fall within a certain range.

Data Range

The range of valid parameter data input is a bit more complicated. The reason is that Fanuc supports three different increment systems (IS). These are identified in the *Parameter Manual* as:

- IS-A
- IS-B
- IS-C

Increment system IS-A has been replaced with increment system IS-B. Although this section covers a specific parameter example (**1430**), it is important to understand the difference between increment systems IS-B and IS-C, and how they relate to a particular parameter. First, let's look at the basics. Fanuc calls this feature *Maximum Stroke*, which defines the minimum and maximum allowed data entry - either in the program or the data settings.

Maximum Stroke

The data input for the three units of input will be different for increment system IS-B and increment system IS-C. Both entries will fit into an 8-digit format. The difference will be the assumed position of the decimal point. The decimal point is not entered, so it is *assumed*.

The following table defines the maximum limits of input the control system - *not the machine* - can handle:

Increment System		Maximum Stroke
IS-B	Metric	±99999.999 millimeters ±99999.999 degrees
	Inch	±9999.9999 inches ±99999.999 degrees
IS-C	Metric	±9999.9999 millimeters ±9999.9999 degrees
	Inch	±999.99999 inches ±9999.9999 degrees

The control system, in order to be effective, must always allow a range that is larger than the largest range the machine tool manufacturer may require. That brings us back to the parameter **1430**, discussed here as a representative example.

Maximum Cutting Feedrate Example

For the maximum cutting feedrate example, Fanuc *16i* and *18i* specify the following range of feed rate:

Input increment	Unit of data	Valid data range	
		IS-A, IS-B	IS-C
Metric	1 mm/min	6-240000	6-100000
Imperial	0.1 inch/min	6-96000	6-48000
Rotary	1 deg/min	6-240000	6-100000

Note that all values fall within the *Maximum Stroke* described earlier. To interpret the ranges, a real machine example may be useful. This particular machine has the following specifications relating to the cutting feedrate:

- Maximum cutting feedrate:
  10000 mm/min or 394 in/min

How would these specifications be listed in the parameter **1430**? Again, it depends on the IS - the increment system. Since the IS-B is far more common than the IS-C, the entry in metric should be 10000.0, but the unit of data is in full units of 1 mm, so the parameter setting will be 10000. For inches, the unit of data is 0.1 in/min, so the input will be 3940 for 394.0 in/min. Cutting feedrate is normally programmed with a single decimal point, for example, in inches, F27.6 = 27.6 in/min.

VALID DATA RANGES

The emphasis in this section on the parameter data entry using the *2-word* or *2-word axis* mode was not coincidental. Virtually all dimensional data stored in the parameters use these formats.

For the complete overview and reference, the following table shows valid ranges for *all* data types:

Data type	Valid data range	Remarks
Bit	0 or 1	
Bit axis	0 or 1	
Byte	-127 to 128	In some parameters, signs are ignored
Byte axis	0 to 255	
Word	-32768 to 32767	In some parameters, signs are ignored
Word axis	0 to 65535	
2-word	-99999999 to 99999999	
2-word axis	-99999999 to 99999999	

One more comment. Parameters of the *axis* type can be set for each available axis separately.

PRACTICAL APPLICATIONS

While all the descriptions, specifications and instructions presented so far form a very valuable background for understanding system parameters, there were only a few practical examples presented. This section will look at practical applications in more detail.

> This section is presented for reference only

Useful parameters presented here are generally aimed at the CNC programmer, operator or supervisor, not necessarily a service technician. Most of these parameters are for reference only, and not to be changed.

PARAMETERS OF SETTING

This initial section contains two parameters that may be useful:

- Parameter 0000
- Parameter 0001

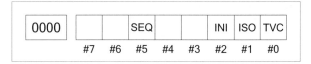

There are three bits that are worth looking into:

ISO   Code used for data output

     0 : EIA code
     1 : ISO code

Bit #1 (ISO) selects the output format. In the old days of tape input, there was a distinction between EIA and ISO format in terms of the number of holes across the tape. Each set of holes represented a single character. EIA had an odd number of holes, ISO (or ASCII as it was called) had an even number of holes across. Today, this distinction is irrelevant, so use the default. ISO is the most common default.

INI   Unit of input

     0 : In mm
     1 : In inches

Bit #2 (INI) sets the measurement units to millimeters or inches. Note that this setting will change automatically when **G20** or **G21** command is used in the program.

Setting (Handy)

Another way - a much simpler one - to activate the bits #1 and #2 is via the screen *Setting (Handy)*, described on *page 234* - the illustration shows the full top portion:

```
SETTING (HANDY)              O1234 N0001

PARAMETER WRITE   = 0   (0 : DISABLE 1 : ENABLE)
TV CHECK          = 0   (0 : OFF 1 : ON)
PUNCH CODE        = 0   (0 : EIA 1 : ISO)
INPUT UNIT        = 0   (0 : MM 1 : INCH)
I/O CHANNEL       = 0   (0-3 : CHANNEL NUMBER)
```

The last bit to be described is **bit #5 (SEQ)**.

SEQ   Automatic insertion of sequence numbers

     0 : Not performed
     1 : Performed

As the label suggests the setting of bit #5 has something to do with block numbers. When a program is entered from the control system keyboard, it has to be entered exactly. This process can be somewhat shortened, if the operator does not have to enter block numbers for each sequence.

Setting the bit #5 of parameter **0000** will input block numbers automatically. That leaves a question of what the increment of the block numbers will be. For that, you have to search for another parameter, **3216**:

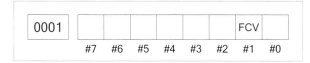

This is a word-type parameter and the allowed range of input is from 0 to 9999. This parameter will be active only if the bit #1 of parameter **0000** is set to 1.

The other parameter of interest is parameter **0001**:

```
0001 |   |   |   |   |   | FCV |   |
      #7  #6  #5  #4  #3  #2   #1  #0
```

There is only one bit to consider - bit #1:

FCV   Tape format

    0 : Series 16 standard format
    1 : Series 15 format

Fanuc *Model 15* has been one of the fully featured control systems for a long time. In spite of fact that this control has some of the most modern features available even today, it also has some that were improved on controls positioned slightly lower. This presents some incompatibility problems.

The main purpose of this parameter is to make programs developed for the *Model 15* control type to work on the *16i* and *18i* controls. This parameter establishes compatibility with older programs.

Setting the bit #1 to 1 establishes compatibility for the following functions:

- Subprogram call M98
- Threading with equal leads G32 (T-series)
- Cycles G90, G92, G94 (T-series)
- Multiple repetitive canned cycle G71 to G76 (T-series)
- Fixed cycles G73, G74, G76, G80 to G89 (M-series)
- Cutter compensation C (M-series)

Some restrictions may apply, so refer to the *Operator's Manual* of your *16i/18i* control.

The most noticeable differences are in multiple repetitive cycles used on CNC lathes. The more modern controls (Models *16i* and *18i* included) use a two block format, while the earlier controls (Model *15* included) used only one block for the same cycles. Fanuc has changed the format to allow more flexibility at the machine, rather than constantly changing system parameters.

Turning and boring commnads are described in a separate chapter - see *page 173*. For the threading cycle **G76** and its differences see *page 214*.

PARAMETERS OF FEEDRATE

The primary parameter in this section is parameter **1401**. It is a bit-type parameter:

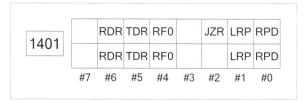

As you can see, this is another bit-type parameter, but with separate settings for the T-series (top row) and the M-series (bottom row).

The part that should be of some interest to the programmers and operators are the bits #1 and #4. Here are bit **#1 (LRP)** descriptions and details:

LRP   Positioning (G00)

    0 : Positioning is performed with non-linear type positioning so that the tool moves along each axis independently at rapid traverse
    1 : Positioning is performed with linear interpolation so that the tool moves in a straight line

Changing this parameter means changing the way a multi axis rapid motion is performed. Earlier controls did not have this option - multi axis motion was always based independently for each axis. The result of this motion was the well known 'dog-leg motion' or 'hockey-stick' motion. Its greatest disadvantages was that the actual tool path did not correspond to the shortest distance between two points. The latest controls can remedy that by setting the bit #1 of parameter **1401** to 1.

Fanuc recommends to set the bit #1 of parameter **1401** to 1 in the following two situations:

- When using a multi-path system, set this parameter to the same value for all paths

- Be sure to set this parameter to 1 when performing three-dimensional coordinate conversion or using the tilted working plane command mode

The other bit, bit **#4 (RF0)** offers these settings:

RF0 When cutting feed rate override is 0% during rapid traverse,

  0 : The machine tool does not stop moving
  1 : The machine tool stops moving

Either setting of this parameter is OK. It's actual setting is more dependent on personal preference. This is how it works:

When you run the first part or are making any other type of step-by-step program evaluation, you typically work in single block mode. Upon each *Cycle Start*, the next block will be processed. If the next block is an axis motion, you would like to know what the motion will be, before the tool starts physically moving. Setting the feed rate override to 0% can do the trick but only by setting the bit #4 of parameter **1401** to 1. This setting will provide time to observe and evaluate the real distance-to-go on the screen. Setting the bit #4 to 0 will start the motion immediately upon *Cycle Start*, at a very low feed rate.

Two other bits **#5 (TDR)** and **#6 (RDR)** refer to dry run. They should be left at their factory defaults.

Another parameter of interest is **1421**.

1421	F0 rate of rapid traverse override for each axis

It is a *word type* parameter and it contains the amount of rapid feed rate (rapid motion or rapid traverse) when the rapid override switch is set to **F0**.

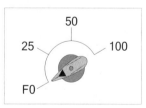

There are certain benefits to explore this parameter.

First the ranges:

Input increment	Unit of data	Valid data range	
		IS-A, IS-B	IS-C
Metric	1 mm/min	30 - 15000	30 - 12000
Imperial	0.1 in/min	30 - 6000	30 - 4800
Rotary	1 deg/min	30 - 15000	30 - 12000

Set the **F0** rate of the rapid override for each axis.

Rapid traverse override signal		Override value
ROV2	ROV1	
0	0	100 %
0	1	50 %
1	0	25 %
1	1	F0

F0: Parameter **1421**

Now for the practical use. Current CNC machines have rapid rates that were unheard of only a few years ago. Rapid rates between 50000 and 75000 mm/min (~2000 and 3000 inches/minute) are not uncommon. Take a short comparison:

In the mid 1980's, 7500 mm/min or 300 inches/min was the top rapid rate of a CNC machining center. Take that rate overridden at the machine to 25% - the machine moved at 1875 mm/min or about 75 inches/min. Now, take the modern CNC machining center with the maximum rapid rate of 'only' 60000 mm/min or about 2400 inches/min. At 25% rapid override, the rapid traverse motion will still be at relatively high 15000 mm/min or 600 inches/min. Setting the parameter **1421** to a more manageable amount might be worth consideration.

Note - some high-speed CNC machining centers also offer a 10% feed rate override, in addition to the common 25%, 50%, and 100% settings

➟ *There are three other feed rate parameters of interest*

- 1420   Rapid traverse rate for each axis
- 1422   Maximum cutting feed rate for all axes
- 1430   Maximum cutting feed rate for each axis

These parameters are based on the machine design and should only be used for reference.

PARAMETERS OF DISPLAY

*Note: Several parameters listed in this section of **Parameter Manual** are related to a particular screen size, so make sure to know the screen size of your control.*

A parameter that may be of some interest especially to CNC operators is **3107**, especially its bit **#4 (SOR)**. Screen size is not important in this case.

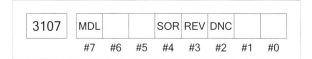

This is a bit-type parameter and parameter #4 is used for displaying program directory:

SOR　Display of the program directory

　　　0 : Programs are listed in the order of registration.
　　　1 : Programs are listed in the order of program number

The description is self-explanatory.

Language Selection

For those user whose language is not English, you can select from a large number of languages using parameters **3102**, **3119**, and **3190**:

- **3102**　Dutch, Spanish, Korean, Italian, Chinese (traditional characters), French, German, Japanese
- **3119**　Portuguese
- **3190**　Russian, Chinese (simplified characters), Czech, Swedish, Hungarian, Polish

For the English language selection, bits for all other languages must be 0. Also note that when the language parameters are set, power has to be turned off before the selected language becomes effective.

Programs Registration and Editing

Parameters **3201** and **3202** are both bit-type parameters and contain a number of settings relating to program registration and editing. Most practical setting are done through the parameter **3202**. The listing is quite large, so lookup the *Parameter Manual* for your control system for more details.

The following summary is a list of four main settings that you should be aware of in relation to parameter **3202**. Although Fanuc refers to the word 'subprogram', it only refers to its most common usage. A 'subprogram' can be a regular program, but in most cases it is a *Fanuc Macro-B* program or at least a special subprogram. This is the summary of what parameter **3202** can be used for:

➪ *Editing of subprograms with numbers O8000 to O8999*

➪ *[CONDENSE] directory function*

➪ *Program number search*

➪ *Editing of subprograms with numbers O9000 to O9999*

Individual setting cover subjects such as:

- Program editing (allowed or disallowed)
- Program deletion (allowed or disallowed)
- Program search and output restrictions
- ... and a few others

Password Protection

A certain range of stored program numbers can be protected by a password. To protect a program, its number must be in the 9000-series. Parameter **3210** can be set to password protect programs in the range of 9000 to 9999.

3210	Password

In most cases, this feature is used for important programs, such as special macros. Look for summary of parameter **3202** above. Settings of these two parameters are often combined. This parameter offers password protection only to the 9000-series programs.

There are four additional parameters that are associated with program protection of *other* program numbers:

3220	Password
3221	Keyword
3222	Program protection range (minimum value)
3223	Program protection range (maximum value)

Parameter **3220** is a 2-word type of parameter and accepts a large range between 0 to 99999999. Use this parameter to set a protective password. Any value other than zero becomes the password, which will not be visible once set. Any program that has been password protected (locked) will not be available for viewing or editing.

In order to set the password, parameter **3220** must be unlocked. Unlocked state exists in two instances:

- When the setting of 3220 is 0
- When the setting of 3221 matches the password

The related parameter **3221** is also a 2-word type of parameter and has the same range from 0 to 99999999.

Range of program numbers that can be password protected is controlled by the two remaining parameters, **3222** and **3223**:

- 3222 Program protection range (minimum value)
- 3223 Program protection range (maximum value)

The range of either parameter is between 0 and 9999.

Suppose, you want to lock programs O8500 to O8599. In this case, parameter **3222** will be set to 8500 and parameter **3223** will be set to 8599.

In order to lock programs of the 9000-series (O9000 to O9999), both parameters **3222** and **3223** must be set to zero (0).

Program Display

Some sensitive, important or proprietary programs, subprograms and macros included, can be hidden from the screen display of the control and work in the background only. This is controlled by parameter **3232**.

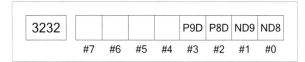

ND8 When a program with a program number from 8000 to 8999 is being executed as a subprogram or macro program, the display of the program on the program screen is:

0 : Not prohibited
1 : Prohibited

ND9 When a program with a program number from 9000 to 9999 is being executed as a subprogram or macro program, the display of the program on the program screen is:

0 : Not prohibited
1 : Prohibited

The bits #2 and #3 are similar - they have the same meaning as bits #0 and #1 but their range of program numbers is between 80000000 to 89999999 (bit #2) and 90000000 to 99999999 (bit #3).

Automatic Block Numbering

Parameter **0000** was described on page 240. Bit **#5 (SEQ)** of this parameter, when set to 1, activates automatic block numbering when the program is input at the control manually.

| 3216 | Increment in sequence numbers inserted automatically |

Parameter **3216** contains the *increment* for each block. If the setting is 1, the result will be *N1, N2, N3*, etc. If the setting is 5, the result will be *N5, N10, N15*, etc.

> Keep in mind that large block number increments have adverse effect on the system memory capacity

Other Parameters

Parameters of display and editing are quite numerous. Those of interest to CNC programmers and CNC operators have been listed in this section.

Parameters **3104** relates to display of machine position and units, **3105** relates to the display of spindle speed and tool number, parameter **3114** controls screen change when a function on the control panel is pressed. The functions keys are **POS, PROG, OFFSET/SETTING, SYSTEM, MESSAGE, CUSTOM** and **GRAPH**. There are two parameters related to the safety of other parameters (parameters **3225** and **3226**) and several others parameter that are worth looking into.

As important as the parameters discussed so far are, most of them were covered for reference and understanding them without the intent of actually changing them. The same approach will apply for next sections, but more parameters will be subject to change.

PARAMETERS OF PROGRAMS

When it comes to CNC programs and programming, there is a large amount of information to be learned from parameters. Just by leafing through the *Parameter Manual* of the control system, you will get a sense of what is what, what parameters can be changed and which to leave alone, their format and structure. As the previous sections have shown, Fanuc provides a great degree of organization when it comes to parameters. The following sections all have some relationship to programs. While the previous format of presenting parameters will be left intact, not every bit or other data will be explained - only those considered by the author as *'more important'*. In this sense, the selection is subjective, but the objective is to provide sufficient details. Of course, all other data you will find in the *Parameter Manual*.

Decimal Point

Parameters of the program start with parameter **3401**.

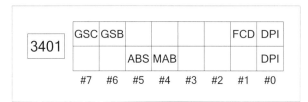

Possibly, only the bit **#0 (DPI)** is of some interest. It sets the rule of how a program data input is interpreted, if no decimal point is given for an address that normally requires it. Take, for example, a simple dimensional address *X1*. Even when the units of measurement are specified, for example, in inches, is *X1* equal to one inch, one hundred thousandths, or less, or more?

This parameter setting provides the answer.

DPI When a decimal point is omitted in an address that can include a decimal point

   0 : The least input increment is assumed
   1 : The unit of mm, inches, or second is assumed

Normal setting of this parameter is *zero*, at least for CNC machining centers and lathes That means *X1* will be interpreted as *X0.0001*, as 0.0001 is the least increment in inches. By the same definition, *X10* in metric will be interpreted as *X0.010*, as the least increment in the metric configuration is one micron or 0.001 mm.

For more detail on addresses without a decimal point, check *page 255, 'No Decimal Point Entry'*.

If we look at the alternative, where the setting of the parameter **3401** is 1 (one), what are the changes? In this case, the control is set to *calculator type of input*, where setting of 1 is the unit currently active - 1 mm, 1 inch or 1 second. The analogy here is how we input numbers into a pocket calculator. For example, 5 will be equivalent 5.0 and 5.3 wil be equivalent to 5.3.

When this setting is active, and metric mode is current (for example), *X45* will be interpreted as 45 mm. If a decimal point is required, the input of (for example) *X45.6*, will be interpreted as 45.6 mm.

Power-On Defaults

Parameter **3402** covers various default settings when the power is turned on. This parameter was used earlier in this chapter to provide an example - see *page 236*.

End of Program Functions

Parameter **3404** also has a number of options, this time with the focus on *program end* functions **M02** and **M30**, as well as the number of M-functions that can be programmed within a single block:

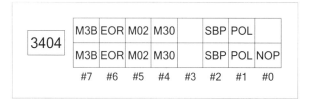

Bits #4 and #5 of parameter **3404** define **how** the *End of Program* function will be interpreted.

M30 When M30 is specified in a memory operation:

   0 : M30 is sent to the machine, and the head of the program is automatically searched for.
      So, when the ready signal FIN is returned and a reset or reset and rewind operation is not performed, the program is executed, starting from the beginning

   1 : M30 is sent to the machine, but the head of the program is not searched for. (The head of the program is searched for by the reset and rewind signal.)

The Fanuc description may be clear to some and a bit vague to others. Experience will widen your sense of interpreting technical texts, including Fanuc manuals. The exact Fanuc definitions are quoted here for one reason only - they are the ones you have to interpret and learn. They are the ones you will use in everyday work. However, for learning purposes, these 'official' definitions are also followed by a more practical explanation, with additional details. How does that affect the bit **#4 (M30)**?

If the bit #4 is set to 0 (zero), when the control encounters the *End of Program* function **M30**, it will automatically return to the first block of the program, ready for another *Cycle Start*. This is the most common setting.

In the unlikely case that #4 is set to 1 (one), the control does *not* return to the top of program.

A function that is very closely related to **M30** is **M02**. When you read the individual Fanuc instructions for **M30** and compare them with those for **M02**, they are identical, *except* for the function itself:

M02 When M02 is specified in memory operation

  0 : M02 is sent to the machine, and the head of the program is automatically searched for.
  So, when the end signal FIN is returned and a reset or reset and rewind operation is not performed, the program is executed, starting from the beginning

  1 : M02 is sent to the machine, but the head of the program is not searched for. (The head of the program is searched for by the reset and rewind signal.)

To understand the difference between **M02** and **M30**, you have to go a bit to the past (not the first time, either). In the days of paper tape, the program was punched on a tape, loaded onto a reel and placed into the reader area of the control. Then, it was fed through a tape reader to the take-up spool. This method was no different as that for a cassette tape, reel-to-reel tape, or - later - VHS tape.

If **M30** was the last function programmed, the tape automatically rewound back to the beginning. The problem was with tapes that were very short. The tape was quite expensive and for reel-to-reel applications, there had to be a leader and a trailer to secure the tape to the reel. The length of each leader and trailer was approximately 1 meter (3 feet), depending on the tape reader design. For long programs (equalling a long tape as well), it was justified.

On the other hand, if a short program required only a foot of tape, for example, the two extra meters or six feet of tape could not be justified. In addition, not all controls were equipped with a reel-to-reel equipment. Many were equipped with only the tape reader and a chute for the tape to fall in. This is where **M02** function became handy.

While the **M30** function was always associated with the end of program and return to the top of program (called *rewind*), **M02** was always associated with end of program but no rewind.

Times have changed, but Fanuc support for the legacy (older) systems has not. This parameter **3404** and its bits #4 and #5 control the behavior of both **M02** and **M30**.

➡ *Author's opinion:*

Leave both bits #4 and #5 set to 0 (zero). There is no need for **M02** *in the part programs anymore, so even a very old program with* **M02** *will behave the same way as newer programs with* **M30**.

Number of M-Functions in a Block

There is one more **3404** parameter bit that relates to the M-functions - it is bit **#7 (M3B)**:

M3B The number of M codes that can be specified in one block:

  0 : One
  1 : Up to three

Traditionally, only one M-function has been allowed in one program block. For programs that require some degree of portability between machines with similar control systems, this *one-M-per-block* rule should always be adhered to. Modern control systems do allow up to three different M-functions in one block, providing they do not conflict with each other.

Bit **#7 (M3B)** of parameter **3404** offers that choice.

> While up to three M-functions in a block may provide a degree of efficiency, allowing only one M-function in a block offers greater portability of programs from one control to another similar control

If the bit #7 of parameter **3404** is set to 0 (zero) (one M-function only), the control will issue an alarm (error condition) for any occurrence of more than one M-function in one block.

PARAMETERS OF SPINDLE CONTROL

There are two spindle related parameters that may be of interest to programmers and operators. One covers the maximum spindle speed and one spindle orientation.

Maximum Spindle Speed

Parameter **3772** is for reference only. It stores the maximum spindle speed the machine has been designed for. The units are revolutions per minute. Valid range is between 0 to 32767 r/min. High speed machining centers would have a different control setting, as their maximum spindle speed exceeds 32767 r/min.

Spindle speed for machining centers is programmed with address *S*, generally followed by the direction of the spindle rotation. For example,

```
S3500 M03
```

The block specifies 3500 r/min at normal clockwise spindle rotation.

If the programmed speed or the speed entered via MDI is higher than the maximum speed, the spindle speed remains at the maximum speed set in parameter **3772**. The same applies if the maximum speed is programmed but the feedrate override switch is set over 100%.

For CNC lathes and tuning centers, the maximum speed cannot be exceeded either, regardless whether cutting speed (**G96**) or spindle speed (**G97**) is in effect.

Spindle Orientation Code

| 4960 | M code specifying the spindle orientation |

During automatic tool change (ATC), the spindle has to stop and orient into a position that is aligned with the arm of the tool changer. This is called *spindle orientation*. The ATC process is automatic, done when tool change function **M06** is activated.

There are times, when the spindle orientation needs to be programmed or entered via MDI. Parameter **4960** specifies the M-function for programmable spindle orientation in the range of 6 to 97. The most common function is **M19** but some machine use **M20**.

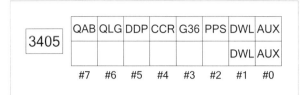

The next parameter that may come useful is **3405**. The one possible bit of interest in this parameter **3405** is the bit **#1 (DWL)**. It refers to the interpretation of **G04** dwell command. **G04** can be programmed as dwell in seconds, milliseconds, or as dwell per rotation. Dwell per second is the normal setting (Bit #1 = 0).

There are several bits to set, applied mainly to turning. Our primary interest is in the bit **#1 (DWL)**:

DWL The dwell time (G04) is:

    0 : Always dwell per second
    1 : Dwell per second in the feed per minute mode, or dwell per rotation in the feed per rotation mode

G04 is a dwell command issued as a separate block (outside of fixed cycles common to CNC machining centers). The main reason for programming a dwell at all is to complete certain activity before another activity starts. Several fixed cycles for drilling operations use dwell, but in this case, its definition is part of the cycle itself as address *P*. For dwell used as a separate block, its input is *X* or *U* in seconds, or *P* in milliseconds. For example, `G04 X0.5` represents one half of a second. For CNC lathes and turning centers only, `G04 U0.5` can be used with the same result of 0.5 of a second dwell. The input version such as `G04 P500` resembles the input in several fixed cycle (**G82**, for example) and represents one half of a second as well. Many CNC programmers like to use the P-address even for **G04**, to be consistent with the fixed cycles that provide no alternative.

Automatic Corner Rounding

For some lathe applications, bit **#4 (CCR)** of parameter **3405** may also be useful. It is associated with the corner handling function on CNC lathes for automatic fillet or chamfer generation. This *Automatic Corner Rounding* is a feature that allows a chamfer or a radius between two perpendicular and orthographic lines. In plain English, an automatically generated chamfer or a radius will be between a face (or a shoulder) and a diameter. There are also some programming rules that have to be observed.

PARAMETERS OF TOOL OFFSET

Offset - or compensation - of tool settings in this section relates to tool length and tool radius. Individual bits of several parameters are useful to know.

Offset Address

Parameter **5001** contains two related bit settings:

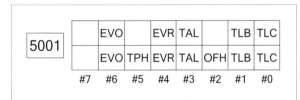

OFH Offset number of tool length compensation, cutter compensation and tool offset

   0 : Specifies the tool length compensation using an H code, and cutter compensation C using a D code. Tool offset conforms to bit 5 (TPH) of parameter No. 5001

   1 : Specifies the tool length compensation, cutter compensation and tool offset using H codes

TPH Specifies whether address D or H is used as the address of tool offset number (G45 to G48)

   0 : D code
   1 : H code

   TPH is valid when bit 2 (OFH) of parameter No. 5001 is set to 0

Normally, tool length offset is programmed as G43 H.. and cutter radius offset as `G41 D..` or `G42 D..`. Bit **#2 (OFH)** controls which letter is programmable for each offset type.

Bit **#5 (TPG)** is listed here only because it is mentioned in bit #2. It relates to *Position Compensation* only, which is the predecessor of cutter radius offset as we know it today. Position compensation used **G45**, **G46**, **G47**, and **G48**. Listing of this type of compensation is strictly for backward compatibility reasons and has no use in modern CNC programming.

Other bits are not important for programming or operation of the CNC machine.

Lathe Offsets

Parameter **5002** is used exclusively by lathe controls.

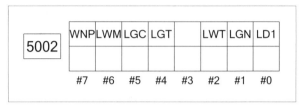

There are a number of options in this parameter, all used for settings of *Geometry* and *Wear* offsets for CNC lathes. No particular bit requires special explanation. This parameter is included here for the purpose of being aware of it, if some changes are required.

Clearing Offsets

Just like the parameter **5002**, parameter **5003** is included here for the same reason - to be aware of it.

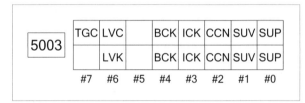

The settings in parameter **5003** relate to the method of offset cancellation and offset behavior with **G28**. Generally, the default settings work quite well.

Radius vs. Diameter

The last of the parameters in this group is **5004**. Two of the bits are important, one for milling applications, the other for turning applications.

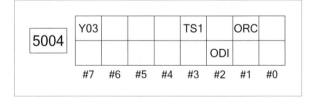

ORC Tool offset value

   0 : Set by the diameter programming
       (Can be set in only the axis under diameter programming)

   1 : Set by the radius programming

ODI A cutter compensation amount is set using:

0 : A radius

1 : A diameter

Bit **#1 (ORC)** is for CNC lathes only. It specifies whether programming the X-axis specifies diameter or radius. Setting the bit to zero (0) represents the common method of programming - *as a diameter*.

A corresponding bit for milling application specifies whether cutter radius offset (compensation) is set as radius or diameter. The common method of setting is zero (0) - *as a radius*.

> *Note - Internally, cutter radius is always offset by the radius, regardless of the parameter setting. For the operator, radius setting makes more sense. Diameter setting is often used by companies that have a special department for computerized tool setup and automatic offset setting using data transfer via a chip.*

Two remaining parameters of interest are related to wear offset.

Wear Offset

Wear offset is an adjustment used for fine-tuning measured dimensions. Regardless of the machine type, this adjustment has a maximum value that can be input. The type of active increment system is also important. Two parameters control the value - **5013** and **5014**.

Both are two-word types:

5013	Maximum value of tool wear compensation

5014	Maximum value of incr. input for tool wear comp.

Parameter **5013** specifies the maximum *absolute* value based on the increment system:

Input Increment	IS-A	IS-B	IS-C
Metric - mm	0.01	0.001	0.0001
Imperial - inch	0.001	0.0001	0.00001

Increment system IS-B is the most common.

Valid data range of parameter **5013** is based on the following table:

Input Increment	IS-A	IS-B	IS-C
Metric - mm	0 - 99999	0 - 999999	0 - 9999999
Imperial - inch	0 - 99999	0 - 999999	0 - 9999999

Parameter **5014** specifies the maximum *incremental* value based on the increment system - the values are identical to parameter **5013**:

Input Increment	IS-A	IS-B	IS-C
Metric - mm	0.01	0.001	0.0001
Imperial - inch	0.001	0.0001	0.00001

Valid data range of parameter **5014** is based on the following table:

Input Increment	IS-A	IS-B	IS-C
Metric - mm	0 - 99999	0 - 999999	0 - 9999999
Imperial - inch	0 - 99999	0 - 999999	0 - 9999999

▪◆ *Example 1:*

Assume that the control uses the common increment system IS-B and the maximum absolute value is 0.5 mm. Offset **5013** will be set to *500*.

▪◆ *Example 2:*

Assume that the control uses the common increment system IS-B and the maximum *incremental* value is 0.1 inches. Offset **5013** will be set to *1000*.

Increment system IS-C is often used by wire EDM machines, where programming coordinates can be up to five decimal place accuracy.

The next section will cover parameter settings for various cycles. Both milling and turning will be explained.

PARAMETERS OF CYCLES

Parameters in this section relate to *fixed cycles* - sometimes called the *canned cycles* - as well as *multiple repetitive cycles,* specifically designed for lathe work.

The word *'cycles'* is a collective description of various toolpath shortcuts, where one or two blocks replace many program block in the so called *'long hand'*. Cycles are used in CNC programming for a number of reasons:

- Shorter programming time
- Shorter program length
- Quick changes at the machine

It is the last item that should be of particular interest to the CNC operators. Even the best efforts by the part programmer do not always guarantee the optimum machining. There are various factors that cannot be foreseen at the time of programming. Being able to optimize the program at the machine is a great benefit.

Fanuc 15 Mode - Compatibility

Parameter **5103** contains several settings that allow compatibility with the older type of programming, using the *Fanuc Model 15* as base.

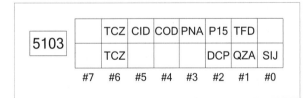

Only one bit is of interest here - *bit #0*:

SIJ   When the tape format for Series 15 is used (with bit 1 (FCV) of parameter No. 0001 set to 1), a tool shift value for the drilling canned cycle G76 or G87 is specified by:

0 : Address Q

1 : Address I, J, or K

On CNC machining centers, there are two fixed cycles that require a shift from the hole centerline - **G76** for fine boring (see *page 124*), and **G87** for backboring (see *page 123*). Normally, the shift is programmed by the address *Q* - for example:

```
G99 G76 X.. Y.. R.. Z.. P.. Q0.01 F..
```

Fanuc model 15 also allowed programming the shift amount for each axis separately, using the *I/J/K* vectors. For the **G17** XY-plane, the vectors are *I* and *J* and represent the shift distance and direction along the X-axis and the Y-axis respectively:

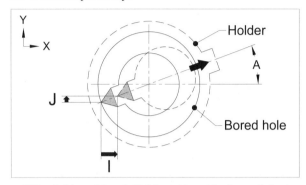

If I = 0.01 and J = 0.005 in **G20** mode (imperial), the **G76** cycle in *Fanuc 15* mode is:

```
G99 G76 X.. Y.. R.. Z.. P.. I0.01 J0.005 F..
```

Remember that *I* and *J* vectors *replace* the Q-amount that is common to both axes.

Spindle Rotation Direction

Other reference-only settings are in two parameters **5112** and **5113** - both are 2-word type parameters:

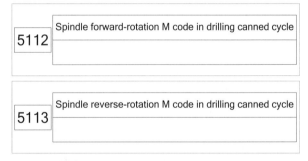

The range of either parameter is from *0 to 255*. Parameter **5112** sets the spindle *forward* rotation M-function in a drilling canned cycle. Parameter **5113** sets the spindle *reverse* rotation M-function in a drilling canned cycle.

- Setting of zero (0) in parameter 5112 will output the M-function as M03

- Setting of zero (0) in parameter 5113 will output the M-function as M04

M03 and **M04** are normal spindle rotation functions and are changed - usually by the machine tool builder - for specific purposes only.

Clearances in G73 / G83 Cycles

In the earlier chapter *Machining Holes*, the illustration for the two peck drilling cycles clearly shows a dimension generally marked in manuals and other documentation by a low case letter *'d'* (see *page 118*):

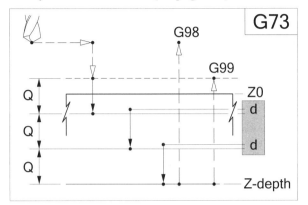

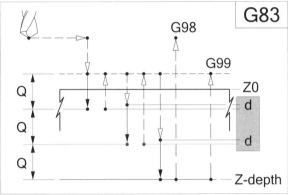

The distance *'d'* is not programmable, it can only be set by a system parameter. For peck drilling, the distance *'d'* is a clearance between one peck and the next. In both cycles, this is the distance that will prevent the tool from hitting the completed depth from a previous peck. There are two setting parameters - **5114** and **5115**:

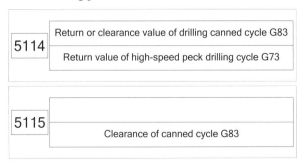

Keep in mind that the upper row refers to CNC turning system, bottom row to CNC milling systems.

As settings for milling systems are far more common, the bottom row will be examined. Setting for the **G83** cycle on the lathes follows the same principles.

The distance *'d'* in **G73** peck drilling cycle is called *return value*, the distance *'d'* in **G83** peck drilling cycle is called *clearance*. Either way, they share the same range of input - between 0 and 32767 (word type). As for other parameters, the increment system is important for correct interpretation of the setting:

Input Increment	IS-A	IS-B	IS-C
Metric - mm	0.01	0.001	0.001
Imperial - inch	0.001	0.0001	0.0001

Increment system *IS-B* is the most common for CNC machining and turning centers. In this case, *IS-C* is the same as *IS-B*. *IS-A* is not normally used.

➥ *Example:*

A particular *'d'* setting in metric units is 0.5 mm and in imperial units 0.02 inches and *IS-B* is active. In this case, the parameters will be set to:

Metric:

- 5114   500   ... *formatted from 00000.500*
- 5115   500   ... *formatted from 00000.500*

Imperial:

- 5114   200   ... *formatted from 0000.0200*
- 5115   200   ... *formatted from 0000.0200*

Note - Since the return value (5114) and the clearance value (5115) use separate parameters, they do NOT have to have the same setting value

For example, you may find that on a vertical machining center, the chip build-up in the hole causes problems with the drill. Increasing the setting in parameter **5115** may help. Parameter **5114** does not have to be changed.

PARAMETERS OF THREADING CYCLE

There are two threading cycles available for CNC lathes - **G92** and **G76**. For all practical applications, **G92** - sometimes called the *'box threading cycle'* for its limitations - is considered obsolete. However, the two parameters discussed in this section apply equally to the old **G92** *and* the top-of-the-line, **G76** cycle (*page 215*).

There are only two parameters related to threading cycles on lathes. Considering how important the threading cycles are - particularly the **G76** - it is extremely important to understand the structure of the **G76** cycle and its related settings. Its format is presented here for reference only, but all its details you can find in a much more comprehensive section, starting on *page 215*.

The two parameters are **5130** and **5131** - they are both of the byte-type format:

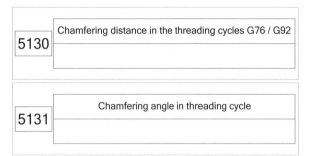

Both parameters use the word *chamfering*. In relation to a threading cycle, the word *chamfering* is not to be confused with the same word used for breaking corners or deburring edges. A better description of this feature would be *gradual pull out* from the thread. The following schematic illustration shows both the ***chamfering distance*** (parameter **5130**) and the ***chamfering angle*** (parameter **5131**) for the lathe threading cycle:

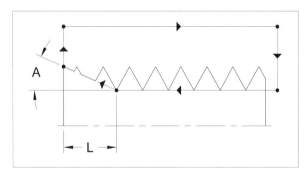

- **Parameter 5130 is measured as the amount of lead *L***
  The range is 0 to 127 in increments of 0.1 pitch

- **Parameter 5131 is measured as an angular dimension**
  The range is 1 to 89 in full degrees

The amount of lead is *programmable* as the second pair of digits in the first block of **G76** cycle. The programmable range is from 0.0 to 9.9 in 0.1 lead increments. The amount of lead *L* is normally programmed in the **G76** cycle. 45° pull-off is the default.

PARAMETERS OF MULTIPLE REPETITIVE CYCLES

Fanuc controls offer a number of cycles for lathes and turning centers. They are called *multiple repetitive cycles*. Unlike fixed cycles for machining holes, where the program processing always continues from the beginning to the end, multiple repetitive cycles are processed in both directions - the cycle evaluation is *repeated* until the final machining is completed.

Available Cycles

All cycles in this group are in the **G70** to **G76** range (using the most common G-code group B)

- **G70**   Finishing cycle used after G70, G71, and G73
- **G71**   Roughing cycle - basic direction is horizontal
- **G72**   Roughing cycle - basic direction is vertical
- **G73**   Pattern repeating cycle
- **G74**   End face peck drilling or grooving (along Z-axis)
- **G75**   Diameter drilling or grooving (along X-axis)
- **G76**   Thread cutting

This is only a brief listing for reference. All these cycles have been described in the following chapters:

- **G70 / G71 / G72 / G73**   Turning and Boring
  ... starts on page 173

- **G74 / G75**   Grooving On Lathes
  ... starts on page 195

- **G76**   Thread Cutting
  ... starts on page 207

Parameters for G71 / G72 Cycles

Programmed depth of cut in **G71** and **G72** is stored in parameter **5132**.

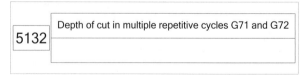

This is the U-address in the first block of the **G71** cycle or the W-address in the first block of the **G72** cycle.

As a 2-word type parameter, its settable range is from 0 to 99999999, with minimum increment of 0.001 mm or 0.0001 inch. For example, depth of 3.5 mm will be set as 3500 and depth of 0.175 inches as 1750.

Parameter **5133** is the retract (escape) amount in the **G71** and **G72** cycles. In the cycle, this is the R-address in the first cycle block of either **G71** or **G72**.

As a 2-word type parameter, its settable range is from 0 to 99999999, with minimum increment of 0.001 mm or 0.0001 inch. For example, 0.5 mm retract will be set as 500, and 0.02 inch retract will be set as 200.

The retract is performed at the end of each roughing pass and the movement away from the cut is at 45°.

Parameters for G73 Cycle

Cycle **G73** is not as commonly used as **G71** or **G72**, but it has its use for roughing castings or forgings where the rough shape is close to the finished shape. The two parameters that set the retract (escape) amount are **5135** and **5136**. The need for two parameters rather than one is the **5135** controls retract in the X-axis, while **5136** controls retract in the Z-axis.

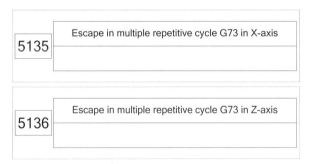

It is also a 2-word type parameter, but its range is from -99999999 to 99999999. Minimum increment is 0.001 mm or 0.0001 inch, same as for the previous cycles.

In the program, the X-amount is set by the U-address in the first cycle block, the Z-amount is set by the W-address of the first cycle block.

Parameter **5137** set the number of pattern repeats (division count). Its range is from 1 to 99999999 and the data entry is also 0.001mm or 0.0001 inch.

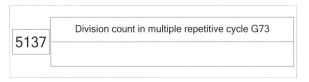

Parameters for G74 / G75 Cycles

Cycles **G74** and **G75** are used for drilling or rough grooving. The direction of cutting in **G74** is along the Z-axis, the cutting direction in **G75** cycle is along the X-axis. Both cuts are interrupted cuts (peck drilling or peck grooving) and the retract (return) clearance between each cut is set by parameter **5139**.

For both cycles, the amount of retract is programmed by the R-address in the first cycle block.

Parameters for G76 Cycle

Multiple repetitive threading is a very complex internal algorithm using a surprisingly easy input of cycle data. There are four parameters associated with the threading cycle **G76**:

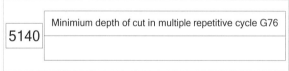

Parameter **5140** controls the setting of minimum cutting depth (minimum threading depth). The allowed range is from 0 to 99999999, using data entry of 0.001 mm or 0.0001 inch, in increment system IS-B.

In the part program, the minimum depth is the Q-address in the first block, and it is a radius value.

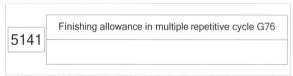

Parameter **5141** controls the setting of the finishing allowance. The allowed range is from 0 to 99999999, using data entry of 0.001 mm or 0.0001 inch, in increment system IS-B.

In the part program, the minimum depth is the R-address in the first block of the **G76** cycle.

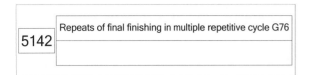

Parameter **5142** controls the number of repetitions of the finial threading cut. The allowed programmable range is from 0 to 99, using an integer as the number of repetitions. The actual parameter range is greater, from 0 to 99999999.

In the part program, the number of repetitions is specified by the first two digits of the P-address in the first block - for example:

P02....

... specifies two repetitions. Normal default is one.

The second pair of the P-address is the pull-out distance in number of leads, for example

P..10..

... specifies one (1.0) lead of gradual pull-out. This setting is controlled by parameter **5130**, already described on *page 252*.

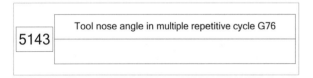

The final parameter for the **G76** threading cycle is **5143**. It controls the value of the thread angle. In the program, this is the third pair of digits in the P-address used in the first block:

P....60

... specifies a 60° thread angle.

☛ *NOTE - Only six angles are normally allowed as input:*

When tape format for Fanuc series 15 is *not* used, the selection is:

00   29   30   55   60   80

When tape format for Fanuc series 15 *is* used, the range is from 0 to 120.

BATTERY BACKUP

Normally, any CNC machine is under power from the main power wall outlet, receiving continuous power supply. There are two special occasions, where the power supply will not be available:

- Power failure
- Machine power off

Having a constant and uninterrupted supply of power to the CNC machine guarantees that all internal settings are preserved. That includes all internally stored programs, offset settings, tool registrations, and system parameters. Several other settings are preserved as well. That bring the inevitable question - what happens if the normal power supply is interrupted for whatever reason?

The answer is simple - use *battery backup*.

This is one area of working with CNC machines that receives rather small attention. Yet, the consequences can be quite severe.

Every CNC machine contains a battery pack that supplies the necessary power, when the main external power supply is not available. This happens virtually every day, when the machine is turned off for the night or weekend. It also happens when the machine is moved from one location to another.

As long as the battery backup is working, there is nothing to worry about. However, batteries do not last forever, and once in a while have to be replaced. Once a year replacement is quite common.

The main point of this section is to emphasize the importance of changing the batteries while the CNC machine is still under external power. Do not wait until the batteries die, that is too late.

This basic rule should always be remembered:

> Always check or change backup batteries while the CNC machine receives external power

PARAMETERS AND G10

A brief mention of setting parameters through the program is listed on *page 232*. This section will go into a few more details. The command **G10** is often used to input various offsets for repetitive jobs and setups. What is less known about this command is that it can also be used to set various parameters of the control system.

As the command **G10** is also used to input offsets through the program, a special label is used along with the **G10** command to identify the entry as parameters, not offsets. The label is *L50* and is used with the **G10**:

```
G10 L50
```

To cancel parameter input, the **G11** command has to be programmed. The format of **G10** program data can be illustrated in this series of blocks:

```
G10 L50           (PARAMETER SETTING MODE ON)
  N.. R..         (FOR NON-AXIS TYPE SETTINGS)
  N.. P.. R..     (FOR AXIS TYPE SETTING ONLY)
G11               (PARAMETER SETTING MODE OFF)
```

➪ where ...

N = *Parameter number*
R = *Parameter setting value*
P = *Axis number*

For example, cutter radius offset setting (M-series) has to be temporarily changed from standard offset radius setting to offset diameter. Parameter **5004** - bit#2 controls the setting, and this is the required data entry:

```
G10 L50              (PARAMETER SETTING MODE ON)
  N5004 R00000100    (BIT 2 OF 5004 SET TO 1)
G11                  (PARAMETER SETTING MODE OFF)
```

Typically, this entry would be followed by the input of various offsets, specified as diameters. Before the **G10** can return to its more common purpose of setting offsets, the **G11** must be programmed to cancel parameter entry mode. Also, do not forget to set the parameter back when the offsets are loaded.

NO DECIMAL POINT ENTRY

As you may have noticed from the numerous examples, parameter data entry generally does *not* use decimal points. Many entries that normally do use a decimal point have to be converted into input values without a decimal point. Of course, the actual value has to remain the same. This section provides some guidelines.

> Explanation presented here can also be used for regular data entry, including offsets and program data

Historical Background

Numbers with decimal points are called *real numbers*. When numerical control technology was originally developed, the technology of the day did not support numbers with decimal points (real numbers). The initial input designation was to fill the field of eight places with the minimum and the maximum of the possible axis travel. This decision has taken into consideration the size and precision of the NC machines available then with a view to the future The decision was to design a control system that could accept minimum and maximum travel within the following ranges - for the imperial system:

IMPERIAL SYSTEM	
Minimum	0000.0001
Maximum	9999.9999

... and also for the metric system:

METRIC SYSTEM	
Minimum	00000.001
Maximum	99999.999

Note that there is no conversion between the two unit systems, only the decimal point has been shifted. Practically, it means that the maximum comparable travel for any machine has been determined as:

➪ *In METRIC system:*

99999.999 mm   is equivalent to   3937.007835 inches

in feet, the number represents 328. 0839862 feet
in meters, the number represents 99. 99999 meters

➪ *In IMPERIAL system:*

9999.9999 inches   is equivalent to   253999.9975 mm

in feet, the number represents 833.3333325 feet
in meters, the number represents 253.9999975 meters

The actual conversion may be interesting as such, but is irrelevant here, as both units provide a maximum travel higher than the largest machine travel available.

Equivalent Numbers

All CNC machines today are set to an old system that is called *leading zero suppression*. There also used be the opposite system - *trailing zero suppression* - that is best left alone and forgotten. Consider this number located in all eight places provided:

00003700

There are four zeros before the first significant number (3) and two zeros after the last significant number (7). Leading zero suppression is a concept that allows the user to ignore all zeros before the first significant digit, and the control will interpret it correctly. In the example,

00003700 = 3700

What does the number represent? The answer depends on which units are in effect. Millimeters or inches? As no decimal point is available - **and** - leading zero suppression is in effect, it means the *least motion increment* will determine the result.

Least Increment

The least increment is the smallest amount of motion the CNC machine is capable of. Again, the current units make a difference:

- *In METRIC system:*

0.001 mm *is the least increment* (00000.001 mm)

- *In IMPERIAL system:*

0.0001 inch *is the least increment* (0000.0001 inch)

Note the numeric equivalent in parentheses, showing all eight digits.

Input Without a Decimal Point

Without decimal point and leading zero suppression and eight place format, these are common equivalents:

- *In METRIC system:*

1.0 = 1.000 = 00001.000 = 00001000 = 1000
0.1 = 0.100 = 00000.100 = 00000100 = 100
0.01 = 0.010 = 00000.010 = 00000010 = 10
0.001 = 00000.001 = 00000001 = 1

- *In IMPERIAL system:*

1.0 = 1.0000 = 0001.0000 = 00010000 = 10000
0.1 = 0.1000 = 0000.1000 = 00001000 = 1000
0.01 = 0.0100 = 0000.0100 = 00000100 = 100
0.0001 = 0000.0001 = 00000001 = 1

From the examples it is easy to see how an input error can happen. This is equally important for entering parameters and for input of offsets or program data.

SUMMARY

It is a difficult task to select and explain parameters for a modern CNC system. As a matter of curiosity, the *Parameter Manual* alone for the *16i/18i* control system has around 800 pages. The objective of this chapter was to make CNC programmers, CNC operators and machine shop supervisors aware of the possibilities parameters, their formats and settings.

In the summary, a few items to remember:

1 View parameters as you wish
   - viewing only does not affect their settings
2 Make a backup of all parameters - the original ones and those changed
3 Thinks twice before a parameter change
4 Ask for permission to change parameters, even if you are qualified to do so
5 If possible, change one parameter at a time, and check effect of the change
6 If in doubt, ask for guidance or let a qualified technician do the work
7 If in doubt, do not make any changes *(see Item 6)*
8 Never change machine related data
9 Many parameters are control based

There are many more parameters available and some that have not been included may also be beneficial to the CNC programmer or CNC operator. Equal consideration applies to those parameters that were included here - some may not be important enough to list. Hopefully, this chapter gives you at least a glimpse into the world of parameters and their effect on everyday machining.

24 PROGRAM OPTIMIZATION

While the ultimate objective of program development is to make the 'perfect program' from the beginning, there are many reasons why the part program sent to the CNC machine will not meet this objective. The first reason that comes to mind are errors caused by poor programming - programmer's errors. While programmer's errors are not uncommon (even when the program is generated by a CAM software), their correction cannot be considered as program optimization, for one good reason. Programming errors prevent machining, therefore they prevent any attempt at making the program work better - and more efficiently.

> Optimizing a part program is to improve its efficiency; in effect, its main purpose is to improve cycle time

For programs that are correct when they reach the machine tool - and that is hopefully a great majority - optimization can take several forms.

PROGRAM REVIEW

The first step for any program optimization is to find how the current program performs. That is usually discovered during part setup, particularly when proving the program itself and the first part run. If any changes are required, they will fall into two categories:

- Simple changes that can be done at the CNC machine
- More complex changes that require re-programming

As part of any program review for the purposes of optimization, both the part program and the part setup will have to be scrutinized. This evaluation also includes any special or related activities. An experienced CNC operator will always take a broad look at many aspects of the program, particularly:

- Cycle time
- Spindle speeds
- Cutting feedrates
- Part zero selection
- Program structure
- Special comments

- Program as it relates to the drawing
- Tool selection
- Work holding method
- Order of tools
- Depth of cut
- Width of cut
- ... and others

These items are not in any particular order. For better organization of the optimization process, they should be separated into more manageable groups - those that will affect the program and those that will affect the setup.

> Program optimization should include both program related changes and setup related changes

PROGRAM RELATED CHANGES

On a daily basis, CNC operators make changes to the part program all the time. Change of spindle speed or cutting feedrate is 'normal' and hardly ever considered real optimization. Even these relatively small and, to some degree, expected changes amount to the beginning of program optimization. Continuing the process should always be a cooperative effort - programmers, operators, supervisors, engineers, tooling experts, and other professionals, they all should be involved, particularly in complex situations.

For large size batches (hundreds, thousands, and more parts in a single batch), the optimization process is even more important. Consider the table below for savings.

As can be expected, the optimization for the large number of parts is especially worth the time.

Number of parts	Time saved per part			
	0.5 sec	1 sec	10 sec	1 min
1 000	8.3 min	16.7 min	2.7 hrs	16.7 hrs
10 000	1.4 hrs	2.8 hrs	27.8 hrs	167 hrs
100 000	13.9 hrs	27.8 hrs	278 hrs	1667 hrs

CHANGES AT THE CONTROL

In addition to the aforementioned changes to speeds and feeds, a qualified CNC operator often does more to improve the cycle time. Before looking at such possibilities in more detail, the issue of speeds and feedrates should not be dismissed too quickly.

Spindle Speeds and Feedrates

It is a well known fact that most programs under utilize the capabilities of modern CNC machines and cutting tools. Even when you look at the recommended cutting speed for certain material and tool, the range can be quite large. For example, it is easy to work with a suggested cutting speed of 200 m/min or even a range of 160-180 m/min, but how do you work with a range that is listed as 100-400 m/min (330-1300 ft/min)?

Most programmers take rather conservative approach to speeds and feeds and leave it to the operator to find the optimum cutting values.

Clearances

Excessive clearances are one of the more common program entries to optimize. In most cases, they are also easy to optimize at the control, by making a few changes to the program. Make sure the changes are constructive.

Clearance is a space allocated for the tool to start cutting. For drilling and related operations, the space will be above the part and below the part in case of through holes. Normal clearances above the part are 2 to 3 mm, or 0.1 inches. Typical metric clearance is often 2.5 mm, which is the practical equivalent of 0.1 inch clearance.

While these amounts are very common, ask yourself whether they are really necessary. Keep in mind that any clearance of this type will be a *cutting* motion - the tool will move through the empty space at a programmed cutting feedrate. For large volumes, the 2.5 mm (0.1") may be a waste of time (see the table at the beginning of this chapter). The same applies for through holes.

Normally, there is a need for the drill (for example) to break through the part with its full diameter. In order to assure that it happens, a certain breakthrough amount is included in the drilling depth. This is another clearance that may be scrutinized and changed if it is unreasonable. 0.5 mm or 0.02" should be enough in most cases.

Another clearance that is often unnecessarily large is the start (and end tool) tool position for contouring. Look at the drawing and the example that follows:

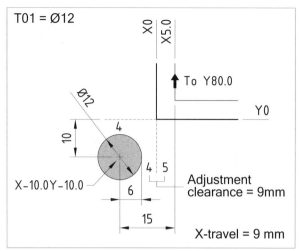

```
N1 G21
N2 G17 G40 G80 T01              (12 MM END MILL)
N3 M06
N4 G90 G54 G00 X-10.0 Y-10.0 S1100 M03 T02
N5 G43 Z-4.5 H01 M08
N6 G41 G01 X5.0 D51 F200.0      (D51 = 6.000)
N7 Y80.0 F150.0
N8 ...
```

The program reflects the illustration above. The three underlined entries are also the critical entries. Together, they define the actual distance-to-go for the X-axis in block N6, from the start point to the edge of the contour:

- **Start point**   = X-10.0
- **Target point**   = X5.0
- **Offset amount**   = 6.000
- **Total distance - no offset**   = 15 mm
- **Total distance - with offset** = 9 mm

In the block N6, the motion is at a cutting feedrate while no cutting takes place - this is not necessary and should be changed from **G01** to **G00**.

Even when rapid motion is programmed, it is really necessary to move 9 mm just to approach the first contour of the part? What should be the optimum distance?

Why not Zero Clearance?

Ideally (that means *mathematically*), the tool cutting edge could be located to *touch* the first contour. With cutter radius offset in effect, its position would be X-1.0 (X5.0 as the target X-location *minus* the cutter radius of 6.000 mm = X-1.0). As tempting as this solution can be, don't try it. The space is very necessary for any radius offset adjustment. Whether it is 1 mm, 2 mm, 3mm makes not much of a difference.

Optimized Program

The two significant changes in the optimized program will be the **G01** to **G00** change as well as the actual start/end XY tool position. First, the updated illustration:

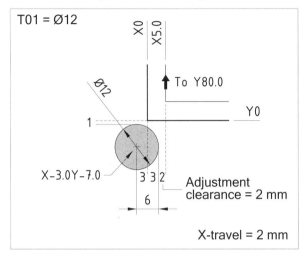

The following program reflects the changes:

```
N1 G21
N2 G17 G40 G80 T01           (12 MM END MILL)
N3 M06
N4 G90 G54 G00 X-3.0 Y-7.0 S1100 M03 T02
N5 G43 Z-4.5 H01 M08
N6 G41 X5.0 D51              (D51 = 6.000)
N7 G01 Y80.0 F150.0
N8 ...
```

The **G00** command and the cutting feedrate in block N6 has been removed, which required the inclusion of **G01** in block N7.

Does that mean the original program was incorrect? In most cases it does not. Keep in mind that optimization of the program itself is a time consuming task. Also keep in mind that most part programs *can be* optimized but they are *not*. Program optimization of any kind must always be justified to avoid any counter effects.

Dwells

Any programmed pause or delay in the part program is called a *dwell*. Dwell is programmed in seconds or milliseconds and its purpose is to delay certain activity. In drilling operations (spot drill, counter bore, countersink, spot face), the purpose of the dwell is to provide a clean surface. Other uses of dwell apply to situations where certain time is required to complete a hardware function, for example, move a tailstock.

Any dwell related to completion of a hardware function should be scrutinized very carefully before any possible changes. These dwells may appear to be too long, but as long as they are justified, they should not be tampered with.

The category of dwell used during machining is different. Consider the following spot drilling cycle entry:

```
...
N4 G90 G54 G00 X134.0 Y88.75 S1300 M03 T02
N5 G43 Z25.0 M08
N6 G99 G82 R2.5 Z-4.3 P1000 F150.0
...
```

The one second dwell (1000 milliseconds) sounds reasonable, unless you have seen how long it actually takes during machining. In many programs, the dwell times are too high, and not necessary.

The purpose of a dwell in the hole operations is to let the tool 'sit' at the hole depth for at least one revolution to clean the bottom. That should be (at least theoretically) enough. Even some sort of rounding the dwell time will be much smaller than many programmed values. To calculate minimum dwell, you have to know the spindle speed (rpm):

Minimum dwell (seconds) = 60 / rpm

For the program example above, the minimum dwell will be 60 / 1300 = 0.046 of a second, or 46 milliseconds. Even rounded to 100 milliseconds, programmed as *P100*, there is quite a significant saving of 0.9 seconds on each hole (!). Multiply the saving by the number of holes, then multiply that by the number of parts and the time savings from the dwell alone can be staggering.

Combined Tool Motions

It does not take any complex calculations to find out that the three motions of X, Y, and Z axes programmed individually will increase the cycle time when compared with a motion of all three axes simultaneously. Consider the last modified program example:

```
...
N4 G90 G54 G00 X134.0 Y88.75 S1300 M03 T02
N5 G43 Z25.0 M08
N6 G99 G82 R2.5 Z-4.3 P100 F150.0
...
```

Now, consider the optimized program version:

```
...
N4 G90 G54 G43 G00 X134.0 Y88.75 Z25.0 H01
   S1300 M03 T02
N5 M08
N6 G99 G82 R2.5 Z-4.3 P100 F150.0
...
```

While the number of blocks have not changed, the N4 block now includes simultaneous motion of all three axes. For a large number of parts, this could be a significant cycle time reduction, providing that the actual setup is clear of obstacles and the three-axis motion can be performed safely.

Also note that block N5 includes only the coolant ON function **M08**. This is a precaution - even necessity - for controls that do not accept more than one M-function in a block. Modern controls accept up to *three* M-functions in the same block, providing they do not clash with each other. The same program segment can be even shorter for the most modern controls:

```
...
N4 G90 G54 G43 G00 X134.0 Y88.75 Z25.0 H01
   S1300 M03 M08 T02
N5 G99 G82 R2.5 Z-4.3 P100 F150.0
...
```

Is this the better way of programming? For general CNC work, most likely not. As an optimized program for a large volume of parts, definitely yes.

SETUP RELATED CHANGES

All these previously described small changes can be done at the control and more often than not produce significant cycle time improvement. In many cases, even the most effective program related changes will not be enough, unless the actual part setup and/or machine setup are also included. Some items that belong to this category are:

- **Change of part design**
- **Change of fixture (work holding)**
- **Changing the number of tools**
- **Change of machine and/or control system**
- *... and other similar changes*

Initial planning of the part and machine setup should always be very thorough. Setup is always categorized as an overhead operation - necessary but unproductive. It is in this particular area that many improvements can be achieved.

Only a limited number of quick setup changes is available at the machine. That is the reason why setup methods, including all tool and work holding devices, should be selected with utmost care.

SUMMARY

Program optimization should always be subject to be evaluated very carefully. For some jobs, it is actually more efficient to use a program that could potentially benefit from optimization than trying to change the program by dedicating too much time to do it.

Optimizing a part program never means degradation of the setup, program or part quality. It never means compromising the safety of machining operations. Optimization is a positive step into improvements at many levels of CNC operations.

The few ideas presented in this chapter are no more than basic suggestions. Whether you are changing program or setup or both for the purpose of optimizing cycle time, keep in mind that this is a huge field that offers virtually unlimited possibilities. Follow the trend in the development of tools, standard fixtures, look for new inventions, and see how they fit your current conditions.

Successful CNC work (programming *and* machining) is not just the quality of the part, but quality of the part combined with efficient productivity.

25 REFERENCES

G-CODES AND M-FUNCTIONS

Milling - G-codes

G00	Rapid positioning
G01	Linear interpolation
G02	Circular interpolation - clockwise
G03	Circular interpolation - counterclockwise
G04	Dwell function (as a separate block)
G07	Hypothetical axis interpolation
G09	Exact stop check - one block only
G10	Programmable data input (Data setting)
G11	Data setting mode cancel
G15	Polar coordinate command cancel
G16	Polar coordinate command
G17	XY plane designation
G18	ZX plane designation
G19	YZ plane designation
G20	English units of input
G21	Metric units of input
G22	Stored stroke check - on
G23	Stored stroke check - off
G25	Spindle fluctuation detection - on
G26	Spindle fluctuation detection - off
G27	Machine zero position check
G28	Machine zero return (reference point 1)
G29	Return from machine zero
G30	Machine zero return (reference point 2)
G31	Skip function
G33	Threading function
G37	Tool length automatic measurement
G40	Cutter radius compensation - cancel
G41	Cutter radius compensation - left
G42	Cutter radius compensation - right
G43	Tool length offset - positive
G44	Tool length offset - negative
G45	Position compensation - single increase
G46	Position compensation - single decrease
G47	Position compensation - double increase
G48	Position compensation - double decrease
G49	Tool length offset cancel
G50	Scaling function cancel
G51	Scaling function
G52	Local coordinate offset
G53	Machine coordinate system
G54	Work coordinate offset 1
G54.1	Additional work coordinate offset (with P)
G55	Work coordinate offset 2
G56	Work coordinate offset 3
G57	Work coordinate offset 4
G58	Work coordinate offset 5
G59	Work coordinate offset 6
G60	Single direction positioning
G61	Exact stop mode
G62	Automatic corner override mode
G63	Tapping mode
G64	Cutting mode
G65	Custom macro call
G66	Modal custom macro call
G67	Modal custom macro call cancel
G68	Coordinate system rotation - active
G69	Coordinate system rotation - cancel
G73	High speed peck drilling cycle
G74	Left hand tapping cycle
G76	Fine (precision) boring cycle
G80	Fixed cycle cancel
G81	Drilling cycle
G82	Drilling cycle with dwell (Spot drilling cycle)
G83	Peck drilling cycle
G84	Right hand tapping cycle
G84.2	Rigid tap cycle - right hand
G84.3	Rigid tap cycle - left hand

G85	Boring cycle
G86	Boring cycle
G87	Back boring cycle
G88	Boring cycle
G89	Boring cycle
G90	Absolute input of dimensional values
G91	Incremental input of dimensional values
G92	Tool position register
G93	Inverse time feed rate
G94	Feed per minute (in/min or mm/min)
G95	Feed per revolution (in/rev or mm/rev)
G98	Return to the initial level in a fixed cycle
G99	Return to the R-level in a fixed cycle

Milling - M-functions

Check your manual for exact listing of M-functions.

M00	Compulsory program stop
M01	Optional program stop
M02	End of program (usually no reset/rewind)
M03	Spindle rotation normal – CW
M04	Spindle rotation reverse - CCW
M05	Spindle rotation stop
M06	Automatic tool change (ATC)
M07	Coolant mist - on
M08	Coolant pump motor - ON
M09	Coolant pump motor - OFF
M13	Spindle rotation normal (CW) + Coolant ON
M14	Spindle rotation reverse (CCW) + Coolant ON
M19	Programmable spindle orientation
M30	End of program (with reset and rewind)
M48	Feedrate override cancel - OFF (not active)
M49	Feedrate override cancel - ON (active)
M60	Automatic pallet change (APC)
M78	B-axis clamp
M79	B-axis unclamp
M87	X-axis mirror ON
M88	Y-axis mirror ON
M89	XY axes mirror cancel
M98	Subprogram call (with program number P)
M99	Subprogram end - Macro end

G-CODES AND M-FUNCTIONS

Turning - G-codes

All G-codes listed belong to *Group A*:

G00	Rapid positioning
G01	Linear interpolation
G02	Circular interpolation - clockwise
G03	Circular interpolation - counterclockwise
G04	Dwell function (as a separate block)
G07	Hypothetical axis interpolation
G09	Exact stop check - one block only
G10	Programmable data input (Data setting)
G11	Data setting mode cancel
G20	English units of input
G21	Metric units of input
G22	Stored stroke check - on
G23	Stored stroke check - off
G25	Spindle fluctuation detection - on
G26	Spindle fluctuation detection - off
G27	Machine zero position check
G28	Machine zero return (reference point 1)
G29	Return from machine zero
G30	Machine zero return (reference point 2)
G31	Skip function
G32	Threading function - constant lead
G34	Threading function - variable lead
G35	Circular threading CW
G36	Circular threading CCW
G40	Cutter radius compensation - cancel
G41	Cutter radius compensation - left
G42	Cutter radius compensation - right
G50	Maximum rpm preset/Tool position register
G52	Local coordinate offset
G53	Machine coordinate system
G54	Work coordinate offset 1
G55	Work coordinate offset 2
G56	Work coordinate offset 3
G57	Work coordinate offset 4

CNC Custom Setup for Milling and Turning
REFERENCES

G58	Work coordinate offset 5
G59	Work coordinate offset 6
G65	Custom macro call
G66	Modal custom macro call
G67	Modal custom macro call cancel
G68	Mirror image for double turrets
G69	Mirror image for double turrets cancel
G70	Profile finishing cycle
G71	Profile roughing cycle - Z-axis direction
G72	Profile roughing cycle - X-axis direction
G73	Pattern repetition cycle
G74	Drilling cycle (end face peck drilling cycle)
G75	Grooving cycle
G76	Multiple repetitive threading cycle
G90	Box type roughing cycle (straight/tapered)
G92	Box type threading cycle (straight/tapered)
G94	Box type face cutting cycle (straight/tapered)
G96	Constant cutting speed mode (surface speed)
G97	Constant spindle speed mode (r/min)
G98	Feed per minute (in/min or mm/min)
G99	Feed per revolution (in/rev or mm/rev)

Turning - M-functions

Check your manual for exact listing of M-functions.

M00	Compulsory program stop
M01	Optional program stop
M02	End of program (usually no reset and
M03	Spindle rotation normal – CW
M04	Spindle rotation reverse - CCW
M05	Spindle rotation stop
M06	Automatic tool change (ATC)
M07	Coolant mist - ON
M08	Coolant pump motor - ON
M09	Coolant pump motor - OFF
M10	Chuck open
M11	Chuck close
M12	Tailstock quill IN
M13	Tailstock quill OUT
M17	Turret indexing forward
M18	Turret indexing reverse
M19	Programmable spindle orientation
M21	Tailstock IN
M22	Tailstock OUT
M23	Thread gradual pull-out ON
M24	Thread gradual pull-out OFF
M30	End of program (with reset and rewind)
M41	Low gear selection
M42	Medium gear selection 1
M43	Medium gear selection 2
M44	High gear selection
M48	Feedrate override cancel - OFF (deactivated)
M49	Feedrate override cancel - ON (activated)
M98	Subprogram call (with program number P)
M99	Subprogram end - Macro end

DECIMAL EQUIVALENTS

Listing in the following table covers fractional, wire gauge (number), letter and metric (mm) values for given decimal equivalents in inches.

Decimal inch	Fraction	Number or Letter	Metric (mm)
0.0059		97	0.15
0.0063		96	0.16
0.0067		95	0.17
0.0071		94	0.18
0.0075		93	0.19
0.0079		92	0.20
0.0083		91	0.21
0.0087		90	0.22
0.0091		89	0.23
0.0095		88	0.24
0.0100		87	0.25
0.0105		86	
0.0110		85	0.28
0.0115		84	
0.0118			0.30
0.0120		83	
0.0125		82	
0.0126			0.32
0.0130		81	
0.0135		80	
0.0138			0.35
0.0145		79	

Decimal inch	Fraction	Number or Letter	Metric (mm)	Decimal inch	Fraction	Number or Letter	Metric (mm)
0.0150			0.38	0.0531			1.35
0.0156	1/64			0.0550		54	
0.0157			0.40	0.0551			1.40
0.0160		78		0.0571			1.45
0.0177			0.45	0.0591			1.50
0.0180		77		0.0595		53	
0.0197			0.50	0.0610			1.55
0.0200		76		0.0625	1/16		
0.0210		75		0.0630			1.60
0.0217			0.55	0.0635		52	
0.0225		74		0.0650			1.65
0.0236			0.60	0.0669			1.70
0.0240		73		0.0670		51	
0.0250		72		0.0689			1.75
0.0256			0.65	0.0700		50	
0.0260		71		0.0709			1.80
0.0276			0.70	0.0728			1.85
0.0280		70		0.0730		49	
0.0292		69		0.0748			1.90
0.0295			0.75	0.0760		48	
0.0310		68		0.0768			1.95
0.0313	1/32			0.0781	5/64		
0.0315			0.80	0.0785		47	
0.0320		67		0.0787			2.00
0.0330		66		0.0807			2.05
0.0335			0.85	0.0810		46	
0.0350		65		0.0820		45	
0.0354			0.90	0.0827			2.10
0.0360		64		0.0846			2.15
0.0370		63		0.0860		44	
0.0374			0.95	0.0866			2.20
0.0380		62		0.0886			2.25
0.0390		61		0.0890		43	
0.0394			1.00	0.0906			2.30
0.0400		60		0.0925			2.35
0.0410		59		0.0935		42	
0.0413			1.05	0.0938	3/32		
0.0420		58		0.0945			2.40
0.0430		57		0.0960		41	
0.0433			1.10	0.0965			2.45
0.0453			1.15	0.0980		40	
0.0465		56		0.0984			2.50
0.0469	3/64			0.0995		39	
0.0472			1.20	0.1015		38	
0.0492			1.25	0.1024			2.60
0.0512			1.30	0.1040		37	
0.0520		55		0.1063			2.70

CNC Custom Setup for Milling and Turning
REFERENCES

Decimal inch	Fraction	Number or Letter	Metric (mm)
0.1065		36	
0.1083			2.75
0.1094	7/64		
0.1100		35	
0.1102			2.80
0.1110		34	
0.1130		33	
0.1142			2.90
0.1160		32	
0.1181			3.00
0.1200		31	
0.1220			3.10
0.1250	1/8		
0.1260			3.20
0.1280			3.25
0.1285		30	
0.1299			3.30
0.1339			3.40
0.1360		29	
0.1378			3.50
0.1405		28	
0.1406	9/64		
0.1417			3.60
0.1440		27	
0.1457			3.70
0.1470		26	
0.1476			3.75
0.1495		25	
0.1496			3.80
0.1520		24	
0.1535			3.90
0.1540		23	
0.1562	5/32		
0.1570		22	
0.1575			4.00
0.1590		21	
0.1610		20	
0.1614			4.10
0.1654			4.20
0.1660		19	
0.1673			4.25
0.1693			4.30
0.1695		18	
0.1719	11/64		
0.1730		17	
0.1732			4.40
0.1770		16	

Decimal inch	Fraction	Number or Letter	Metric (mm)
0.1772			4.50
0.1800		15	
0.1811			4.60
0.1820		14	
0.1850		13	4.70
0.1870			4.75
0.1875	3/16		
0.1890		12	4.80
0.1910		11	
0.1929			4.90
0.1935		10	
0.1960		9	
0.1969			5.00
0.1990		8	
0.2008			5.10
0.2010		7	
0.2031	13/64		
0.2040		6	
0.2047			5.20
0.2055		5	
0.2067			5.25
0.2087			5.30
0.2090		4	
0.2126			5.40
0.2130		3	
0.2165			5.50
0.2188	7/32		
0.2205			5.60
0.2210		2	
0.2244			5.70
0.2264			5.75
0.2280		1	
0.2283			5.80
0.2323			5.90
0.2340		A	
0.2344	15/64		
0.2362			6.00
0.2380		B	
0.2402			6.10
0.2420		C	
0.2441			6.20
0.2460		D	
0.2461			6.25
0.2480			6.30
0.2500	1/4	E	
0.2520			6.40
0.2559			6.50

Decimal inch	Fraction	Number or Letter	Metric (mm)	Decimal inch	Fraction	Number or Letter	Metric (mm)
0.2570		F		0.3480		S	
0.2598			6.60	0.3504			8.90
0.2610		G		0.3543			9.00
0.2638			6.70	0.3580		T	
0.2556	17/64			0.3583			9.10
0.2657			6.75	0.3594	23/64		
0.2660		H		0.3622			9.20
0.2677			6.80	0.3642			9.25
0.2717			6.90	0.3661			9.30
0.2720		I		0.3680		U	
0.2756			7.00	0.3701			9.40
0.2770		J		0.3740			9.50
0.2795			7.10	0.3750	3/8		
0.2810		K		0.3770		V	
0.2812	9/32			0.3780			9.60
0.2835			7.20	0.3819			9.70
0.2854			7.25	0.3839			9.75
0.2874			7.30	0.3858			9.80
0.2900		L		0.3860		W	
0.2913			7.40	0.3898			9.90
0.2950		M		0.3906	25/64		
0.2953			7.50	0.3937			10.00
0.2969	19/64			0.3970		X	
0.2992			7.60	0.4040		Y	
0.3020		N		0.4062	13/32		
0.3031			7.70	0.4130		Z	
0.3051			7.75	0.4134			10.50
0.3071			7.80	0.4219	27/64		
0.3110			7.90	0.4331			11.00
0.3125	5/16			0.4375	7/16		
0.3150			8.00	0.4528			11.50
0.3160		O		0.4531	29/64		
0.3189			8.10	0.4688	15/32		
0.3228			8.20	0.4724			12.00
0.3230		P		0.4844	31/64		
0.3248			8.25	0.4921			12.50
0.3268			8.30	0.5000	1/2		12.70
0.3281	21/64			0.5118			13.00
0.3307			8.40	0.5156	33/64		
0.3320		Q		0.5312	17/32		
0.3346			8.50	0.5315			13.50
0.3386			8.60	0.5469	35/64		
0.3390		R		0.5512			14.00
0.3425			8.70	0.5625	9/16		
0.3438	11/32			0.5709			14.50
0.3445			8.75	0.5781	37/64		
0.3465			8.80	0.5906			15.00

PARAMETERS CLASSIFICATIONS

The following list is for Fanuc series 16i, 18i, 160i, 180i, 160is, 180is - all Models B. Other control models will have a very similar listing. Except as noted, the listing is related to all machine types these models support. Typical Fanuc series designations are:

- M-series   ... Milling
- T-series   ... Turning
- L-series   ... Laser
- P-series   ... Puncher

Wire EDM machines usually use controls that are basically designed for milling and adapted for EDM.

Keep in mind that the purpose of the following list is to give you an idea not only about how parameters are classified, but also of the various groups covered.

As you see, the list is quite long:

1. Parameters of Setting
2. Parameters of Reader/Puncher Interface
   Or Remote Buffer
   - Parameters Common to all Channels
   - Parameters of Channel 1 (I/O CHANNEL=0)
   - Parameters of Channel 1 (I/O CHANNEL=1)
   - Parameters of Channel 2 (I/O CHANNEL=2)
   - Parameters of Channel 3 (I/O CHANNEL=3)
3. Parameters of Dnc1/Dnc2 Interface
4. Parameters of M-Net Interface
5. Parameters of Remote Diagnosis
6. Parameters of Dnc1 Interface #2
7. Parameters of Memory Card Interface
8. Parameters of Factolink
9. Parameters of Data Server
10. Parameters of Ethernet
11. Parameters of Power Mate CNC Manager
12. Parameters of Axis Control/Increment System
13. Parameters of Coordinates
14. Parameters of Stored Stroke Check
15. Parameters of Chuck And Tailstock Barrier (T Series)
16. Parameters of Feedrate
17. Parameters of Acceleration/Deceleration Control
18. Parameters of Servo (1 of 2)
19. Parameters of Di/Do
20. Parameters of Display And Edit (1 of 2)
21. Parameters of Programs
22. Parameters of Pitch Error Compensation
23. Parameters of Spindle Control
24. Parameters of Tool Compensation

Decimal inch	Fraction	Number or Letter	Metric (mm)
0.5938	19/32		
0.6094	39/64		
0.6102			15.50
0.6250	5/8		
0.6299			16.00
0.6406	41/64		
0.6496			16.50
0.6562	21/32		
0.6693			17.00
0.6719	43/64		
0.6875	11/16		
0.6890			17.50
0.7031	45/64		
0.7087			18.00
0.7188	23/32		
0.7283			18.50
0.7344	47/64		
0.7480			19.00
0.7500	3/4		
0.7656	49/64		
0.7677			19.50
0.7812	25/32		
0.7874			20.00
0.7969	51/64		
0.8071			20.50
0.8125	13/16		
0.8268			21.00
0.8281	53/64		
0.8438	27/32		
0.8465			21.50
0.8594	55/64		
0.8661			22.00
0.8750	7/8		
0.8858			22.50
0.8906	57/64		
0.9055			23.00
0.9062	29/32		
0.9219	59/64		
0.9252			23.50
0.9375	15/16		
0.9449			24.00
0.9531	61/64		
0.9646			24.50
0.9688	31/32		
0.9843			25.00
0.9844	63/64		
1.0000	1		25.40

25. Parameters of Wheel Wear Compensation
26. Parameters of Canned Cycles
    - Parameters of Canned Cycle for Drilling
    - Parameters of Threading Cycle
    - Parameters of Multiple Repetitive Canned Cycle
    - Parameters of Small-hole Peck Drilling Cycle
27. Parameters of Rigid Tapping
28. Parameters of Scaling And Coordinate System Rotation
29. Parameters of Single Direction Positioning
30. Parameters of Polar Coordinate Interpolation
31. Parameters of Normal Direction Control
32. Parameters of Index Table Indexing
33. Parameters of Involute Interpolation
34. Parameters of Exponential Interpolation
35. Parameters of Flexible Synchronous Control
36. Parameters of Straightness Compensation (1 of 2)
37. Parameters of Inclination Compensation
38. Parameters of Custom Macros
39. Parameters of One Touch Macro
40. Parameters of Pattern Data Input
41. Parameters of Positioning By Optimal Acceleration
42. Parameters of Skip Function
43. Parameters of Automatic Tool Offset (T Series) And Automatic Tool Length Measurement (M Series)
44. Parameters of External Data Input
45. Parameters of Fine Torque Sensing
46. Parameters of Manual Handle Retrace
47. Parameters of Graphic Display
    - Parameters of Graphic Display / Dynamic Graphic Display
    - Parameters of Graphic Color
48. Parameters of Run Hour And Parts Count Display
49. Parameters of Tool Life Management
50. Parameters of Position Switch Functions
51. Parameters of Manual Operation And Automatic Operation
52. Parameters of Manual Handle Feed, Manual Handle Interruption And Tool Direction Handle Feed
53. Parameters of Manual Linear/Circular Function
54. Parameters of Reference Position Setting With Mechanical Stopper
55. Parameters of Software Operator's Panel
56. Parameters of Program Restart
57. Parameters of High-Speed Machining (High-Speed Cycle Machining / High-Speed Remote Buffer)
58. Parameters of Rotary Table Dynamic Fixture Offset
59. Parameters of Polygon Turning
60. Parameters of External Pulse Input
61. Parameters of Hobbing Machine And Simple Electric Gear Box (EGB)
62. Parameters of Axis Control By PMC
63. Parameters of Two-Path Control
64. Parameters of Interference Check Between Two Tool Posts (Two-Path) (For Two-Path Control)
65. Parameters of Synchronous/Composite Control And Superimposed Control
66. Parameters of Angular Axis Control
67. Parameters of B-Axis Control
68. Parameters of Simple Synchronous Control
69. Parameters of Sequence Number Comparison And Stop
70. Parameters of Chopping
71. Parameters of High-Speed And High-Precision Contour Control By RISC (M Series)
    - Parameters of Acceleration/Deceleration before Interpolation
    - Parameters of Automatic Speed Control
72. Parameters of High-Speed Position Switch (1 of 2)
73. Other Parameters
74. Parameters of Trouble Diagnosis
75. Parameters of Maintenance
76. Parameters of Embedded Macro
77. Parameters of High-Speed Position Switch (2 of 2)
78. Parameters of Superimposed Command Function In Binary Operation
79. Parameters of Servo Speed Check
80. Parameters of Manual Handle Functions
81. Parameters of Manual Handle For 5-Axis Machining
82. Parameters of Manual Handle Feed
83. Parameters of Multi-Path Control
84. Parameters of Acceleration Control
85. Parameters of External Deceleration Positions Expansion
86. Parameters of Operation History
87. Parameters of Display And Edit (2 of 2)
88. Parameters of Tool Management Functions
89. Parameters of Straightness Compensation (2 of 2)
90. Parameters of Interpolation Type Straightness Compensation
91. Parameters of Machining Condition Selecting Screen
92. Parameters of Dual Check Safety
93. Parameters of Servo (2 of 2)
94. Parameters of Servo Guide Mate
95. Parameters of Interference Check For Rotary Area
96. Parameters of Slide Axis Control For Link-Type Presses
97. ...

This list may be longer or shorter, depending on what features the particular control system offers.

COMMON ABBREVIATIONS

Regardless of any specific industry, most technical fields use various short forms, abbreviations and acronyms. CNC technology is no different - it has its share of short forms as well. Some of them are related strictly to machining, others to CNC work - in most cases, they overlap. The following list is a reference that identifies only some of the most common abbreviations you will find in CNC machine shop environment, including some common abbreviations found in engineering drawings. Listing of the common abbreviations is in alphabetical order and by no means comprehensive:

APC	Automatic Pallet Changer
ASCII	American Standard Code for Information Interchange - electronic standard
ATC	Automatic Tool Changer
CAD	Computer Aided Design - design part of CAD/CAM
CAD/CAM	Computer Aided Design - Computer Aided Manufacturing - a concept of using computers from design to manufacturing
CAM	Computer Aided Manufacturing - also referred to CNC programming software
CAPP	Computer Aided Process Planning - a computer software used for scheduling various shop activities based on previous history, availability of material and machining capacity, and requirements of the customer
CBORE	Counterbore - also C'BORE - used in drawings
CHFR	Chamfer
CIM	Computer Integrated Manufacturing - all elements of manufacturing are controlled by a host computer
CMM	Coordinate Measuring Machines - used to inspect part features using electronic probes - involves a stand-alone equipment
CNC	Computer Numerical Control
CRT	Cathode Ray Tube - an industry term for computer screen (as opposed to LCD)
CS	Cutting speed - also known as **Surface speed**
CSS	Constant Surface Speed - same as **Cutting Speed**
CSINK	Countersink - also C'SINK - used in drawings
DIA	Diameter - used in drawings - also as a symbol ⌀
DNC	Direct Numerical Control or **Distributed Numerical Control** - method of using an external computer to send continuous blocks of data to the CNC machine
DP	Depth - used in drawings, particularly for holes
EDM	Electrical Discharge Machine - known as Wire EDM
EIA	Electronic Industry Association - electronic standard
ft/min	Feet per minute - cutting speed - also FPM or SFM or SFPM
FMS	Flexible Manufacturing System
GD&T	Geometric Dimensioning and Tolerancing - also (incorrectly) GDT - method of identifying relationship between two or more features in a drawing
HBM	Horizontal Boring Mill - includes a fourth axis, typically the B-axis as well as W-axis parallel to Z-axis
HMC	Horizontal Machining Center - includes a fourth axis, typically the B-axis
HOME	Informal description of machine reference point - officially known as **Machine Zero**
IPM	Inches per minute - feedrate designation - also **ipm** or **in/min**

IPR	Inches per revolution - feedrate designation - also **ipr** or **in/rev** - mainly used on CNC lathes		PLC	Programmable Logic Control (Process Logic Control) - electronic signals used to control a process to begin a new action
IPT	Inches per tooth - material load also referred to as 'chip load' or **inches per flute (ipf)**		R	Radius - also RAD - used in drawings
ISO	International Standards Organization - an international body that sets various standards		RPM	Revolutions per minute - Spindle Speed as compared to Cutting Speed (CS) - also written as **r/min**
JIT	Just In Time - inventory reduction system based on the principle manufacturing and shipping only when required by a customer		RS	Recommended Standard - for example, RS-232
			SC or S/C	Over sharp corners - drawing designation of measurement
LCD	Liquid Crystal Display - display screen that replaced CRT type		SF or S/F	Spot face - used in drawings as a feature identification
LEAD	Amount of thread motion along an axis, in 360° turn		SFM	Surface Feet Per Minute - Cutting Speed (CS) designation in Imperial units- same as CS or CSS - also SFPM
MDI	Manual Data Input - control feature allowing input of program data, generally for a temporary use			
m/min	Meters per minute - cutting speed - also **MPM** or **mpm**		SI	Système international d'unités - International System of Units - a French language abbreviation identifying modernized metric based standards
mm/min	Millimeters per minute - feedrate designation			
mm/rev	Millimeters per revolution - feedrate mainly used on CNC lathes		SPC	Statistical Process Control - a system using random sampling "to ensure a process is within its operating limits allowing for corrections to be made before scrapping part and eliminat- ing 100% inspection"
MPG	Manual Pulse Generator - Fanuc name for the setup 'handle'			
NC	Numerical Control - means numerical control technology in general, including CNC - or - it may refer to old pre-CNC hard wired control systems			
			TIR	Total Indicator Reading
			TPF	Taper per foot - used in drawings as a feature identification
ORIGIN	Setting equivalent to X0 Y0 Z0 - used for Machine Zero (Home) and/or Part Zero (Program Zero)		TPI	Threads per inch - number of threads over 1 inch length
PC	Personal Computer - a desktop or laptop computer for everyday use		TYP or TYP.	Typical - used in drawing to identify common dimensions
PITCH	Distance between two corresponding points of adjacent threads			

TAP DRILL SIZES

Metric Threads - Coarse

Tap size	Tap drill - mm	Inch equivalent
M1x0.25	0.75	0.0295
M1.2x0.25	0.95	0.0374
M1.4x0.3	1.10	0.0433
M1.5x0.35	1.15	0.0453
M1.6x0.35	1.25	0.0492
M1.8x0.35	1.45	0.0571
M2x0.4	1.60	0.0630
M2.2x0.45	1.75	0.0689
M2.5x0.45	2.05	0.0807
M3x0.5	2.50	0.0984
M3.5x0.6	2.90	0.1142
M4x0.7	3.30	0.1299
M4.5x0.75	3.75	0.1476
M5x0.8	4.20	0.1654
M6x1	5.00	0.1969
M7x1	6.00	0.2362
M8x1.25	6.75	0.2657
M9x1.25	7.75	0.3051
M10x1.5	8.50	0.3346
M11x1.5	9.50	0.3740
M12x1.75	10.20	0.3937
M14x2	12.00	0.4724
M16x2	14.00	0.5512
M18x2.5	15.50	0.6102
M20x2.5	17.50	0.6890
M22x2.5	19.50	0.7677
M24x3	21.00	0.8268
M27x3	24.00	0.9449
M30x3.5	26.50	1.0433

Metric Threads - Fine

Tap size	Tap drill - mm	Inch equivalent
M3x0.35	2.65	0.1043
M3.5x0.35	3.15	0.2283
M4x0.5	3.50	0.1378
M4.5x0.5	4.00	0.1575
M5x0.5	4.50	0.1772
M5.5x0.5	5.00	0.1969
M6x0.75	5.25	0.2067
M7x0.75	6.25	0.2461
M8x1	7.00	0.2756
M9x1	8.00	0.3150
M10x0.75	9.25	0.3642
M10x1	9.00	0.3543
M10x1.25	8.75	0.3445
M11x1	10.00	0.3937
M12x1	11.00	0.4331
M12x1.25	10.75	0.4232
M12x1.5	10.50	0.4134
M13x1.5	11.50	0.4528
M13x1.75	11.25	0.4429
M14x1.25	12.75	0.5020
M14x1.5	12.50	0.4921
M15x1.5	13.50	0.5315
M16x1	15.00	0.5906
M16x1.5	14.50	0.5709
M17x1.5	15.50	0.6102
M18x1	17.00	0.6693
M18x1.5	16.50	0.6496
M18x2	16.00	0.6299
M20x1	19.00	0.7480
M20x1.5	18.50	0.7283
M20x2	18.00	0.7087
M22x1	21.00	0.8268
M22x1.5	20.50	0.8071

Tap size	Tap drill - mm	Inch equivalent
M22x2	20.00	0.7874
M24x1	23.00	0.9055
M24x1.5	22.50	0.8858
M24x2	22.00	0.8661
M25x1.5	23.50	0.9252
M27x2	25.00	0.9843
M28x2	26.00	1.0236
M30x2	28.00	1.1024
M30x3	27.00	1.0630

Imperial Threads - UNC/UNF

Tap size	Tap drill inch	Decimal inch	Alternative mm
#0-80	3/64	0.0469	
1/16-64	3/64	0.0469	
#1-64	#53	0.0595	
#1-72	#53	0.0595	
#2-56	#50	0.0700	
#2-64		0.0709	1.80
3/32-48	#49	0.0730	
#3-48	#47	0.0785	1.90
#3-56	#45	0.0820	
#4-32	#45	0.0820	
#4-36	#44	0.0860	
#4-40	#43	0.0890	
#4-48	#42	0.0935	
#5-40	#39	0.0995	
#5-44	#37	0.1040	
1/8-40	#38	0.1015	
#6-32	#36	0.1065	
#6-36	#34	0.1110	
#6-40	#33	0.1130	
5/32-32	1/8	0.1250	
5/32-36	#30	0.1285	

Tap size	Tap drill inch	Decimal inch	Alternative mm
#8-32	#29	0.1360	
#8-36	#29	0.1360	
#8-40	#28	0.1405	
3/16-24	#26	0.1470	
3/16-32	#22	0.1570	
#10-24	#25	0.1495	
#10-28	#23	0.1540	
#10-30	#22	0.1570	
#10-32	#21	0.1590	
#12-24	#16	0.1770	
#12-28	#14	0.1820	
#12-32	#13	0.1850	4.70
7/32-24	#16	0.1770	
7/32-32	#12	0.1890	4.80
#14-20	#10	0.1935	
#14-24	#7	0.2010	5.10
1/4-20	#7	0.2010	5.10
1/4-28	#4	0.2090	
1/4-32	7/32	0.2188	5.50
5/16-18	F	0.2570	6.50
5/16-20	17/64	0.2656	
5/16-24	I	0.2720	6.90
5/16-32	9/32	0.2813	7.10
3/8-16	5/16	0.3125	8.00
3/8-20	21/64	0.3281	
3/8-24	Q	0.3320	8.50
3/8-32	11/32	0.3438	
7/16-14	U	0.3680	9.40
7/16-20	25/64	0.3906	9.90
7/16-24	X	0.3970	10.00
7/16-28	Y	0.4040	
1/2-13	27/64	0.4219	
1/2-20	29/64	0.4531	11.50

Tap size	Tap drill inch	Decimal inch	Alternative mm
1/2-28	15/32	0.4688	
9/16-12	31/64	0.4844	
9/16-18	33/64	0.5156	13.00
9/16-24	33/64	0.5156	13.00
5/8-11	17/32	0.5313	13.50
5/8-12	35/64	0.5469	
5/8-18	37/64	0.5781	
5/8-24	37/64	0.5781	
11/16-12	39/64	0.6094	
11/16-16	5/8	0.6250	
11/16-24	41/64	0.6406	
3/4-10	21/32	0.6563	16.50
3/4-12	43/64	0.6719	17.00
3/4-16	11/16	0.6875	17.50
3/4-20	45/64	0.7031	17.50
3/4-28	23/32	0.7188	
13/16-12	47/64	0.7344	
13/16-16	3/4	0.7500	
7/8-9	49/64	0.7656	19.50
7/8-12	51/64	0.7969	20.00
7/8-14	13/16	0.8125	
7/8-16	13/16	0.8125	
7/8-20	53/64	0.8281	
15/16-12	55/64	0.8594	
15/16-16	7/8	0.8750	
15/16-20	57/64	0.8906	
1-8	7/8	0.8750	
1-12	59/64	0.9219	
1-14	15/16	0.9375	
1-20	61/64	0.9531	
1 1/16-12	63/64		
1 1/16-16	1.0		

Straight Pipe Taps - NPS

Tap size	Tap drill (inches)	Metric alternative
1/8-27	S	8.80
1/4-18	29/64	11.50
3/8-18	19/32	15.00
1/2-14	47/64	18.50
3/4-14	15/16	23.75
1-11½	1 3/16	30.25
1 1/4-11½	1 33/64	38.50
1 1/2-11½	1 3/4	44.50
2-11½	2 7/32	56.00

Pipe size	TPI	Tap drill	Decimal size
1/16	27	1/4	0.2500
1/8	27	11/32	0.3438
1/4	18	7/16	0.4375
3/8	18	37/64	0.5781
½	14	23/32	0.7188
3/4	14	59/64	0.9219
1.0	11-1/2	1-5/32	1.1563
1-1/4	11-1/2	1-1/2	1.5000
1-1/2	11-1/2	1-3/4	1.7500
2.0	11-1/2	2-7/32	2.2188

Taper Pipe Taps - NPT

Tap size	Tap drill (inches)	Metric alternative
1/16-27	D	6.30
1/8-27	R	8.70
1/4-18	7/16	11.10
3/8-18	37/64	14.50
1/2-14	45/64	18.00
3/4-14	59/64	23.25
1-11½	1 5/32	29.00
1 1/4-11½	1 1/2	38.00
1 1/2-11½	1 47/64	44.00
2-11½	2 7/32	56.00
2 1/2-8	2 5/8	67.00
3-8	3 1/4	82.50

Taper Pipe Taps - NPT - Continued

Pipe Size	TPI	Drilled only		Taper reamed	
		Tap Drill	Decimal Size	Tap Drill	Decimal Size
1/16	27	D	0.2460	15/64	0.2344
1/8	27	Q	0.3320	21/64	0.3281
1/4	18	7/16	0.4375	27/64	0.4219
3/8	18	37/64	0.5781	9/16	0.5625
1/2	14	45/64	0.7031	11/16	0.6875
3/4	14	29/32	0.9062	57/64	0.8906
1.0	11-1/2	1-9/64	1.1406	1-1/8	1.1250
1-1/4	11-1/2	1-31/64	1.4844	1-15/32	1.4688
1-1/2	11-1/2	1-47/64	1.7344	1-23/32	1.7188
2.0	11-1/2	2-13/64	2.2031	2-3/16	2.1875

60-DEGREE THREAD FORMS

The three thread forms below show standard dimensions of *Unified National* (UN) and *metric* threads, plus the older *American National* standard. All threads have 60° included angle.

Each of the three threads can be cut with the same partial profile insert. The *American National* form is considered obsolete and not used for new part designs.

Depth of thread formulas are listed in the next section.

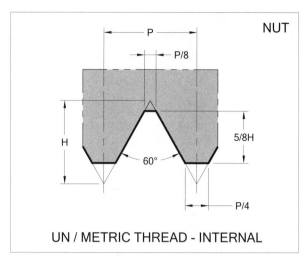

UN / METRIC THREAD - INTERNAL

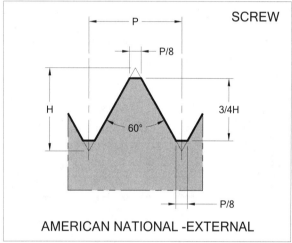

AMERICAN NATIONAL - EXTERNAL

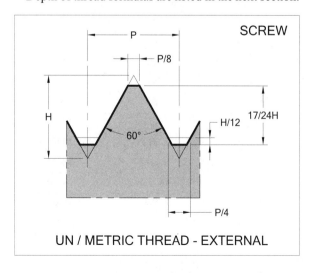

UN / METRIC THREAD - EXTERNAL

USEFUL FORMULAS

This section contains various formulas, useful for both CNC programmers and operators. Some formulas are part of chapters, others had been added for reference.

Speeds and Feeds

$$r/min = \frac{m/min \times 1000}{\pi \times D_{mm}}$$

$$r/min = \frac{ft/min \times 12}{\pi \times D_{inch}}$$

- **METRIC FEEDRATE** - *in mm/min*:

  $F = r/\min \times mm/fl \times N$

  where ...
  F = Programmed cutting feedrate in millimeters per minute
  r/min = Programmed spindle speed (S)
  mm/fl = Chip load per flute in millimeters
  N = Number of cutting flutes (teeth)

- **IMPERIAL FEEDRATE** - *in in/min*:

  $F = r/\min \times in/fl \times N$

  where ...
  F = Programmed cutting feedrate in inches per minute
  r/min = Programmed spindle speed (S)
  in/fl = Chip load per flute in inches
  N = Number of cutting flutes (teeth)

  > Note that for lathe work, feedrate is normally programmed in *mm/rev* or *in/rev*

- **TAPPING FEEDRATE**

Tapping feedrate must always synchronize the pitch of the tap and the programmed spindle speed:

$F = r/\min \times Pitch$

Pitch is the distance between threads - for example, 1/2-20 tap has 20 threads per inch (TPI). To calculate the pitch, use the following formula:

$Pitch = 1/TPI$

Pitch for the **1/4-20** tap is 0.05 inches.

For metric threads, the pitch is part of the thread drawing definition - for example, **M10x1.5** is a 10 mm diameter thread with 1.5 mm pitch.

- **CUTTING TIME**

Cutting time in minutes can be calculated from the current feedrate and the length of cut:

$T = L/F$

where ...
T = Cutting time in minutes
L = Length of cut in current units (mm or inches)
F = Feedrate in current units (mm/min or in/min)

Machining Holes - Spot Drilling

Formulas for machining holes are generally related to spot drilling and drilling of through holes and blind holes. The basic elements of the spot drilling formulas:

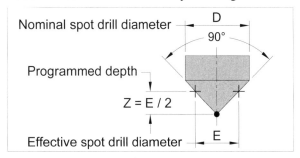

- **SPOT DRILLING - Small Holes**

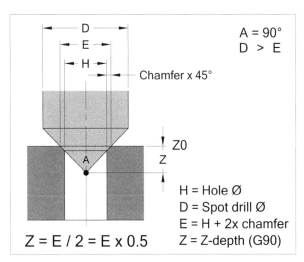

- **SPOT DRILLING - Large Holes**

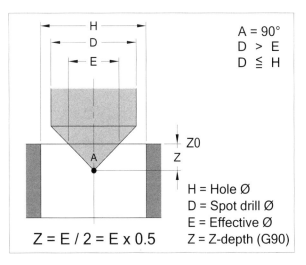

Machining Holes - Drilling

The most important part of Z-depth calculation is to calculate the drill point length. This is the 'P' length in the formulas. For the common drills, the length P is calculated using a constant. The next illustration shows a formula to calculate P length for *any* drill point angle:

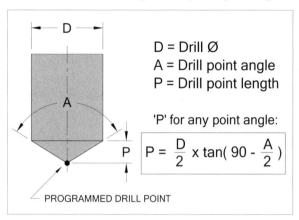

D = Drill Ø
A = Drill point angle
P = Drill point length

'P' for any point angle:

$$P = \frac{D}{2} \times \tan\left(90 - \frac{A}{2}\right)$$

➡ CONSTANTS FOR COMMON DRILL ANGLES

For the majority of drill depth calculations, mathematical constants are easier to use while providing sufficient accuracy for most of drilling operations.

60°	P = D x 0.866
75°	P = D x 0.652
80°	P = D x 0.596
82°	P = D x 0.575
90°	P = D x 0.500
100°	P = D x 0.420
110°	P = D x 0.350
118°	P = D x 0.300
120°	P = D x 0.289
135°	P = D x 0.207
150°	P = D x 0.134
180°	P = 0.000

➡ BLIND HOLES

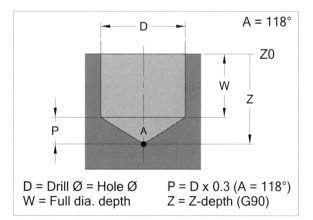

D = Drill Ø = Hole Ø
W = Full dia. depth
P = D x 0.3 (A = 118°)
Z = Z-depth (G90)

➡ THROUGH HOLES

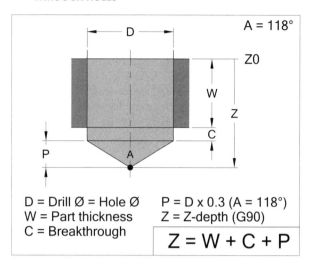

D = Drill Ø = Hole Ø
W = Part thickness
C = Breakthrough
P = D x 0.3 (A = 118°)
Z = Z-depth (G90)

$$Z = W + C + P$$

Tapers

A taper is a conical shape that can be machined on CNC lathes. A taper is created by a uniform change in diameter, over a given length. Tapers are common for many lathe applications, and an important part of CNC knowledge. There is a difference between imperial tapers and metric tapers. The reason is the way each taper is defined in the drawing.

- **In metric units taper is defined as a RATIO**
- **In imperial units taper is defined as a TAPER PER FOOT**

Typical examples are:

➡ 3 TPF ...taper defined as 3 inches taper per foot
1 : 10 ...taper defined as a ratio to a diameter

> A taper is the change in diameter per unit of length

The above general definition applies equally to imperial and metric tapers.

Specific definitions are:

- Imperial (as a Taper Per Foot - TPF)

> Taper per foot is the change in diameter in inches over one foot of length

- Metric (as a ratio - 1 : X)

> Taper is the change in diameter by 1 mm over the length of X

Control Memory Capacity

The CNC memory capacity is limited. Initially, all program were stored on a punched tape. The punched holes of the tape were spaced from each other by an established standard EIA RS-227-A. The spacing between punched holes was exactly 0.1000 inches.

1 foot = 0.3048 meters *and* **1 meter = 3.28083 feet**

Many modern CNC units show the memory capacity in characters, meters or feet. A typical program file is a plain text file, and the number of bytes is approximately the number of characters. Memory capacity specified in meters or feet refers to the length of obsolete tape and converting one unit to another is necessary.

◆ *To find the meters of tape equivalent,*
if the memory capacity is known in characters:

> Number of characters x 0.00254
> = Memory storage capacity in meters
> ... is the same as ...
> (Number of characters x 2.54) / 1000

◆ *To find the feet of tape equivalent,*
if the memory capacity is known in characters:

> Number of characters / 120
> = Memory storage capacity in feet

◆ *To find the memory capacity in characters*
if the memory capacity is known in meters:

> Meters / 0.00254
> = (Meters x 1000) / 2.54
> = Memory storage capacity in characters

◆ *To find the memory capacity in characters*
if the memory capacity is known in feet:

> Feet x 120 = Memory storage capacity in characters

◆ *EXAMPLES of memory capacity:*

Meters	Feet	Characters
15	49.21	5 905
20	65.62	7 874
40	131.23	15 748
80	262.47	31 496
320	1049.87	125 984

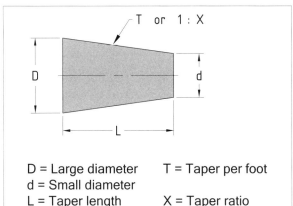

D = Large diameter   T = Taper per foot
d = Small diameter
L = Taper length   X = Taper ratio

◆ **METRIC TAPERS**

$$1 \div X = \frac{D - d}{L}$$

$$X = \frac{L}{D - d}$$

$$L = (D - d) \times X$$

$$d = \frac{D - L}{X}$$

$$D = \frac{L}{X} + d$$

◆ **IMPERIAL THREADS**

$$TPF = \frac{(D - d)}{L \times 12}$$

$$L = \frac{(D - d) \times 12}{TPF}$$

$$d = \frac{D - (L \times TPF)}{12}$$

$$D = \frac{L \times TPF}{12} + d$$

Trigonometric Chart

This simple trigonometric chart includes all formulas to solve a right angle triangle.

		$a/c = \sin A = \cos B$
		$b/c = \cos A = \sin B$
		$a/b = \tan A = \cot B$
		$b/a = \cot A = \tan B$
$a = c \times \sin A$	$b = c \times \cos A$	$c = a / \sin A$
$a = c \times \cos B$	$b = c \times \sin B$	$c = a / \cos B$
$a = b \times \tan A$	$b = a \times \tan B$	$c = b / \sin B$
$a = b / \tan B$	$b = a / \tan A$	$c = b / \cos A$
$a = \sqrt{c^2 - b^2}$	$b = \sqrt{c^2 - a^2}$	$c = \sqrt{a^2 + b^2}$
$A = 90° - B$	$B = 90° - A$	$C = 90°$

Index

Symbols

/ function 219

Numerics

60-degree thread forms 274

A

Absolute data input 86
Absolute mode 227
Address keyboard 25
Adjustments 83
ALTER key 41
Alternating flank infeed 211
ANSI 154
Arc radius 102
Arc vectors 102
ATC 55
ATC process 60
Auto operation 28
Auto power 36
Automatic block numbering 244
Automatic corner rounding 247
Automatic Tool Change (ATC) 55
Axis selection 41

B

Back boring 123
Battery backup 254
Blind holes 126
Block delete 219
Block skip 34, 219
Block skip switch 219
Boring bar extension 166
Boring cycles 123
BT toolholders 53
Button switches 33

C

Calculator input 104
CAM programs and machining 108
CAN key 26
CAT tool holders 53
Center drills 125
Center finder 77
Changing tool numbers 166
Chatter 113
Chip-to-Chip time 56
Chuck clamp and unclamp 160
Chuck pressure 160
Clearances 179, 194, 212
Climb milling 112
CNC lathe tools 151
   Corner radius change 158
   Cutting inserts 156
   Gage insert 154
   Identification 153
   ISO and ANSI 154
   Machine zero 161
   Right and left hand 153
   Selection of jaws 160
CNC machine specifications 5
   Horizontal Machining Center 6
   Lathes and Turning centers 7
   Vertical Machining Center 5
CNC process 1
COM work offset
   Common work offset 85
Common abbreviations 269
Contour - External cutting 176
Contour - Internal cutting 177
Contour points 174
Control panel 20, 22
   ON-OFF buttons 23
Control panel symbols 45
Control system 19
   Address keyboard 25
   Control Panel 22
   Control panel 20
   Display screen 23
   Edit Keys 26
   EOB - CAN - INPUT keys 26
   Help key 26
   Memory capacity 21
   Navigation keys 23
   Numeric keyboard 25
   Operation Panel 20
   Overview 19
   Page and cursor keys 26
   Reset key 26
   Selection keys 23
   Shift key 25
   Soft keys 23
   System memory 20
Conventional milling 112
Conversational and macro programming 108
Coolant functions 37
Coolants 37, 113
   Benefits 114
   Working with coolants 114
Corner grooves 195

279

INDEX

Counterboring 117
Countersinking 117
CRC error 179
Crest of thread 213
Cursor keys 26
CUSTOM key 25
Cutter radius compensation error 179
Cutter radius offset 12, 102, 135, 184
    Contour start and end 148
    Cutting direction 136
    D-address 135
    Equidistant toolpath 136
    G40-G41-G42 135
    Geometry and Wear 142
    Initial considerations 139
    Insufficient clearance 145
    Look-ahead type 138
    Offset adjustment 140
    Offset memory types 142
    Programs from computer 144
    Radius too large 147
    Solving problems 145
    Tool size change 149
Cutter radius too large 147
Cutting condition 2
Cutting direction 136
Cutting direction change 193
Cutting inserts 156
Cutting tools 54
Cycle end 29
Cycle start 28, 64

D

D-address 135, 137
Data input 86
Data without decimal point 255
Datum shift with G10 226
Decimal equivalents table 263
Decimal point 17, 104, 245
Deep hole drilling 118
DELETE key 41
Depth of thread 214
Diameter grooves 195
Difficulty in part measurement 224
Digit 0 - Letter O 104
Dimensional errors 104
Dimensional tolerances 101
Display screen 23
Distance-To-Go 78
Distance-to-Go 88
DNC 22
DNC mode 37, 40, 106
D-offset in subprograms 229
D-offset number 137
Down milling 112
Drawing evaluation 1, 101
    Tolerances 101

Drilling 276
Drilling example 116
Drilling formulas 125
Drilling holes 115
Dry run 36
Dummy tool 64
Dwell time 106, 117, 247, 259
    Minimum dwell calculation 117

E

Edge finder 77, 78
Edit key (protection) 45
Edit keys 26
Edit mode 40
Emergency stop 45
Empty tool pockets 59
EOB key 26
Equidistant toolpath 136
Errors 104
Errors / Alarms / Faults 31
E-stop 45
EXT offset 133, 134
EXT work offset 85, 98, 100
Extended work offsets 100
External grooves 200
External work offset 85, 86

F

F dimension 179
F0 selection 41
Face grooves 195
Facing cuts 185
Facing to centerline 192
Fanuc 15 mode 250
Fanuc system specifications 7
    Controlled axes 7
    Data Input / Output 10
    Editing functions 9
    Feedrate functions 8
    Interpolation functions 8
    Operation functions 8
    Options 10
    Program input 9
    Setting and display functions 9
    Spindle functions 8
    Tool functions 9
Feedhold 28
Feedrate for tapping 120
Feedrate optimization 43
Feedrate override 42
Feedrate per minute 42
Feedrate per revolution 42
F-feedrate in subprograms 229
Finding a center 71
Fine boring cycle G76 124
First part run 4

Fixed cycles 115
   Boring 123
   Clearances in G73/G83 cycles 251
   Depth 117
   Drilling 116
   Fine boring 124
   Initial level 117
   Peck drilling 118
   Rigid tapping 122
   R-level 117
   Spot drilling 117
   Tapping 120
   Tapping feedrate 120
Fixed type ATC 57
Fixture for tool assembly 93
Fixture setup 4
Flank infeed 210
Formulas
   Drilling 276
   Speeds and feeds 274
   Spot drilling 275
   Tapers 276
   Trigonometry 278
Formulas - drilling 125
Full thread profile 211

G

G00 83
G01 34, 42
G02 34, 42
G03 34, 42
G04 247
G10 127, 129, 131, 226
G10 and parameters 232
G10 for roughing and finishing 229
G10 format 226
G10 format for lathes 230
G17 12
G20 11
G21 11
G28 38, 89, 161
G30 38
G32 - Long hand threading 213
G40 135
G41 135, 164, 190, 193
G42 135, 164, 190
G43 87, 88, 89, 96, 135
G44 87, 88
G49 87, 89
G52 66, 127, 129, 131, 225
G54 37, 75, 76, 82, 83, 88
G54 P1 to P48 100
G54-G59 75, 85, 100, 127
G55 37
G65 127
G70 173
G70-G71-G72-G73 Summary 188

G71 173, 178
G71 Lathe cycle 177, 186
   One block format 186
   Two block format 177
G71-G72 Cycle - Parameters 252
G71-G72 Lathe cycles - Type I and Type II 187
G72 173, 178
G72 Lathe cycle 185, 187
   One block format 187
   Two block format 185
G73 118, 173
G73 Cycle - Parameters 253
G73 Lathe cycle 186, 187
   One block format 187
   Two block format 186
G74 120, 122
G74-G75 Cycles - Parameters 253
G74-G75 grooving cycles 205
G76 124
G76 - Threading cycle 214
G76 cycle - P-address 216
G76 Cycle - Parameters 253
G76 Thread cutting cycle 215
   Examples 216
   One block format 215
   Two block format 215
G81 118
G82 106, 117, 118
G83 118
G84 120, 122
G85 123
G86 123
G87 123
G88 123
G89 124
G90 35, 74, 105, 129
G91 35, 105
G92 - Box threading 213
G96 44
G97 44
G97G76 216
G98 116
G99G81 116
G-codes 75
G-codes list - Milling 261
G-codes list - Turning 262
Geometry offset 142, 163, 204
GRAPH key 25
Groove dimensions 198
Groove machining 195
Grooving
   Depth control 199
   External and Internal 200
   Position and Width control 204
   Position control 201
   Width control 203
Grooving - insert reference point 201
Grooving cycles 205
Grooving insert 196

Grooving on lathes 195
Grooving tool 196

H

H-address 87, 135
Hand of thread 213
Handle 49
Handle mode 37, 39
Handy setting 234
Hard jaws 160
Height of part 84
Help key 26
Holes 115
Home mode 37
Home position 38, 161
HSK tool holders 54

I

I and J vectors 102
Imaginary tool point 189
Increment system 239
Incremental data input 86
Incremental jog 45
Incremental mode 227
Infeed methods 210
Initial level 117
Initial program settings 12
INPUT and +INPUT keys 86, 90
INPUT key 26
Input of real numbers 238
INSERT key 41
Insert orientation 208
Insert reference point 201
Insert shapes 156
Insufficient clearance 145
Internal grooves 200
ISO 154

J

Jog mode 37, 39

L

Lathe chuck 160
Lathe cycles G70-G71-G72 173
Lathe drilling cycles G74 and G75 205
Lathe geometry 161
Lathe offsets 161
   Changing tool numbers 166
   Geometry offset 163
   R and T columns 164
   Tip numbers 164
   Tool reference point 162
   Wear offset 165

Work setup 163
Lathe tools 151
Lathes - part zero
   CNC lathes 71
Lead 212
Lead vs. Pitch 212
Left hand tools 153
Length offset 87
Letter O - Digit 0 104
Local coordinate system 129, 130, 225
Logical errors 17, 31
Logical program structure 11
Longest tool 97
Look-ahead offset type 138
Lubrication alarm 32
Lubrication error 31

M

M00 29, 30, 32, 34, 56
M00 with a comment 221
M01 29, 30, 32, 33, 34
M02 29, 30
M03 34, 44, 118
M04 34, 44
M05 34
M06 32, 33, 56, 60, 62, 63, 64
M08 37, 106, 108
M09 37
M30 29, 30, 32
Machine geometry 92
Machine lock 36
Machine origin 38
Machine zero 73
Machine zero return 38
Machining a part 101, 111
   Coolants 113
   Drawing evaluation 101
   Machining objectives 101
   Maintaining surface finish 111
   Material inspection 103
   Part inspection 108
   Preventing a scrap 109
   Process of machining 101
   Production machining 107
   Program evaluation 103
   Program input 106
   Program verification 106
   Stock allowance 103
   Surface finish 103
Machining direction 112
Machining holes 115, 275
Machining objectives 101
Major thread diameter 213
Managing tolerances 166
Manual absolute 36, 52
Material check 3
Material identification 1

Material inspection 103
Maximum tool diameter 58, 59
Maximum tool length 58
Maximum tool weight 58, 59
MDI input 38
MDI mode 37, 39, 234
Memory capacity 21
Memory mode 37, 40, 106
MESSAGE key 25
M-functions - Milling 262
M-functions - Turning 263
Milling tools 53
Minimum dwell 117
Minor thread diameter 213
Mode selection 37
Modified flank infeed 210
MPG 49
M-S-T Lock 35
Multi start threads 218
Multipart setup 127
   Distances known 129
   Distances unknown 127
   G10 setting 130
   G52 setting 130
   Setting by macro 131
   Z-axis control 132
Multiple offsets 203, 205
Multiple repetitive cycles 173
Multiple work offsets 82

N

Navigation keys 23
Neck grooves 195
Negative stock allowance 182
Nose radius change 158
Numeric keyboard 25

O

OFFSET / SETTING key 24
Offset adjustment 91, 140
Offset change 193
Offset change by program 225
Offset errors 18
Offset memory types 89
   Type A 90
   Type B 90
   Type C 90
Offset screen 191
Offsets numbering 203
ON/OFF switches 32
Operation panel 20, 27
   Buttons and switches 28
   Layout 28
   Status lights 29
Optional program stop 30
Optional stop 33

Oriented spindle stop (OSS) 124
Origin 65
OSS (Oriented spindle stop) 124
Overcutting error 147
Overrides
   Feedrate override 42
   Rapid override 42
     F0 selection 41
   Spindle Override 44

P

P and Q contour identification 183
P-address 106
P-address in G76 cycle 216
Page keys 26
Parameter types 235
Parameters 172, 232
Parameters - Classifications 267
Parameters and G10 254
Parameters of ...
   Cycles 249
   Display 243
   Feedrate 241
   Multiple repetitive cycles 252
   Programs 245
   Setting 240
   Spindle control 247
   Threading cycle 251
   Tool offsets 248
Part holding 2
Part inspection 4, 108
Part orientation 68
Part setup evaluation 73
Part stopper 67
Part zero 65, 73, 162
Part zero location 80
Part zero shift 225
Part zero symbols 65
Partial thread profile 211
Password protection 243
Peck drilling 118
   Q-depth 119
Pitch 212
Pitch Diameter 213
Pitch vs. Lead 212
Planes 12
POS key 24
Positive stock allowance 182
Pot numbers 56
Power ON 28
Precision boring cycle G76 124
Preparatory commands 75
Preventing a scrap 109
Process of machining 101
Production machining 107
Production run 4
PROG key 24

Program documentation 3
Program end 30
Program errors 17
   Incorrect units 18
   Logical errors 17
   Offset errors 18
   Syntax errors 17
Program evaluation 3, 103
Program input 106
Program interpretation 11, 15
   Dwell time 16
   Initial settings 12
   Program structure 11
   Program structure consistency 14
   Speeds and feeds 15
   Status block 13
   Type of operation 15
   Units 15
Program loading 4
Program modification 111
Program optimization 4, 257
Program Stop 30
Program testing 37
Program testing and setup mode 37
Program transfer 3
Program verification 2, 106
Program writing 2
Program zero 65

R

R and T columns 164
Radial infeed 210
Radius from I and J 102
Radius vs. Diameter 248
Random memory type ATC 57
Rapid mode 37, 39
Rapid override 41, 42
Ready light indicator 28
Reference points 73
   Machine zero 73
   Part zero 73
Registering tools 58
Reset 26
Right hand tools 153
Rigid tapping 122
R-level 117
Root of thread 213
Rotary switches 37

S

S-address 44
Safety block 13
Safety issues 4, 52
Scrap 109
   Causes 109
   Steps to eliminate 109

   Working with offsets 109
Scrap prevention 140
Selection keys 23
Setting offsets 4
Setting parameters by G10 232
Setting part zero 65
Setup handle 49
   Axis selection 50
   Dial wheel 50
   Mode selection 50
   Range selection 50
Setup mode 37
Setup related changes 260
Shift Key 25
Shifting part zero 225
Single block 33, 34
Soft jaws 160
Soft keys 23
Software verification 107
Speed and Feed relationship 218
Speeds and feeds 15, 274
Spindle function 44
Spindle orientation 124, 247
Spindle override 44
Spindle rotation direction 250
Spindle rotation in threading 208
Spindle speed limitation 218
Spot drilling 117, 275
Spot drills 125
Spring passes 214
Standard grooves 195
Stand-by tool position 56
Status block 13
Status indicator lights 29
Step-by-step ATC 60
Stock allowance 103, 182, 183
Surface finish 103, 111
   Chatter and vibration 113
   Coolants 113
   Drawing specifications 111
   Effect of cutting tools 111
   Machining direction 112
   Selection of cutting data 112
Surface finish in threading 210
Symbols 45
   Modes 46
   Operations 47
   Switches 46
Synchronized tapping 122
Syntax errors 17, 31, 104
SYSTEM key 25
System memory 20
System parameters 233
   Access 233
   Backup 234
   Bit type 236
   Fanuc classification 233
   Parameter types 235
   Parameters for operators 233

Valid data ranges 240

T

Tap drill 120
Tap drill sizes
   Imperial UN 272
   Metric coarse 271
   Metric fine 271
   Straight pipe taps NPS 273
   Taper pipe taps NPT 273
Tapers 276
Tapping 120
   Troubleshooting 120
Tapping feedrate 120
Test cuts 221
Thread cutting 207
   Clearances 212
   Hand of tool 208
   Infeed methods 210
   Methods of programming 213
   Spindle rotation 208
   Spindle speed limitation 218
   Threading data 214
   Threading defaults 218
   Threading process 207
   Threading tools 207
Threading bars 208
Threading cycle G76 214
Threading data
   Angle of thread 214
   Depth of thread 214
   Feedrate 214
   First pass depth 214
   Spring pass 214
Threading definitions 212
Threading feedrate 212, 214
Threading insert 211
   Full profile 211
   Partial profile 211
Through holes 126
TNR 189
Toggle switches 33
Tolerances 101, 166
Toll tip numbers 164
Tool assembly fixture 93
Tool change clearance 180
Tool change position 56
Tool changing 159, 180
   ATC 55
   ATC process 60
   Empty pockets 59
   Fixed type ATC 57
   Magazine labeling 57
   Manual tool change 55
   Programming format 64
   Random memory type ATC 57
   Stand-by position 56
   Tool magazine (carousel) 56

Tool magazine capacity 56
Tool waiting 58
Tool holders - milling 53
Tool length offset 87, 100
   Adjustments 91
   Applying 89
   Commands 87
Tool length setting 92
   Preset method 93
   Reference tool method 97
   Touch-off method 95
Tool magazine 56
Tool magazine capacity 56
Tool magazine labeling 57
Tool nose radius offset 189
Tool numbering - lathes 151
Tool numbers 15
Tool pockets (pots) 56
Tool repetition 62
   Different tool 62
   Same tool 62
Tool size change 149
Tool specifications 58
   Maximum tool diameter 58
   Maximum tool length 58
   Maximum tool weight 58
Tool waiting 58
Tool waiting position 56
Tooling 53
   Cutting tools 54
   Tooling system 54
Tooling fixture 93
Tooling preparation 3
Tooling selection 2
Tooling setup 3
Tooling system 54
Tools
   Milling 53
Tool-to-Tool time 56
TPI - Threads per inch 213
Trial cuts 221
   Milling applications 222
   Turning applications 223
Trigonometric calculations table 278
Turning and boring 173
Turret 152
Twist drills 126

U

Units 15, 83, 150
   Incorrect units 18
Units conversion 150
Units of input 239
Up milling 112
Using a vise 67

V

Vibration 113
Vise orientation 69
Vise setup 67
Vises for CNC work 67
V-threads 274

W

Wear offset 142, 165, 171, 249
Wear offset parameters 172
Work completion 3
Work offset change 226
Work offsets 73, 83
    Calculations 79
    Data input 86
    Extended set 132
    Z-axis setting 81
Work offsets - multiple 82
Work shift 72
W-stock allowance 183

Z

Z-axis lock 35
Z-axis neglect 35
Zero return 29, 38